Frontiers of Nuclear Medicine

Aktuelle Nuklearmedizin

Frontiers of Nuclear Medicine

Aktuelle Nuklearmedizin

Editor: Wolfgang Horst

Springer-Verlag Berlin Heidelberg New York 1971

Professor Dr. med. Wolfgang Horst

Direktor der Universitätsklinik und Poliklinik
für Nuklearmedizin und Radiotherapie,
Kantonsspital, CH-8006 Zürich

With 180 Figures

ISBN-13: 978-3-642-65094-9 e-ISBN-13: 978-3-642-65092-5
DOI: 10.1007/978-3-642-65092-5

Softcover reprint of the hardcover 1st edition 1971

Library of Congress Catalog Card Number 71-152 727

Preface

That nuclear medicine has advanced so far and so fast is due in no small measure to GEORGE VON HEVESY. His work on radioactive indicator technique laid the foundation on which this young branch of medicine was able to develop in the decade which followed World War II.

In the intervening years the second generation has grown up in nuclear medicine. Some of them were still exposed to VON HEVESY's influence, for instance, his address to the 1957 meeting of the Italian Society of Nuclear Medicine in Turin, or his Marie Curie Memorial Lecture at the Pittsburg meeting of the North American Society of Nuclear Medicine in 1961. Others again will remember that he helped to found the European Society of Nuclear Medicine in 1962—1963 and became its honorary president.

It was VON HEVESY who, together with HEILMEYER, insisted that this be a completely open society, having neither national nor geographical attributes, its Europeanness being reflected in the variety of languages spoken at its congresses. Its members exhibited a similar variety, including in addition to those medically qualified—specialists in internal medicine, radiologists and laboratory research workers—physicists, chemists and engineers.

A group of young second-generation scientists from eleven countries have dedicated these papers to the memory of the great pioneer of nuclear medicine. This book contains new results reported by doctors, physicists, chemists and computer specialists—results so far-ranging as to push the frontiers of nuclear medicine still further forward.

Zurich, April 1971

W. HORST

Vorwort

Die Nuklearmedizin verdankt ihre heutige schnelle Entwicklung in besonderem Maße GEORG VON HEVESY. Seine Arbeiten über die radioaktive Indikatortechnik waren ein wichtiges Fundament, auf dem sich dieser junge Zweig der Medizin im ersten Jahrzehnt nach dem 2. Weltkrieg entwickelte.

Inzwischen ist bereits die zweite Generation von Nuklearmedizinern herangewachsen. Sie hat das Wirken VON HEVESYS z. T. noch miterlebt, so z. B. in Italien seinen Festvortrag anläßlich der Tagung der Italienischen Gesellschaft für Nuklearmedizin 1957 in Turin oder seine Gedächtnislesung für Marie Curie 1961 in Pittsburg anläßlich der dortigen Tagung der Gesellschaft für Nuklearmedizin von Nordamerika. Andere werden sich an VON HEVESY erinnern, als er 1962 und 1963 die europäische Gesellschaft für Nuklearmedizin mitbegründete und ihr Ehrenpräsident wurde.

Er war es, der damals mit HEILMEYER zusammen diese Gesellschaft als offene Gesellschaft verstanden wissen wollte, die weder nationale noch geographische Attribute tragen sollte und die ihre europäische Vielfalt auch in der Zahl der Kongreßsprachen widerspiegelt. Die gleiche Vielfalt findet sich unter ihren Mitgliedern: Ärzte — Internisten, Radiologen, Laboratoriumsmediziner usw. —, Physiker, Ingenieure und Chemiker.

Eine Gruppe von jungen Wissenschaftlern der zweiten Generation hat nun wissenschaftliche Arbeiten dem Andenken des großen Pioniers der Nuklearmedizin gewidmet. Es sind Ärzte, Physiker, Chemiker und Computerfachleute aus 11 Ländern, die in diesem Buch ihre neuen Ergebnisse mitteilen — Ergebnisse, die in ihrer Vielgestaltigkeit die wissenschaftliche Front der heutigen Nuklearmedizin aufreißen.

Zürich, April 1971

W. HORST

George von Hevesy
in memoriam

* 1. VIII. 1885 Budapest
† 5. VII. 1966 Freiburg i. Br.

George von Hevesy

GEORGE VON HEVESY was a pioneer of nuclear medicine. He received the Nobel Prize in 1943 for his studies in the field of radioactive indicator technique and was one of the founders and honorary president of the European Society of Nuclear Medicine as well as honorary member of the North American Society of Nuclear Medicine.

GEORG VON HEVESY war ein Pionier der Nuklearmedizin. Er erhielt 1943 für seine Arbeiten über die radioaktive Indikatortechnik den Nobelpreis und war Mitbegründer und Ehrenpräsident der europäischen Gesellschaft für Nuclearmedizin sowie Ehrenmitglied der Gesellschaft für Nuclearmedizin von Nordamerika.

The *George von Hevesy Prize for Nuclear Medicine* was awarded for the first time at the 7th annual meeting of the European Society of Nuclear Medicine, held in Zurich in 1969. The recipient was

Der *Georg von Hevesy Preis für Nuklearmedizin* wurde erstmals anläßlich der 7. Jahrestagung der europäischen Gesellschaft für Nuclearmedizin 1969 in Zürich verliehen an

H. SAUL WINCHELL, M.D., Donner Laboratory, University of California, Berkeley.

On the occasion of the 8th annual meeting of the European Society of Nuclear Medicine, held in Hanover in 1970, the first *George von Hevesy Memorial Lecture* was read by

Anläßlich der 8. Jahrestagung der europäischen Gesellschaft für Nuclearmedizin 1970 in Hannover wurde die erste *Georg von Hevesy Gedächtnisvorlesung* gehalten von

Professor E. ERIC POCHIN, M.A., M.D., F.R.C.R., President emeritus I.C.R.P., Medical Research Council, University of London.

GEORGE VON HEVESY was born in Budapest on 1st August, 1885. He studied chemistry, mathematics and physics in Budapest, Berlin and Freiburg im Breisgau, where he also obtained his doctorate. He then worked for two years with LORENZ and WILLSTÄTTER at the ETH Zurich, in 1911 was with HABER in Karlsruhe and until 1913 with RUTHERFORD in Manchester. In 1913, with PANETH in Vienna, he carried out the earliest work on radioactive indicators. After a short period at the University of Budapest, he joined NIELS BOHR in Copenhagen until 1926; in 1922 he discovered hafnium and in the following year did his first work on radioactive indicators in biology. From 1926 to 1934 he was Director of the Physical-Chemical Institute of the University of Freiburg i. Br., returning to Copenhagen until 1943. In 1935 he and CHIEWITZ carried out the first work with an artificial radionuclide, ^{32}P, with the support of O. H. LAWRENCE and the collaboration of A. KROGH. He was awarded the Nobel Prize in 1943. From then until 1961 he worked in Stockholm, in the THEORELL and VON EULER Institutes and the Radiumhemmet. His interests now embraced clinical, physiological and biological problems and radiobiology. GEORGE VON HEVESY died on 5th July, 1966, in Freiburg i. Br., surrounded by his family and tended by his friend and physician, L. HEILMEYER.

He published more than 400 articles in the course of his career, in addition to a number of world-famous books which had been translated into many languages.

GEORG VON HEVESY wurde am 1. August 1885 in Budapest geboren. — Studium von Chemie, Mathematik und Physik in Budapest, Berlin und zuletzt in Freiburg im Breisgau, wo er auch promovierte. Anschließend zwei Jahre bei LORENZ und WILLSTÄTTER an der ETH Zürich; 1911 bei HABER in Karlsruhe und bis 1913 bei RUTHERFORD in Manchester. In Wien 1913 erste chemische Radioindikatorarbeit mit PANETH. — Nach kürzerer Tätigkeit an der Universität Budapest bis 1926 bei NIELS BOHR in Kopenhagen. 1922 Entdeckung des Hafniums; 1923 erste biologische Radioindikatorarbeit. 1926—1934 Direktor des Physikalisch-chemischen Instituts der Universität Freiburg im Breisgau. 1934 bis 1943 wieder bei NIELS BOHR in Kopenhagen. 1935 gemeinsam mit CHIEWITZ erste biologische Arbeit mit einem künstlichen Radionuklid (^{32}P); Unterstützung durch O. H. LAWRENCE und Zusammenarbeit mit A. KROGH. 1943 Nobelpreis. 1943—1961 in Stockholm im Institut THEORELL und VON EULER sowie im Radiumhemmet. Seine Arbeitskraft galt jetzt klinischen, physiologischen und biologischen Problemen und der Radiobiologie. GEORG VON HEVESY starb am 5. Juli 1966 in Freiburg im Breisgau, umsorgt von seiner Familie und seinem Freund und Arzt L. HEILMEYER.

Sein Lebenswerk umfaßt mehr als 400 Veröffentlichungen neben einer Zahl weltweit bekannter Bücher, die in viele Sprachen übersetzt wurden.

Honorary degrees:
Ehrendoktor der Universitäten von:

1929 Cape Town, Sc.D.
1945 Uppsala, Phil.D.
1948 Freiburg im Breisgau, Phil.D.
1950 Ghent, D.Sc.
1950 Copenhagen, Phil.D.
1954 São Paulo, Med.D.
1955 Rio de Janeiro, Med.D.
1957 Turin, Med.D.
1957 Burlington, Jur.D.
1959 Freiburg im Breisgau, Med.D.
1959 Liège, Sc.D.
1960 London, Sc.D.
1964 Cambridge, Sc.D.
1965 Budapest, Dr. Ing.

Membership of academies:
Mitglied der Akademien:

1925 Danish Academy of Sciences
1926 Akademie der Wissenschaften in Heidelberg
1928 Gothenburg Academy of Sciences
1939 Royal Society, London
1942 Swedish Academy of Sciences
1945 Hungarian Academy of Sciences
1948 Rome Academy of Sciences (Lincei)
1948 Genoa Academy of Sciences
1948 National Academy, Boston
1950 Brussels Academy of Sciences
1950 National Academy of India
1961 Pontificial Academy
1965 Österreichische Akademie der Wissenschaften

Medals and prizes:
Auszeichnungen und Preise:

1929 Cannizzaro Prize, Rome
1942 Danish Cancer Research Prize
1944 Nobel Prize of 1943
1949 Copley Medal, Royal Society
1950 Faraday Medal, Chemical Society
1950 Brussels Academy Medal
1951 Bailey Medal, College of Physicians, London
1956 Silvanus Thompson Medal, British Society of Radiology
1957 Medal of the City of Paris
1957 Mitglied des Ordens Pour le Mérite der BRD
1958 Cotius-Medaille der Akademie der Wissenschaften in Halle
1959 Atoms for Peace Award
1961 Rosenberger Medal of Chicago University
1961 Niels Bohr Medal of the Danish Engineering Society

Honorary member of learned societies:
Ehrenmitglied der wissenschaftlichen Gesellschaften:

1939 Chemical Society, London
1945 Chemical Society, Helsingfors
1948 Royal Institution
1951 Deutsche Bunsen-Gesellschaft
1952 Swedish Medical Society
1955 Biochemical Society, Helsingfors
1955 Austrian Society of Physics and Chemistry
1958 Japan Chemical Society
1958 Society of Nuclear Medicine, U.S.A.
1958 British Society of Radiology
1960 Deutsche Physiologische Gesellschaft
1960 Akademie Halle
1963 Gesellschaft für Nuclearmedizin, Mitbegründer, Ehrenpräsident
1965 Swedish Society of Radiobiology

Lectures:
Akademische Gedächtnis- und Festansprachen:

1930–1931 Baker Lectures, Cornell University
1931 Hugo Müller Lecture, London Chemical Society
1935 Scott Lectures, Cambridge University
1940 Durham Lecture, Harvard University (prevented from delivering by the outbreak of World War II)
1944 Nobel Lecture
1949–1950 Franqui Lectures, Ghent University
1950 Faraday Lecture, Edinburgh
1951 Schoenheimer Memorial Lecture, New York (prevented by illness)
1953 Aschoff Gedächtnis-Lesung, Universität Freiburg im Breisgau
1955 Evening Lecture, Geneva (at the invitation of the Secretary General of the United Nations)
1956 Sylvanus Thompson Lecture, British Institute of Radiology, London
1957 Italian Society of Nuclear Medicine, Turin: Historical Progress of Isotopic Methodology and its Influence on the Biological Sciences
1958 Gedächtnislesung für Heinrich Wieland, Bonn
Jahrestreffen der Mitglieder des Ordens Pour le Mérite der BRD
1960 First Durham Foundation Lectures, Cambridge University
1961 Lecture on Marie Curie and her contemporaries at the invitation of the Society of Nuclear Medicine, Pittsburg
1962 Internationales Symposium für Nuklearmedizin, Freiburg im Breisgau, Festvortrag: Historische Übersicht über einige Anwendungen radioaktiver Isotope in der Hämatologie.
1966 Papal Academy of Sciences, Vatican, Opening Address: Radiation Haematology

George von Hevesy was honored in the following countries:

Austria, Belgium, Brazil, Denmark, Finland, France, Germany, Great Britain, Hungary, India, Italy, Japan, South Africa, Sweden, United States of America, Vatican.

Georg von Hevesy erfuhr Ehrungen unter anderem in folgenden Ländern:

Belgien, Brasilien, Dänemark, Deutschland, Finnland, Frankreich, Großbritannien, Indien, Italien, Japan, Österreich, Schweden, Südafrika, Ungarn, Vatikan, Vereinigte Staaten von Amerika.

W. Horst, Zurich, Switzerland

Contents – Inhaltsverzeichnis

Authors — Autoren

ALTENBRUNN, H.-J., Dr. med. habil., Isotopen-Abteilung, Robert-Rössle-Klinik, DDR-1115 Berlin-Buch, Lindenberger Weg 80

BEHSCHAD, H., Dr. med., Czerny-Krankenhaus, D-69 Heidelberg, Voss-Str. 3

BEKIER, A., Dr. med., Röntgendiagnostisches und radiotherapeutisches Zentralinstitut, Kantonsspital, CH-9006 St. Gallen

BESCH, W., Sektion für Nuklearmedizin, Zentrum für Innere Medizin und Kinderheilkunde der Universität Ulm, D-79 Ulm, Steinhövelstr. 9

BIANCHI, C., M. D., Professor, Clinica Medica Generale, Università di Pisa, I-56100 Pisa

BITTER, F., Dipl.-Phys., Sektion für Nuklearmedizin, Zentrum für Innere Medizin und Kinderheilkunde der Universität Ulm, D-79 Ulm, Steinhövelstr. 9

COLI, A., M. D., Clinica Medica Generale, Università di Pisa, I-56100 Pisa

EATON, Jr., S. BOYD, M. D., Massachusetts General Hospital, Boston, Massachusetts, U.S.A.

FORTH, W., Dr. med., Professor, Institut für Pharmakologie und Toxikologie der Universität des Saarlandes, D-665 Homburg (Saar)

GALASKO, C. S. B., Ch. M., F.R.C.S., F.R.C.S.E., The Nuffield Orthopaedic Centre, Headington, Oxford OX3 7LD, U.K.

GIANNOTTI, P., M. D., Professor, Clinica Chirurgica, Università di Pisa, I-56100 Pisa

HEHN, G., Dr., Institut für Kernenergetik der Universität Stuttgart, D-7 Stuttgart 80, Pfaffenwaldring 31

HERNÁDY, T., M. D., Radiological Clinic, Medical University of Budapest, Budapest, Hungary

ISHIKAWA, D., M. D., Department of Radiology, Faculty of Medicine, University of Tokyo, 3-28-6 Mejirodai, Bunkyo-ku, Tokyo, Japan

JAMES, Jr., A. EVERETTE, Sc. M., M. D., Assistant Professor, Department of Radiology, The Johns Hopkins Hospital, 601 N. Broadway, Baltimore, Maryland 21205, U.S.A.

KAUL, A., Dr. phil. nat., Professor, Nuklearmedizinische Abteilung, Klinikum Steglitz der Freien Universität Berlin, D-1 Berlin 45, Hindenburgdamm 30

MCEWAN, A. C., National Radiation Laboratory, P.O. Box 1456, Christchurch, New Zealand

MACK, J. F., Department of Radiology, St. Mary's Hospital, Long Beach, California, U.S.A.

MÁRTONFFY, KATALIN, M. D., Radiological Clinic, Medical University of Budapest, Budapest, Hungary, now: Universitätsklinik und Poliklinik für Nuklearmedizin und Radiotherapie, Kantonsspital Zürich, Rämistr. 100, CH-8006 Zürich

MEURET, G., Dr. med., Nuklearmedizinische Abteilung, Medizinische Universitätsklinik, D-78 Freiburg (Breisgau), Hugstetter Str. 55

MÖRSDORF, D., Dr. med., Dipl.-Phys., Firma Dr. Karl Thomae, D-795 Biberach

NGO, T. M., M. D., Ph. D., Assistant Professor, Department of Nuclear Medicine and Radiobiology, University Medical Center, Sherbrooke, P. Q., Canada

PALLA, R., M. D., Clinica Medica Generale, Università di Pisa, I-56100 Pisa

PAVEL, D., Dr., I.N.S.E.R.M. (U 90), Hôpital Necker, 151, Rue de Sèvres, F-75 Paris-15e

PFANNENSTIEL, P., Dr. med., Privatdozent, Sektion Nuklearmedizin, Deutsche Klinik für Diagnostik AG, D-62 Wiesbaden, Aukammallee 33

PIXBERG, H. U., Dr. med., Privatdozent, Institut für Nuklearmedizin und spezielle Biophysik, Krankenhaus Oststadt, D-3 Hannover, Podbielskistr. 380

RAYNAUD, J. P., Professor, Centre de Recherches Roussel-Uclaf, 111, route de Noisy, F-93 Romainville

RHODES, B. A., Ph. D., Assistant Professor, Division of Nuclear Medicine, School of Hygiene 2101, The Johns Hopkins Medical Institutions, Baltimore, Maryland 21205, U.S.A.

SCHÄFER, N., Dr. med., Sektion für Cardiologie, Zentrum für Innere Medizin und Kinderheilkunde der Universität Ulm, D-79 Ulm, Steinhövelstr. 9

SCHENK, P., Dr. med., Privatdozent, Czerny-Krankenhaus, D-69 Heidelberg, Voss-Str. 3

SCHLEHE, H., Dr. med., Zentrum für Innere Medizin und Kinderheilkunde der Universität Ulm, D-79 Ulm, Steinhövelstr. 9

SCHWALL, H., Dr. rer. nat., Dipl.-Chem., Firma C. H. Boehringer Sohn, D-6507 Ingelheim

SEIFERT, J., Dr. med., Röntgeninstitut und Strahlenklinik, Klinikum Essen der Ruhr-Universität Bochum, D-43 Essen-Holstenhausen, Hufelandstr. 55

SIGMUND, EVA, Dr. med., Sektion für Nuklearmedizin, Zentrum für Innere Medizin und Kinderheilkunde der Universität Ulm, D-79 Ulm, Steinhövelstr. 9

SODD, V. J., Ph. D., Deputy Chief, Nuclear Medicine Laboratory, Bureau of Radiological Health, Cincinnati General Hospital, 3231 Burnet Avenue, Cincinnati, Ohio 45229, U.S.A.

TÖRÖK, ÉVA, M. D., Radiological Clinic, Medical University of Budapest, Budapest, Hungary

WELLMAN, H. N., M. D., Chief, Nuclear Medicine Laboratory, Bureau of Radiological Health, Cincinnati General Hospital, 3231 Burnet Avenue, Cincinnati, Ohio 45229, U.S.A.

WELLNER, U., Dr. rer. nat., Dipl.-Phys., Nuklearmedizinische Abteilung, Universitätskliniken Köln, D-5 Köln-Lindenthal, Josef-Stelzmann-Str. 9

WINCHELL, H. S., M. D., Ph. D., Donner Laboratory, University of California, Berkeley, California 94720, U.S.A.

WINSTEAD, M. B., Ph. D., Bucknell University, Lewisburg, Pennsylvania 17837, U.S.A.

WOLF, K.-J., Dr. med., Medizinisches Strahleninstitut der Universität Tübingen, D-74 Tübingen, Roentgenweg 11

WUNDER, S., Assistent, Czerny-Krankenhaus, D-69 Heidelberg, Voss-Str. 3

YASUKOCHI, H., M. D., Professor, Department of Radiology, Faculty of Medicine, University of Tokyo, 3-28-6 Mejirodai, Bunkyo-ku, Tokyo, Japan

Computer Diagnosis of Thyroid Scintigrams

H. YASUKOCHI and D. ISHIKAWA

Department of Radiology, Faculty of Medicine, University of Tokyo, Tokyo, Japan

Radioisotope diagnosis of the thyroid diseases is one of the most popular and developed techniques in the field of Nuclear Medicine and many techniques concerning radioisotopes occupy a large part of thyroid diagnosis accompanied with the various radioactive nuclides, ^{125}I, ^{132}I and ^{131}I.

Morphological examinations using radioisotopes are also developed to detect thyroid nodules. In this field, ^{32}P [1] and ^{131}Cs [2] were reported as a positive deposit agent, but the most frequently used technique is the defect detection by scintigraphic techniques using ^{131}I. Usually the detection capacity in scintigraphic techniques depends on two main factors, they are the capacity of the apparatus and that of radioisotopes. On the point of apparatuses, the detection capacity depends on the sensitivity and the resolution of the scintigraphic equipment. The development of a scintillation camera using a pinhole collimator shows better resolution and sensitivity in thyroid scintigrams than by conventional scanners [3], but because of the anatomical position of the thyroid being on the surface of the body and the figure being plane, the dependence of thyroid scintigram on apparatuses is not seriously important in the interpretation of the scintigrams [4]. In the point of radiopharmaceuticals the capacity depends upon the energy of the radioisotope. Some low- or medium-energy radioisotopes such as ^{125}I [5] or ^{99m}Tc [6] are also used for thyroid scintigrams; this shows an improved scintigram mainly depending on the resolution capacity of collimators but, with the same reason, the dependence is also not so remarkable as seriously limited in other organs.

For these reasons thyroid scintigrams by scanner using ^{131}I are selected in this report for analysis as the most popular technique.

Materials

For this research, 1000 consecutive thyroid scintigrams were taken from the routine thyroid examinations in our department and 336 cases were selected for analysis after the following criteria. First of all 400 cases of non-nodular diseases are excluded because this method is for determining the diagnosis of nodular goiters, and then 200 unconfirmed cases are excluded and among the remaining scintigrams 64 cases are also excluded because of incalculable factors, such as having received some medications or some procedures before the examination.

The selected 336 cases consist of three categories: there are 133 cases of malignant tumor, 136 cases of benign tumor and 67 cases of chronic thyroiditis.

Furthermore other 48 cases are added from the other department for comparison.

Scintigrams

On the selected cases, the scintigrams are analysed statistically before the computing on the point of sex, age, uptakes etc., and as a result the following factors are selected as diagnostically significant.

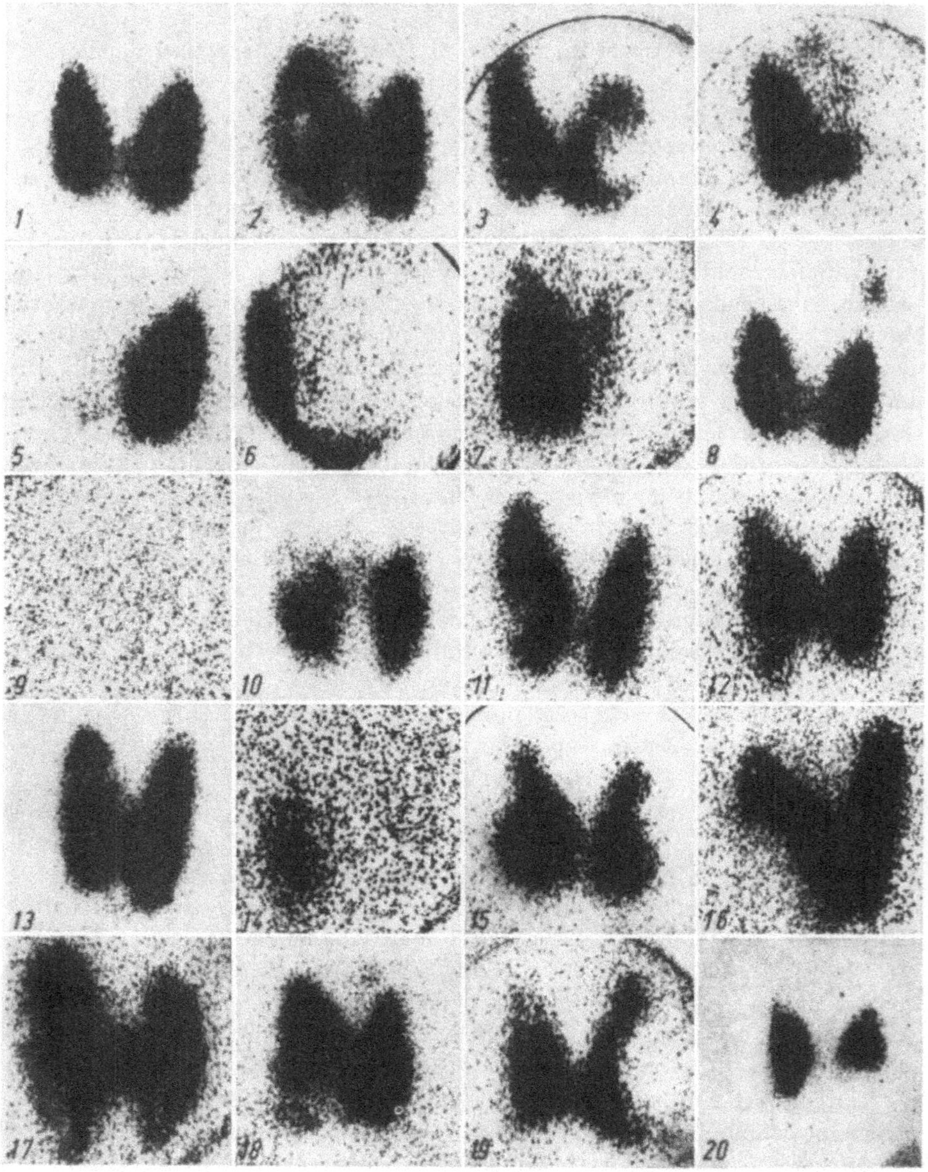

Fig. 1. Patterns of thyroid scintigram in each factor.

1. x_1, 2. x_2, 3. x_3, 4. x_4, 5. x_5, 6. x_6, 7. x_7, 8. x_8, 9. x_9, 10. u_1 (upper), 11. u_1 (lateral), 12. u_2 (lower), 13. u_2 (median), 14. u_3, 15. u_4, 16. v_1, 17. v_2, 18. w_1, 19. w_2, 20. w_3

They are: general pattern of the thyroid glands (x), mainly depending on the deposit patterns of radioisotope and on the defect types of x; the position of the defect in the gland (u); the continuity of the deposit between the defect area and surrounding normal area (v) and the figure of the defect (w) are also interpreted [7]. The frequency of these factors is determined in each disease (y). The details of the factors are as follows.

As to the diseases (y), y_1 means a malignant tumor, y_2 a benign tumor and y_3 chronic thyroiditis. In the pattern of thyroid scintigrams (x), x_1 means the normal (non-defect) type, x_2 small defect (smaller than $\frac{1}{3}$ of one lobe) type, x_3 medium defect (between $\frac{1}{3}$ and $\frac{2}{3}$ of one lobe) type, x_4 large defect (larger than $\frac{2}{3}$ of one lobe but limited in one lobe) type, x_5 one lobe (but opposite lobe is normal) type, x_6 both lobes damaged (defect larger than one lobe) type, x_7 hot nodule type, x_8 metastatic type and x_9 low uptake (including organic, functional and premedication) type. As to the position of the defect (u), u_1 means the defect is situated in the upper and/or lateral side of the lobe, u_2 lower and/or median side, u_3 circular and u_4 multiple. On the continuity between the defect and the surrounding tissue (v), v_1 means clearly defined border between them and v_2 diffuse border. On the figure of the defect (w), w_1 means non-expansive defect where the deposit area is dominant into the defect area, w_2 expansive type where the defect extends to the deposit and makes the border of the deposit area convex, w_3 intermediate type in which the definition of w_1 and w_2 is impossible.

Characteristic scintigrams showing each factor of x, u, v and w are demonstrated in Fig. 1 (x: 1—9, u: 10—15, v: 16—17, w: 18—20).

Analysis

The frequency of each factor mentioned above is obtained in each disease and listed in Table 1. From these data the probabilities of the diagnosis in each scintigram are calculated by Bayes' formula [8] as in other reports [9—11].

$$P y_i(x_j, u_k, v_m, w_n) = \frac{P y_i \cdot P x_j/y_i \cdot P u_k/y_i \cdot P v_m/y_i \cdot P w_n/y_i}{\sum P y_i \cdot P x_j/y_i \cdot P u_k/y_i \cdot P v_m/y_i \cdot P w_n/y_i}.$$

In the formula, $P y_i(x_j, u_k, v_m, w_n)$ shows the probability of the disease (y_i) from a scintigram with the factors (x_j, u_k, v_m, w_n). $P y_i$ means the probability of the disease y_i in the series. $P x_j/y_i$ means the probability of scintigram factor x_j when the disease is y_i. $P u_k/y_i$, $P v_m/y_i$, $P w_n/y_i$ are the same notations as $P x_j/y_i$ but with other scintigram factors. According to the formula, every scintigram is calculated for the probability on each disease. For instance, probability of diagnosis of malignant tumor from the scintigram shown in Fig. 1-2 was calculated as follows,

$$P y_1(x_2, u_1, v_2, w_2) = \frac{0.39 \cdot 0.173 \cdot 0.382 \cdot 0.474 \cdot 0.164}{\begin{array}{c}0.396 \cdot 1.173 \cdot 0.382 \cdot 0.474 \cdot 0.164 + 0.405 \cdot 0.250 \cdot 0.260 \\ \cdot 0.600 \cdot 0.546 + 0.199 \cdot 0.194 \cdot 0.286 \cdot 0.800 \cdot 0.353\end{array}}$$

$$= 0.13,$$

Table 1. *Frequency of each factor*

		y_1	y_2	y_3
1. Px_j/y_i	x_1	0.105	0.081	0.284
	x_2	0.173	0.250	0.194
	x_3	0.075	0.184	0.030
	x_4	0.026	0.159	0.269
	x_5	0.053	0.007	0.015
	x_6	0.026	0.007	0.015
	x_7	—	0.007	—
	x_8	0.026	—	—
	x_9	0.068	0.088	0.156
2. Pu_k/y_i	u_1	0.382	0.260	0.286
	u_2	0.402	0.600	0.439
	u_3	0.123	0.010	0.057
	u_4	0.093	0.130	0.228
3. Pv_m/y_i	v_1	0.526	0.400	0.200
	v_2	0.474	0.600	0.800
4. Pw_n/y_i	w_1	0.374	0.169	0.294
	w_2	0.164	0.546	0.353
	w_3	0.462	0.285	0.353
5. $Py_1 = 0.398$, $Py_2 = 0.405$, $Py_3 = 0.199$				

because the scintigram has a small defect (x_2), the position of the defect is in lateral (u_1), the continuity between the defect and surrounding deposit is diffuse (v_2) and the defect is expanding into the deposit (w_2). With the same computing procedure the probability of benign tumor in this scintigram $Py_2(x_2, u_1, v_2, w_2)$ was 0.57 and that of chronic thyroiditis $Py_3(x_2, u_1, v_2, w_2)$ was 0.30. After computing the probabilities in each disease and comparing the values of the calculated probability, the value of probability of benign tumor was the highest among the diseases. By comparison, the disease of the scintigram was suggested as benign nodule.

In this way the probability values of each disease are calculated in every scintigram of the series.

Results

The results of the calculation are shown in Fig. 2a, b, c. In Fig. 2a the white and the black columns show the frequency of malignant tumor correctly and falsely diagnosed, the dotted and the lined columns show that of benign tumor and chronic thyroiditis falsely diagnosed as malignant tumor. The numbers in the longitudinal axis in the figure mean the calculated probability for malignant tumor (Py_1). For instance, a white column in 0.7 shows the frequency of malignant tumor with correct diagnosis and the probability being between 0.70 and 0.79, and the lined column in 0.7 shows the frequency of chronic thyroiditis which is falsely diagnosed as malignant tumor with the probability between 0.70 and 0.79.

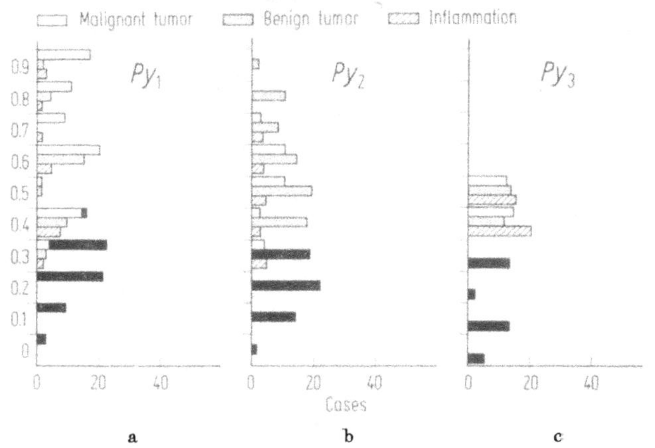

Fig. 2. The probabilities of each scintigram (our series)

In Fig. 2b, the dotted and the black columns show frequency of benign tumors diagnosed correctly and falsely, and the white and the lined columns show the malignant tumor and the chronic thyroiditis which are falsely diagnosed as benign tumor. In Fig. 2c, the lined and the black columns show the frequency of chronic thyroiditis diagnosed correctly and falsely, and the dotted and the white columns benign tumor and malignant tumor falsely determined as chronic thyroiditis.

According to the figures, the accuracy of scintigram reading and the frequency of diagnosable scintigrams are calculated as follows,

The accuracy of the scintigram reading

$$= \frac{\text{the number of correct diagnoses with the probability higher than a certain value}}{\text{the number of scintigrams with the probability higher than a certain value}},$$

The frequency of diagnosable scintigrams

$$= \frac{\text{the number of scintigrams with the probability higher than a certain value}}{\text{the total number of scintigrams analysed}}.$$

The calculations were done separately in each disease and as total, and the results are listed in Table 2. The accuracy was about 60 percent overall, but in combination with the value of the probability, for instance, the criteria of correct diagnosis are determined as value of probability higher than 0.70, the accuracy then becomes over 70 percent. But in this case a number of nondiagnosed scintigrams exist.

The first result shows an accuracy equivalent to that of a trained radiologist after a resident course, the second to that of a specialist on thyroid diseases.

To determine the factors, the frequency may generally depend on the equipment of the special department and the administered dose. To check the significance of these criteria, another series of scintigrams is also analysed. The series is

Table 2. *The accuracy of automatic scintigram reading (our series)*

Probability	Accuracy (%)				Frequency (%)
	malig. t.	benign t.	chr. in.	total	
over 0.90	100	60	—	78	8
over 0.80	100	62	—	75	17
over 0.70	93	74	0	72	25
over 0.60	79	62	0	54	45
over 0.50	61	60	36	56	70
total	51	56	47	54	100

obtained by courtesy of Chiba University (Prof. KAKEHI). There are 48 cases of specially selected thyroid scintigrams with confirmed diagnosis which were used to check the accuracy of the capacity of Japanese radiologists [12]. They comprise 31 cases of malignant tumor, 11 cases of benign tumor and 6 cases of chronic thyroiditis. For these scintigrams the probability of each disease is calculated using the same frequency because the frequency of diseases is not determined randomly.

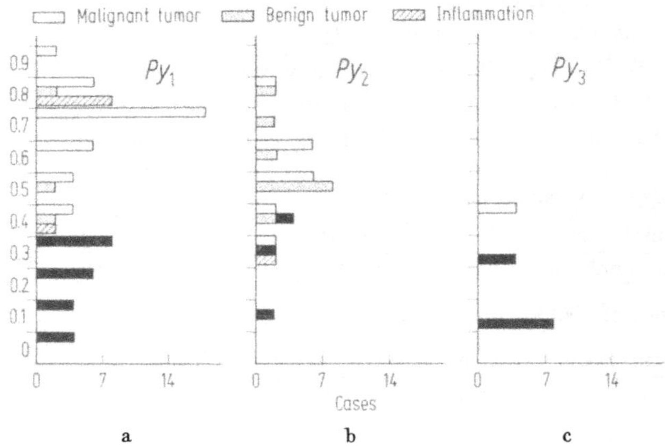

Fig. 3. The probabilities of each scintigram (comparison series)

The results are shown in Fig. 3a, b, c and the degree of accuracy was as listed in Table 3. In each series the accuracy of diagnosing chronic thyroiditis is lower than those of other diseases. This shows the difficulty of the diagnosis of thyroiditis from one scintigram. But if a series of scintigrams is obtained from one patient at considerable intervals, the variety of the scintigrams will reveal more information than the case of one scintigram, and the accuracy of diagnosing chronic thyroiditis may be higher.

Table 3. *The accuracy of automatic scintigram reading (comparison series)*

Probability	Accuracy (%)				Frequency (%)
	malig. t.	benign t.	chr. in.	total	
over 0.90	100	—	—	100	2
over 0.80	80	100	0	84	13
over 0.70	93	67	0	73	46
over 0.60	81	75	0	69	60
over 0.50	73	78	0	67	81
total	68	73	0	60	100

Furthermore the frequency of each factor, specially the factor x_9, will differ depending on the administered dose and/or on the sensitivity of the equipment as noticed before. Even considering these suggestions, the fact that the results of the two series show the same order of accuracy, will encourage us to pursue the possibility of the quantitative diagnosis of thyroid diseases according to the scintigram reading, even in the present status.

Discussion

According to the computer diagnosis of thyroid scintigrams of present cases, we conclude that the scintigram reading has a considerable capacity for diagnosing thyroid diseases. Of course, we do not claim that the ability of the computer diagnosis may surpass that of well trained doctors at any time, but this result suggests to us that the thyroid scintigram analysis will also give quantitative information on clinical procedures not only the defect detection.

In the calculation of these series we have selected only the significant factors because of our economic capacity, but, if possible, we will develop the program further by combining the other information, such as the course of the disease, clinical status, laboratory data [13] and other information on scintigrams, and extend it to other diseases. After developing the analytical program as shown in Fig. 4, we hope to improve the accuracy of the thyroid scintigram reading compared to the present series so that the diagnosis of the thyroid may become more comfortable.

On this program, the reading capacity depends on the doctor's interpretation, but, in future, the automatic readings will be introduced, as is already starting in the facsimile scanning in mammography [14]. The normalization of the lobe figure will be easier than with unilateral organs because the thyroid has similar bilateral lobes.

Because the composition factors of the scintigram are more simple and vague compared with the other morphological examinations, such as X-ray films or histological specimens, the introduction of computer diagnosis will be easier and more reasonable than these other complex and accurate morphological examinations. On this viewpoint, we are now trying to introduce this technique to the hepatoscintigram readings which are the next most frequent scintigrams in our clinic.

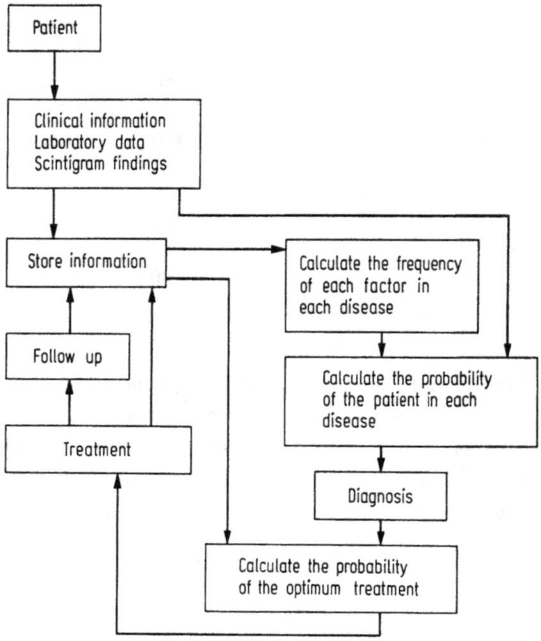

Fig. 4. Future program of thyroid diagnosis and treatment

One year after this paper was written, we have succeeded in identifying the thyroid scintigram automatically using wide field smoothing and border line determination by computer. Details have been presented at the last annual meeting of Nippon Societas Radiologica and will be published in the near future.

References

1. ACKERMAN, N. B.: The diagnosis of thyroid cancer with radioactive phosphorus. Surg. **47,** 615 (1960).
2. KAKEHI, H., UCHIYAMA, G.: The detection of thyroid cancer using cesium 131. Presented at annual meeting of American Society of Nuclear Medicine, 1969.
3. YASUKOCHI, H., ISHIKAWA, D., YAMAZAKI, T., MIYAMAE, T., LIN, S.: A comparison of thyroid scintigrams between by scanner and by camera. Radioisotopes **17,** 294 (1968).
4. — — — — —: Comparison of the diagnostic values of the scintigrams by scanner and by camera. In: Medical Radioisotope Scintigraphy, Vol. 1, p. 529. IAEA, Vienna 1969.
5. HARPER, P. V., ENDLICH, H., LATHROP, K. A., SIEMENS, W.: Clinical experience with iodine 125. J. Nucl. Med. **2,** 127 (1961).
6. — BECK, R., CHARLESTON, D., LATHROP, K. A.: Optimization of a scanning method using technesium 99m. Nucl. **22** (1), 50 (1964).

7. ISHIKAWA, D., YASUKOCHI, H.: Readings on thyroid scintigrams. Nippon Acta Radiologica **28**, 1039 (1968).
8. LEDLY, R. S., LUSTED, L. B.: Reasoning foundations of medical diagnosis. Science **130** (3366), 9 (1959).
9. WARNER, H. R., TORONTO, A. F., VEASEY, L. G., STEPHANSON, R.: A mathematical approach to medical diagnosis, Application to congenital heart disease. J. A. M. A. **177**, 177 (1961).
10. LODWICK, G. S., HAUN, C. L., SMITH, W. E., KELLER, R. F., ROBERTSON, E. D.: Computer diagnosis of primary bone tumors, A preliminary report. Radiology **80**, 273 (1963).
11. — KEATS, T. E., DORST, J. P.: The coding of roentgen images for computer analysis as applied to lung cancer. Radiology **81**, 185 (1963).
12. TATENO, M., KAKEHI, H.: The accuracy of thyroidscintigram reading. Presented at annual meeting of Nippon Societas Radiologica, 1969.
13. BENDER, C. E., FITZGERALD, L. T., WILLIAMS, C. M.: Probability values for protein bound iodine, thyroid iodine 131 uptakes and triodothyronin resin uptakes for hypothyroidism, euthyroidism and hyperthyroidism. Amer. J. Roent. **103**, 886 (1968).
14. WINSBERG, F., ELKIN, M., NACY Jr., J., BORDAZ, W., WEYMOUTH, W.: Detection of radiographic abnormality in mammograms by means of optical scanning and computer analysis. Radiology **89**, 211 (1963).

Anwendung kerntechnischer Berechnungsmethoden für Gammastrahlen bei der Bestrahlungsplanung

Application of Nuclear Calculation Methods to the Planning of Radiation Treatment with Gamma Rays

G. Hehn

Institut für Kernenergetik der Universität Stuttgart

Summary

Computer programs have been developed in recent years for the design of nuclear reactors with improved shielding; these can be applied in radiology to provide a theoretical basis for earlier experimental work and semi-empirical methods.

The Stuttgart PLK program, originally developed to determine heat sources and equivalent doses of mixed neutron and gamma radiation, was adapted to enable calculation by computer of the production of secondary electrons by gamma radiation in the tissues. Where penetration of secondary electrons can be neglected (X-ray and ^{60}Co radiation), the desired local distribution of the energy dose is found directly; where it cannot (betatron gamma radiation), the necessary information about the source of secondary electrons is obtained. Three examples are given to illustrate the efficiency of the method: a layer of homogeneous tissue, thigh and head irradiation. Hard and soft X-rays, cobalt gamma rays and betatron bremsstrahlen were programmed at energy maxima of 10, 30 and 50 MeV.

Kurzfassung

Mit Hilfe individueller und genauer Vorausberechnungen von Strahlungsfeldern läßt sich die Planung für Tumorbestrahlungen effektiver gestalten. Zur Berechnung von Kernreaktoren und vor allem zur optimalen Auslegung von Reaktorabschirmungen sind in den letzten Jahren große Computerprogramme entwickelt worden, die sich auch für radiologische Probleme mit Erfolg einsetzen lassen. Damit stehen auch der Radiologie neben der rein experimentellen Bestimmung von Strahlungsfeldern in Phantomen und den halbempirischen Interpolationsmethoden im größeren Maße als bisher rein theoretische Strahlentransportprogramme zur Verfügung, von deren Anwendung man neue Ergebnisse für die Bestrahlungsplanung erwarten kann. Das Stuttgarter Strahlentransportprogramm PLK, das primär zur Bestimmung der Wärmequellen und der Äquivalentdosis von gemischten Neutronen- und Gammastrahlenfeldern für Reaktorabschirmungen entwickelt wurde, ist so abgeändert worden, daß möglichst viele Informationen über die Erzeugung von Sekundärelektronen durch Gammastrahlen in Gewebe durch den Computer berechnet werden. In allen Fällen, in denen die Reichweite der erzeugten Sekundärelektronen vernachlässigbar ist, d.h. bei Röntgen- und Kobalt-Gammastrahlen, erhält man sofort die gewünschte Ortsverteilung der Energiedosis. In den Fällen, in denen die Wanderung der Sekundärelektronen nicht vernachlässigbar ist, d.h. bei Betatron-Gammastrahlung, bekommt man die benötigten Informationen über die Quellverteilung der Sekundärelektronen. Die Leistungsfähigkeit der Rechenmethode wird an drei Beispielen demonstriert. Rechnungen wurden für eine homogene Gewebeschicht und zwei inhomogene Probleme, eine Oberschenkel- und eine Kopfbestrahlung, durchgeführt. Die Bestrahlungsenergie wurde hierbei zwischen weichen und harten Röntgenstrahlen, Kobalt-Gammastrahlen und Betatronbremsstrahlen mit 10 MeV, 30 MeV und 50 MeV Maximalenergie variiert.

Einleitung

Parallel zur Entwicklung immer größerer Computeranlagen sind in den letzten Jahren die Berechnungsmöglichkeiten für Neutronen- und Gammastrahlenfelder in der Kerntechnik sprunghaft angestiegen. Neutronenfelder werden heute mit großer Genauigkeit in den sogenannten Kritikalitätsrechnungen für die Spaltzone von Kernreaktoren vorausberechnet. Gemischte Neutronen- und Gammastrahlenfelder werden durch die dicke Strahlenabschirmung rechnerisch verfolgt, denn man möchte schon bei der Planung von Kernreaktoren sicher sein, daß die ionisierende Strahlung, die die Abschirmung durchdringt, die maximal zulässige Dosis für das Betriebspersonal nicht übersteigt. Gerade für Abschirmungsprobleme wurden große Computerprogramme entwickelt, die z.B. die Flußdichte von Gammastrahlen und die Dosisrate in Abhängigkeit vom Ort, der Energie der Strahlung und ihrer Ausbreitungsrichtung bestimmen.

Die am häufigsten verwendeten Methoden sind die P_L- und S_N-Approximation der Boltzmannschen Strahlentransportgleichung sowie die Monte-Carlo-Methode. Die Monte-Carlo-Methode wird vor allem am Kernforschungszentrum Oak Ridge/USA intensiv weiterentwickelt. Programme dieser Art sind das OGRE-P1 und OGRE-G [1]. Die S_N-Methode stammt aus dem Kernforschungszentrum Los Alamos/USA. Dazu gehören die Programme DTF IV und DTF X-RAY [2]. Leistungsfähige P_L-Programme sind am Institut für Kernenergetik in Stuttgart entwickelt worden, die die Bezeichnungen PLE und PLK tragen [3].

All diese Methoden und Rechenprogramme haben ihre speziellen Vor- und Nachteile. Die Monte-Carlo-Methode ist besonders für geometrisch komplizierte Gebilde geeignet, während die anderen Methoden ihre Vorteile bei geschichteten Medien besitzen. Da die Methoden sich bei gleichzeitiger Anwendung gut ergänzen, wird man in Zukunft mehr und mehr kombinierte Rechnungen durchführen.

Im folgenden Bericht möchte ich die Anwendung des P_L-Programms PLK für Bestrahlungen in der Medizin zeigen. Zunächst möchte ich eine kurze Information über die Rechenmethode und das Rechenprogramm geben. Anschließend sollen Ergebnisse für eine homogene Gewebeschicht bei Bestrahlung mit weichen und harten Röntgenstrahlen sowie Kobalt-Gammastrahlen betrachtet werden.

Schließlich sollen zwei spezielle Fälle der Karzinombestrahlung, eine Oberschenkel- und eine Kopfbestrahlung mit dem Computer simuliert werden. Für Röntgenstrahlen und Kobalt-Gammastrahlen wird der Ortsverlauf der Energiedosis angegeben und mit der Quellverteilung der Sekundärelektronen verglichen, die man für verschiedene Bremsstrahlungsspektren bei Betatronbestrahlungen erhält.

P_L-Approximation und Computerprogramme

Die Strahlenausbreitung von Gammastrahlen in Materie wird exakt von der sogenannten Boltzmannschen Strahlentransportgleichung beschrieben, wobei zusätzliche Randbedingungen das spezielle Problem eindeutig festlegen. Die Transportgleichung lautet:

$$\begin{array}{lll} \boldsymbol{\Omega}\nabla\Phi(\boldsymbol{X},\boldsymbol{\Omega},E) & +\,\Sigma_T(\boldsymbol{X},E)\,\Phi(\boldsymbol{X},\boldsymbol{\Omega},E) & = S(\boldsymbol{X},\boldsymbol{\Omega},E) \\ \text{Nettoausströmung} & +\ \text{Absorption} & = \text{Quellen} \end{array}$$

mit $\Phi(\boldsymbol{X}, \boldsymbol{\Omega}, E)$, der Photonenflußdichte,

$\Sigma_T(\boldsymbol{X}, E)$, dem totalen Wirkungsquerschnitt

und $S(\boldsymbol{X}, \boldsymbol{\Omega}, E)$, den gesamten Quellen einschließlich Streuquellen.

Jede dieser Größen kann eine Funktion des Ortes $\boldsymbol{X}$, der Ausbreitungsrichtung $\boldsymbol{\Omega}$ und der Photonenenergie E sein. Obige Gleichung läßt sich aus einer einfachen Erhaltungsbedingung für Photonen ableiten, die besagt, daß die Nettoausströmung aus einem Phasenraumelement zusammen mit der Absorption genau so groß sein muß wie alle Photonenquellen in dem gleichen Phasenraumelement. Der entscheidende Punkt der P_L-Approximation obiger Transportgleichung besteht darin, daß die Winkelabhängigkeit der Flußdichte und der Quellen in eine endliche Reihe von Legendre-Polynomen entwickelt wird. Auf diese Weise wird in der P_L-Approximation die Richtungsabhängigkeit aller Größen analytisch behandelt im Gegensatz zur S_N-Approximation, bei der der Winkelbereich in einzelne Intervalle unterteilt wird und damit rein numerisch behandelt wird. Zur Behandlung der Energieabhängigkeit wird die übliche Multigruppenstruktur verwendet.

Die auf diese Weise erhaltenen Multigruppen-P_L-Differentialgleichungen werden mit Hilfe des Differenzenverfahrens anschließend numerisch gelöst.

Für ebene Geometrie und Kugelgeometrie sind die Formeln programmiert worden. Die Programme heißen PLE und PLK. Beide Rechenprogramme sind in Algol 60 für den Telefunken-Rechner TR 4 geschrieben worden. Die P_L-Approximation der Gruppenflüsse und Gruppenquerschnitte, die Zahl der Energiegruppen, die Zahl der Maschenpunkte und die Anzahl verschiedener Gewebearten sind nur durch die Speicherkapazität des Computers begrenzt. Für eine typisch radiologische Aufgabe mit sieben verschiedenen Gewebeschichten, bei P_5-Approximation der Flußdichte und Wirkungsquerschnitte sowie 28 Energiegruppen, beträgt die Rechenzeit 10 Minuten für die Erzeugung der makroskopischen Wirkungsquerschnitte und 20 Minuten für die Flußdichte- und Energiedosisberechnung und den Ausdruck der umfangreichen Ergebnisse. In naher Zukunft werden die P_L-Programme für den Stuttgarter Control Data Computer 6600 zur Verfügung stehen, wodurch eine erhebliche Reduzierung der benötigten Rechenzeit zu erwarten ist.

Bestrahlung einer homogenen Gewebeschicht

Wir berechneten zunächst die Photonenflußdichte und die Energiedosisrate in einer 30 cm dicken Schicht, in der alle menschlichen Gewebearten in den entsprechenden Anteilen homogen vermischt enthalten sein sollen. Die Zusammensetzung dieses zur Vereinfachung der Rechnung gebildeten Standardgewebes ist in

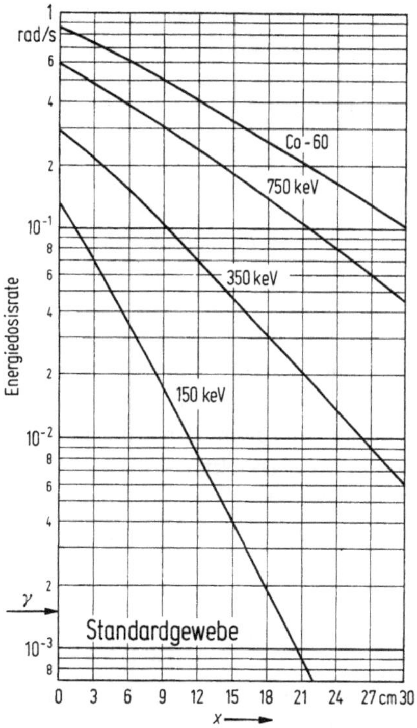

Abb. 1. Energiedosisrate in Standardgewebe für eine Punktquelle mit 1.68 kCi

Tab. 1 [4] gegeben. Alle Rechnungen wurden für eine Punktquelle mit einer Aktivität von 1.68 kCi durchgeführt, die einer Photonenemissionsrate von 6.22 × 10^{13} Photonen pro Sekunde entspricht. Die Entfernung zwischen der Quelle und der zu bestrahlenden Schicht wurde generell auf 60 cm festgelegt. Die Dichte des Standardgewebes wurde auf 1 g/cm³ normiert.

In Abb. 1 sind vier Kurvenverläufe eingetragen, die die Energiedosisrate in rad pro Sekunde für vier verschiedene Quellenergien angeben, und zwar für

150 keV,
350 keV,
750 keV
und Co-60-Energien.

Tabelle 1. *Anzahl der Atome in einem Gramm verschiedener Gewebearten sowie prozentuale Zusammensetzung nach ORNL-TM-2079*

Elemente	Standardgewebe		Muskel		Knochen		Gehirn		Mark	
	%	Atome/g × 10^{24}	%	Atome/g × 10^{24}	%	Atome/g × 10^{24}	%	Atome/g × 10^{24}	%	Atome/g × 10^{24}
H	10	5.98 (−2)*	10	5.98 (−2)	7.1	4.50 (−2)	10	5.98 (−2)	10	5.98 (−2)
O	60	2.26 (−2)	75	2.82 (−2)	39.5	1.49 (−2)	71	2.67 (−2)	74	2.79 (−2)
C	24	1.20 (−2)	11	5.52 (−3)	22.4	1.13 (−2)	15	7.52 (−3)	12.5	6.27 (−3)
N	2.9	1.25 (−3)	2.6	1.12 (−3)	4.7	2.05 (−3)	1.9	8.17 (−4)	3.2	1.38 (−3)
Ca	1.2	1.77 (−4)	3.1 (−3)	4.66 (−7)	12.37	1.87 (−3)	8.6 (−3)	1.29 (−6)	—	—
P	1.1	2.14 (−4)	0.18	3.50 (−5)	13.16	2.57 (−3)	0.34	6.61 (−5)	—	—
S	0.24	4.51 (−4)	0.24	4.51 (−5)	0.41	7.69 (−5)	0.17	3.19 (−5)	—	—
K	0.20	3.08 (−5)	0.30	4.62 (−5)	—	—	0.30	4.62 (−5)	—	—
Na	0.20	5.24 (−5)	0.16	4.19 (−5)	0.08	2.08 (−5)	0.18	4.72 (−5)	—	
Cl	0.20	3.40 (−5)	0.18	3.06 (−5)	—	—	0.23	3.91 (−5)	0.12	2.04 (−5)
Mg	0.03	7.43 (−6)	1.9 (−2)	4.71 (−6)	0.28	6.87 (−5)	0.015	3.72 (−6)	—	—

* Lies $5.98 \cdot 10^{-2} \times 10^{24} = 5.98 \cdot 10^{22}$.

In allen vier Beispielen kann der Energietransport durch Sekundärelektronen vernachlässigt werden, so daß die Quellverteilung für Sekundärelektronen mit dem Verlauf der Energieübertragung auf das Gewebe übereinstimmt. Am Schichtanfang ist die Energiedosis nahezu proportional zur Energie der einfallenden Photonen. Sie ist am kleinsten für 150-keV-Strahlen und am größten für Co-60-Strahlen. Die 150-keV-Strahlen werden im Gewebe stark absorbiert. Die Dosiskurve fällt entlang 30 cm Gewebe um drei Zehnerpotenzen ab, während bei Co-60-Strahlen der Abfall nur eine Dekade beträgt. 750-keV-Strahlen zeigen nahezu das gleiche Verhalten wie Co-60-Strahlen. 350-keV-Strahlen liegen mit einem Abfall von zwei Dekaden halbwegs zwischen den beiden extremen Beispielen.

Entsprechende Rechnungen für Betatronbremsstrahlen mit Maximalenergien von 10 MeV, 30 MeV und 50 MeV zeigen einen ähnlichen Quellverlauf für Sekundärelektronen wie im Falle von Co-60-Strahlen, nur sind die Kurven nach oben verschoben, weil die energiereicheren Photonen größere Energiemengen übertragen können. Hierbei ist der Energietransport durch die Sekundärelektronen erheblich. Vor allem die Compton-Elektronen zeigen eine starke Bevorzugung der Vorwärtsrichtung, was zur Folge hat, daß die Energiedosiskurve gegenüber dem Quellverlauf in Vorwärtsrichtung verschoben wird und das Maximum der Kurve vom Schichtanfang in innere Bereiche der Schicht verlegt wird.

Zur Bestrahlungsplanung benötigt man von der physikalischen Seite primär den Ortsverlauf der Energiedosisrate. Informationen über diesen Ortsverlauf kann man einerseits aus Messungen an entsprechenden Phantomen erhalten oder schneller und billiger durch Benutzung vorhandener Rechenprogramme auf dem Computer.

Nun ist die Energiedosisrate eine sehr integrale Größe. Es gibt weitere physikalische Informationen, die man am besten über die Rechnung erhält. Wie sieht z. B. das Energiespektrum der Gammastrahlen bei verschiedenen Eindringtiefen im Gewebe aus? Welche Ausbreitungsrichtungen hat man nach der Streuung im Gewebe zu erwarten? Diese Informationen über das Feld der Gammastrahlen fallen bei jeder Berechnung an und können auf Wunsch ausgedruckt werden. Ein Vergleich des Energiespektrums der Gammastrahlen im Gewebe zeigt, daß in allen berechneten Beispielen die Energie-Gruppen mit den Quellenergien dominieren. Dieser Effekt ist am kleinsten bei niederen Quellenergien und verstärkt sich mit steigender Bestrahlungsenergie. Ebenso ist die Winkelverteilung stark in Richtung des einfallenden Strahls kollimiert. Das bedeutet, daß die ungestreuten Photonen und Photonen mit kleinen Winkelstreuungen dominieren und daß die Rechenmethode diesen Anteil der Strahlung besonders sorgfältig behandeln muß. Andererseits sind aber auch alle Großwinkelstreuungen interessant, weil sie zur Verbreiterung des Strahlenkegels führen und die unerwünschte seitliche Streustrahlung abgeben. Mit Hilfe vorhandener zweidimensionaler Rechenprogramme lassen sich diese Effekte sauber studieren.

Eine weitere wichtige Information, die ebenfalls am besten über eine Berechnung zu erhalten ist, ist der Ortsverlauf der Quellen der Sekundärelektronen, die bei der Wechselwirkung von Gammastrahlen mit den Gewebeatomen entstehen. Durch Stöße wird die Photonenenergie zunächst hauptsächlich auf Sekundärelektronen übertragen. Die Energie- und Richtungsverteilung dieser Elektronen ist deswegen eine wichtige Bestimmungsgröße. In Tab. 2 haben wir aus unseren

Rechnungen die Energieübertragung auf Sekundärelektronen in der Schichtmitte für sieben verschiedene Bestrahlungsenergien miteinander verglichen. Die Werte sind für die drei wichtigsten Wechselwirkungsarten getrennt angegeben, und zwar für die Compton-Streuung, die Paarbildung und den Photoeffekt. Man sieht, daß die Comptonelektronen in allen Beispielen dominieren mit Ausnahme des Betatrons bei 50 MeV Maximalenergie. Bei der Betatronbremsstrahlung wird die Paarbildung bedeutend. Die gesamte Energie, die hierbei auf die Elektron-Positron-Paare übertragen wird, ist etwa so groß wie die Energie, die von den Comptonelektronen übernommen wird.

Tabelle 2. *Energieübertragung auf Sekundärelektronen in der Mitte einer 30 cm dicken Schicht aus Standardgewebe der Dichte 1 g/cm³; Abstand zwischen Schicht und Punktquelle 60 cm; Emissionsrate der Quelle: 1 Photon/s*

Quelle	Comptonelektronen (MeV/cm³ s)	Paarelektronen + Positronen	Photoelektronen (MeV/cm³ s)
150 keV	3.54 (−9)*	—	4.01 (−10)
350 keV	4.61 (−8)	—	1.09 (− 9)
750 keV	1.81 (−7)	—	2.39 (− 9)
Co-60	3.27 (−7)	5.97 (−11)	3.78 (− 9)
10-MeV-Betatron	7.86 (−7)	1.15 (− 7)	7.95 (− 9)
30-MeV-Betatron	9.33 (−7)	6.57 (− 7)	9.40 (− 9)
50-MeV-Betatron	1.01 (−6)	1.36 (− 6)	1.20 (− 8)

* Lies $3.54 \cdot 10^{-9}$.

Die kleinste Energiemenge entfällt auf die dritte Gruppe, die Photoelektronen; sie beträgt zehn Prozent bei der 150-keV-Quelle, fällt dann steil ab auf ein Prozent bei der Co-60-Quelle und erreicht den kleinsten Wert beim 50-MeV-Betatron mit einem halben Prozent. Da die einzelnen Photoelektronen im Vergleich zu den anderen Sekundärelektronen eine wesentlich kleinere kinetische Energie erhalten, ist ihre Anzahl nicht so klein, wie es aus der Summe aller Energien zunächst scheinen mag.

Die bisherigen Angaben bezogen sich auf eine verhältnismäßig dicke Schicht aus homogenem Gewebe. In der Regel werden jedoch stark heterogene Systeme bestrahlt mit recht unterschiedlichen Gewebeschichten.

Bestrahlung des Oberschenkels

Bei der Bestrahlung des Oberschenkels wurden folgende Gewebeschichten von rechts nach links angenommen:

0,75 cm Haut- und Fettgewebe,
1,25 cm Muskel,
0,25 cm Knochen,
3,50 cm Mark,
0,25 cm Knochen,
1,25 cm Muskel,
0,50 cm Haut- und Fettgewebe.

Die verwendeten Atomzahldichten sind in Tab. 1 gegeben. Für Haut- und Fettgewebe wurden die Werte des Standardgewebes genommen. Die Strahlenquelle ist wieder eine Punktquelle. Die Entfernung zwischen Quelle und Haut

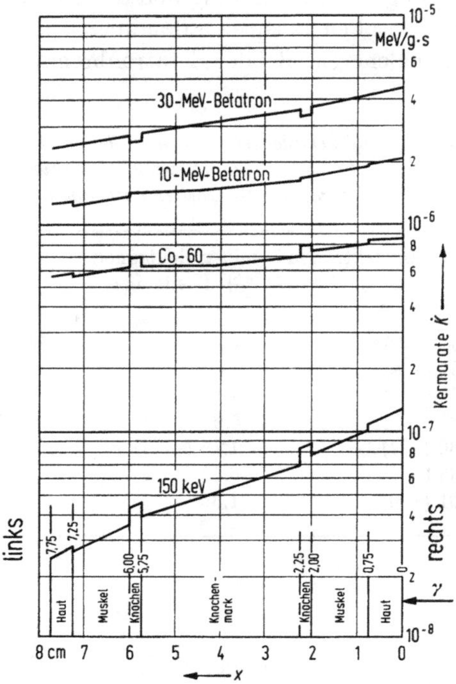

Abb. 2. Energieübertragung auf Sekundärelektronen bei Oberschenkelbestrahlung für eine Punktquelle mit 1 Photon/s

beträgt einheitlich 60 cm. In Abb. 2 sind die Ergebnisse unserer Berechnungen mit dem Programm PLK gegeben. Aufgetragen ist die Kermarate $\dot{K}$ (*Kerma* = *k*inetic *e*nergy *r*eleased in *ma*tter) in MeV pro Gramm und Sekunde für vier verschiedene Strahlenquellen

150 keV,
Co-60,
10-MeV-Betatron
und 30-MeV-Betatron.

In den beiden ersten Beispielen der 150-keV-Strahlen und Co-60-Strahlen kann der Energietransport durch Sekundärelektronen vernachlässigt werden. In diesem Fall kann man obiges Produkt, das der Energieübertragungsrate von Gammastrahlen auf Sekundärelektronen entspricht, mit der Konstanten $1{,}604 \cdot 10^{-8}$ rad/MeV g^{-1} multiplizieren, und man erhält dann wie in Abb. 1 die Energiedosisrate in rad pro Sekunde. Man beachte jedoch, daß alle Werte in Abb. 2 auf eine Quellstärke von 1 Photon pro Sekunde bezogen sind, während in Abb. 1 die Quellstärke $6{,}22 \cdot 10^{13}$ Photonen pro Sekunde betrug, was einer 1,68-kCi-Kobaltquelle entspricht.

In den beiden Beispielen der Betatronstrahlung entsprechen die aufgetragenen Kurven dem Ortsverlauf der Quellstärken der Sekundärelektronen. Diese Elektronen starten überwiegend in einem engen Richtungskegel um die Fortpflanzungsrichtung der sie erzeugenden Photonen. Durch diese Bündelung der Anfangsimpulse in Vorwärtsrichtung entsteht eine Drift der Sekundärelektronen in Vorwärtsrichtung, die eine geringe Verschiebung der Energiedosiskurve gegenüber der Quellkurve zur Folge hat. Die Berechnung dieser Verschiebung erfordert eine größere Abänderung des vorhandenen Rechenprogramms. Die Vernichtungsstrahlung der entstehenden Positronen wurde berücksichtigt. Das verwendete Quellspektrum für die Betatronstrahlung wurde aus neuen Messungen [5] entnommen.

Bestrahlung des Kopfes

Als dritte Anwendung des Rechenprogramms PLK wählten wir eine Gehirnbestrahlung, von oben nach unten nahmen wir folgende Schichten an:

0,15 cm Hautgewebe,	6,00 cm Gehirn,
1,00 cm Knochen,	1,00 cm Knochen,
0,50 cm Liquor,	0,15 cm Hautgewebe.

Die Atomzahldichten wurden aus Tab. 1 genommen. Für Haut und Liquor sind die Werte des Standardgewebes verwendet worden. Wir wählten die gleichen Quellenergien und den gleichen Abstand zwischen Quelle und Haut wie bei der

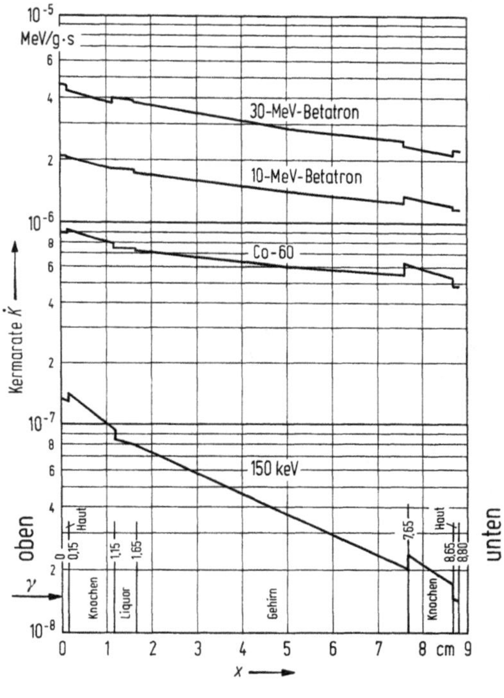

Abb. 3. Energieübertragung auf Sekundärelektronen bei Kopfbestrahlung für eine Punktquelle mit 1 Photon/s

Oberschenkelbestrahlung. Die Emissionsrate der Punktquelle ist ebenfalls auf ein Photon pro Sekunde normiert. In Abb. 3 sind die Ergebnisse der Berechnungen aufgetragen. Gegeben ist wieder die Energieübertragung auf die Sekundärelektronen als Kermarate $\dot{K}$ für die vier verschiedenen Strahlenquellen.

Im Falle der 150-keV-Strahlen und der Co-60-Strahlen kann man die Kurvenwerte mit der Konstanten $1{,}604 \cdot 10^{-8}$ rad/MeV g^{-1} multiplizieren und erhält dann die Zahlenwerte für die Energiedosisrate in rad pro Sekunde.

Für die beiden Betatronbeispiele bekommt man aus den Kurven den Ortsverlauf der Quelldichte für Sekundärelektronen, indem man die angegebenen Werte mit der Massendichte multipliziert.

Vergleicht man Abb. 2 und Abb. 3 miteinander, so sieht man, daß die Strahlenabsorption im Knochengewebe bei der weichen 150-keV-Röntgenstrahlung am ausgeprägtesten ist. Der Effekt ist klein bei Kobaltstrahlen. Er verschwindet vollständig bei den Betatronstrahlen mit einer Maximalenergie von 10 MeV und schlägt um in den entgegengesetzten Effekt bei Betatronstrahlen mit 30 MeV Maximalenergie. Das ist vor allem eine Folge der unterschiedlichen Atomzusammensetzung der Gewebearten. Solche Heterogenitätseffekte lassen sich am besten durch Simulierung des Strahlenfeldes mit dem Computer studieren. Massendichte, atomare Zusammensetzung und geometrische Konfigurationen sind wie die Quellenergie der einfallenden Strahlung Eingabeparameter, die man individuell wählen kann.

Mit den gezeigten Ergebnissen sollte darauf aufmerksam gemacht werden, daß heute Rechenprogramme zur Verfügung stehen, mit denen man die physikalische Seite der Bestrahlungsplanung weiter verbessern kann. Strahlentransportprogramme dieser Art erfordern jedoch sehr große Computeranlagen. Die wachsende Verfügbarkeit solcher Anlagen läßt erwarten, daß mehr und mehr detaillierte Optimalisierungsstudien nicht nur für technische Großprojekte wie Kernreaktoren, sondern ebenso für die Anwendung ionisierender Strahlen in der Medizin durchgeführt werden.

Herrn Prof. Dr. med. E. Szirmai danke ich für viele Diskussionen.

Literatur

1. Penny, S. K., Trubey, D. K., Emmett, M. B.: OGRE. A Monte Carlo System for Gamma Ray Transport Studies. ORNL-3805, 1966.
2. Lathrop, K. D.: DTF-IV. A Fortran IV Program for Solving the Multigroup Transport Equation with Anisotropic Scattering. LA-3373, 1965.
3. Hehn, G., Siegert, H. J.: The Development of a Multigroup Spherical Harmonics Code and its Application to Slab Geometry Shielding Problems. AERE-R 5773, 1968, S. 261–297.
4. Ritts, J. J., Solomito, E., Stevens, P. N.: Calculation of Neutron Fluence-to-Kerma Factors for the Human Body. ORNL-TM-2079, 1968.
5. Brune, D., Mattsson, S., Liden, K.: Application of a Betatron in Photonuclear Activation Analysis. AE-333, 1968.

Production and Clinical Development of a New Ideal Radioisotope of Iodine — Iodine-123

H. N. WELLMAN, V. J. SODD and J. F. MACK

Nuclear Medicine Laboratory, Division of Medical Radiation Exposure, Bureau of Radiological Health, Cincinnati General Hospital, Cincinnati, Ohio, U.S.A.

The use of radioiodine for thyroid tests still comprises some 57% of the procedures performed in Nuclear Medicine Laboratories according to the recent Stanford Research Institute Survey conducted by the Bureau for Radiological Health [1]. Wide diagnostic use of Iodine-131 for thyroid tests results in what has been considered a tolerable absorbed radiation dose to the thyroid. Certainly there has been no direct study indicating a significant deleterious effect from Iodine-131 in such use and probabilities are against any study ever conclusively demonstrating measurable permanent radiation effects. However, a recent review by HEMPLEMAN has raised the possibility that adenomas are induced by as little as a 20 rad dose of external radiation to the thyroid of young individuals [2]. As an approximate figure, an adult with a normal uptake and gland and a 100 μCi administered dose of Iodine-131 might be expected to receive 150 rad to the thyroid. Younger individuals would, of course, receive relatively larger doses because of smaller glands, but similar uptake values. Absorbed doses associated with scanning procedures with Iodine-131 then may approach the levels questioned by HEMPLEMAN as inducing adenomas. Thus, only marginal evidence would prompt one to suspect the diagnostic use of Iodine-131 as having permanent deleterious radiation effects on the thyroid, except perhaps in children. Certainly, at this time, a dose reduction motive alone would probably not warrant abandoning the convenience and economics of Iodine-131.

On the other hand, if significant advantages were inherent in the use of another radioisotope of iodine while still resulting in considerable dose reduction, then it is probable that use of the other radioisotope would be fostered. In the past, both Iodine-132 and Iodine-125 have been proposed as substitutes for Iodine-131, especially because of the inherent dose reduction associated with their use. Apparently for a number of reasons, neither radioisotope has gained wide usage, although occasional instances make the use of Iodine-125 desired over Iodine-131. MYERS, in recent years, has proposed the use of Iodine-123 as having superior theoretical characteristics for thyroid function studies, in addition to the concomitant dose reduction in its use [3, 4, 5]. The theoretical advantages of Iodine-123 are a short half-life of 13.3 hours, an average life of 19.2 hours [10], which approaches the 24 hour peak uptake of the thyroid, no beta particle emission, a 28 keV X-ray, a 159 keV gamma ray and an absorbed thyroid dose of 1—3% that of Iodine-131. Because of the potential theoretical advantages, a program of development of Iodine-123 production techniques and of clinical evaluation of Iodine-123 was undertaken to make the widespread use of Iodine-123 practical.

Methods

a) Production

Numerous techniques for the production of Iodine-123 have been used in our laboratories (Fig. 1). Iodine-123 used in the study was principally produced by the direct $^{121}Sb(\alpha, 2n)\,^{123}I$ reaction. The mean Iodine-124 contamination at time

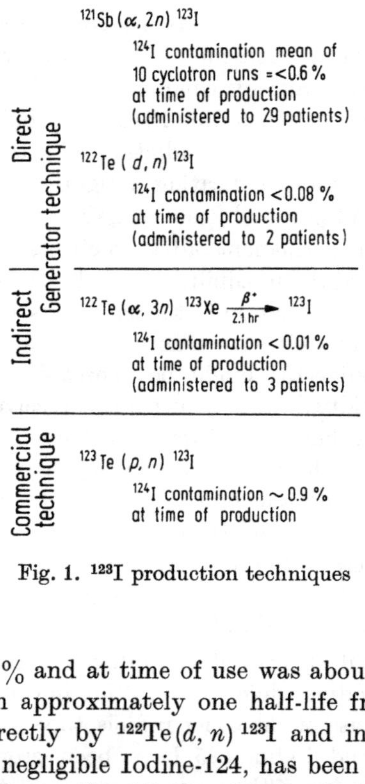

Fig. 1. ^{123}I production techniques

of production was <0.6% and at time of use was about 1.5%. The Iodine-123 was administered within approximately one half-life from time of production. Iodine-123 produced directly by $^{122}Te(d, n)\,^{123}I$ and indirectly by $^{122}Te(\alpha, 3n)$ $^{123}Xe \rightarrow {}^{123}I$, containing negligible Iodine-124, has been given to a few patients. Contamination with the longer lived Iodine-124 has been a serious consideration as, with increasing shelf life, many of the advantages of Iodine-123 can be negated, as discussed in a recent article by RHODES [6]. The indirect production technique utilizing pure Xenon-123 can eliminate this problem.

Most target materials do not have an isotopic purity to eliminate significant contamination when direct cyclotron production of Iodine-123 is used. To eliminate this problem, a unique continuous flow cyclotron generator has been developed (Fig. 2), [7]. In this generator, Helium is pumped through the target holder and then through a carbon dioxide trap which freezes out all contaminant iodine isotopes but allows Xenon-123 to pass. The pure Xenon-123, however, is further cycled through a liquid nitrogen trap where the Xenon-123 is frozen out. The trap is then removed and the Xenon-123 is allowed to decay 3—4 hours to Iodine-

123 which is then leached out in a basic solution and neutralized for patient oral administration. The Iodine-123 yield of the $^{122}Te(\alpha, 3n)\,^{123}Xe \xrightarrow{\beta+} {}^{123}I$ reaction is approximately 0.5 mCi/µA-h.

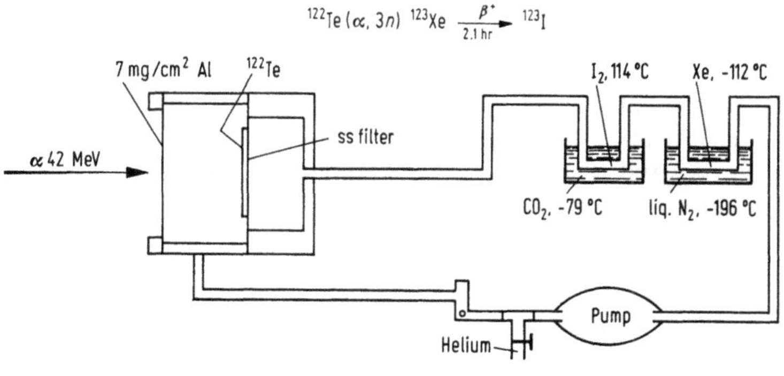

Fig. 2. Continuous flow ^{123}I generator

b) Instrument Development

In order to take advantage of the dose reduction capabilities of Iodine-123 by administering small amounts of the radioiodine for thyroid uptake studies, a special ultrasensitive dual-crystal uptake system was developed in the Radioisotope Laboratory at Cincinnati General Hospital. This system consists of two $\frac{1}{4}'' \times 3''$ sodium iodide (Tl) crystals that utilize the photopeak/scatter ratio to correct for thyroid depth (Fig. 3). The thyroid-neck phantom standardization curve for Iodine-123 (Fig. 4) shows a linear depth response with Iodine-123 as has been previously reported with Iodine-131 [8]. This system compares well with the standard thyroidal uptake systems in accuracy and allows much greater sensitivity.

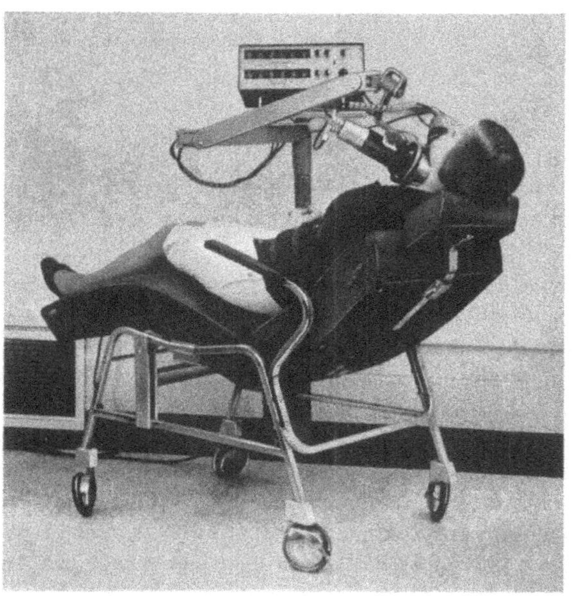

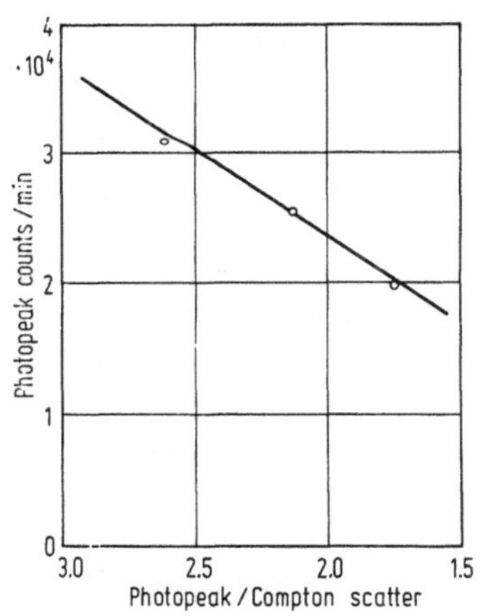

Fig. 4. Thyroid-neck phantom standardization curve (0.1 µCi ^{123}I source)

← Fig. 3. Dual-crystal thyroid uptake system

Clinical Evaluations

With Iodine-123 in hand, protocols were formulated to intercompare the clinical use of Iodine-123 with that of Iodine-131.

a) Protocol I

Protocol I (Fig. 5) was first undertaken to take advantage of the maximum dose reduction capabilities of Iodine-123. Only 50 μCi of Iodine-123 was administered to each patient so that the sensitive dual crystal uptake system could be used to advantage and, also, because the 1045-hole collimator, in conjunction with the Picker Magnascanner III, resulted in similar count rates from 100 μCi of Iodine-131 and 50 μCi of Iodine-123 administered doses. That is to say, Iodine-131 and Iodine-123 were compared on an isocount basis so as to demonstrate the

Day I administer 50 μCi ^{123}I

	at 6 hr	at 24 hr
1. } Rectilinear scan with	{ 28 keV	28 keV
2. } Brookhaven collimator	{ 159 keV	159 keV
3. Gamma scintiphoto	159 keV	159 keV
4. Uptakes		
A 1" × 1" crystal	integral count	integral count
B 1/4" × 3" dual crystal	P/S count	P/S count
C 1/4" × 3" dual crystal, separate	P/S count	P/S count
D 1" × 3" dual crystals in WBC	P/S count	P/S count

End of day II administer 100 μCi ^{131}I

	at 24 hr
1. Rectilinear scan - 31 hole picker collimator	364 keV
2. Uptake	integral count

Summary of each patient

1. Total of seven (7) scintigraphic techniques.
2. Total of nine (9) uptake procedures.

Fig. 5. Protocol I for clinical intercomparison of ^{123}I (5% ^{124}I) with ^{131}I

Systems intercompared *

^{123}I vs. ^{131}I	Correlation coefficient r	n
^{123}I RIL vs. ^{131}I RIL	0.95	18
^{123}I		
WBC vs. HIDL	0.97	17
WBC vs. RIL	0.93	17
WBC vs. HIDLS	0.93	15
HIDL vs. RIL	0.92	15
HIDLS vs. RIL	0.80	14
HIDLS vs. HIDL	0.78	15

*Key: RIL = 1" × 1" crystal
HIDL = 1/4" × 3" dual crystals
HIDLS = 1/4" × 3" dual crystals, separate signals
WBC = 1" × 3" dual crystals

Fig. 6. Correlation of ^{123}I and ^{131}I thyroid uptakes (24 hr) with separate systems

maximum absorbed dose reduction feasible with Iodine-123 thyroid imaging. Thirty patients were studied under this protocol which was composed of seven scans and nine uptake procedures on each patient.

Iodine-123 scans were performed at 6 and 24 hours with the dual 28 and 159 keV low-energies. Gamma camera scintiphotos were taken at 159 keV at both 6 and 24 hours. Iodine-123 uptakes were performed at 6 and 24 hours with four different modes of uptake systems available in the laboratory. A routine 24 hour Iodine-131 uptake and scan were initiated immediately following the Iodine-123 studies. The resultant 184 scintigraphs were then read individually by a panel of the four physicians in as blind a technique as could be devised. The individual scintigraphs on each patient were coded and arranged in random order by a "third party" to be read objectively without benefit of any information about the studies or patients. The same 1 mm light pipe was used with all rectilinear scans, such that no evident photographic difference was identifiable in the scans. The scans were then ranked in order of preference and interpreted by each of the four observers independently.

Results of the 24 hour thyroid uptakes with the previously described systems are tabulated in Fig. 6. The correlation coefficients are all high and probably not significantly different except where the HIDLS mode was used. This suggests that all systems measured the same thing, and that similar values can be used for assessing uptakes as normal or abnormal between the systems.

Results of the objective reading of the scintigraphs are tabulated in Fig. 7. The analysis of variance of the profile ranked studies with each observer's scoring reveals that there was an insignificant difference between the ranking by the four

Order of preference	1	2	3	4	5	6	7
Treatment / Observer	24 hr 364 keV ^{131}I	6 hr 159 keV ^{123}I	24 hr 159 keV ^{123}I	6 hr Gamma camera ^{123}I	24 hr Gamma camera ^{123}I	6 hr 28 keV ^{123}I	24 hr 28 keV ^{123}I
I	1.8	3.8	4.0	3.6	3.5	4.4	5.0
II	2.1	3.6	3.8	3.4	4.0	4.8	4.8
III	1.7	3.3	3.8	4.0	3.9	4.6	5.0
IV	1.4	3.4	4.0	4.1	4.4	4.3	4.9

Results:

A Observers $F(n_1=3,\ n_2=18) = 0.067$

No significant difference between observers

B Treatments $F(n_1=6,\ n_2=18) = 54.4$

Significant difference <<< $p = 0.01$

Fig. 7. Analysis of variance by two-way classification for Protocol I

observers and, furthermore, there is a highly significant difference between the seven treatments that are intercompared. All observers preferred the Iodine-131 studies with the decreasing order of preference from column 1—7; the 24 hour Iodine-123, 28 keV, studies being the least desirable. In addition, all observers were consistent in assessing the homogeneity of scans and scan diagnosis.

These data would suggest that the Iodine-131 scans were preferred. However, several points bias in favor of the Iodine-131 scans despite attempts to equalize all bias. First of all, all observers are more accustomed to reading scans utilizing the technique developed for Iodine-131. Furthermore, the "art" of scanning on the technologist's part consists of a "feel" for how to obtain good scans. This knack would have to be acquired by all technologists performing the Iodine-123 scans, and would require more than 30 cases.

More likely, however, is a degradation of scans obtained with low energy—high resolution collimators, such as the Brookhaven type (1045-hole), due to septal penetration and scattering secondary to the higher energy gamma ray emitting Iodine-124 (602 and 722 keV). At the time of administration, Iodine-124 usually was present as 1—2% of the total activity present. Analogously, at 24 hours, Iodine-124 would represent about 4—8% of the total activity.

Of note is the panel's second preference for 159 keV, 6 hour scans, when relative Iodine-124 contamination is at a minimum.

b) Protocol II

Because of the results of the first protocol, a second protocol was undertaken to elucidate the factors involvde in attempting optimal thyroid scanning with Iodine-123. The protocol (Protocol II, Fig. 8) departed from that previously reported in that a higher mean dose of about 300 μCi of Iodine-123 was employed, and the 31-hole high energy collimator utilized for Iodine-131 was also employed with Iodine-123—in addition to a 1045 low energy—high resolution collimator.

Day I administer 300 μCi ^{123}I p.o.

Rectilinear scans and collimators	at 6 hr	at 24 hr
1. 31 and 1045 hole	28 keV 159 keV	28 keV 159 keV
2. Optional gamma camera	159 keV	159 keV
3. 24 hr uptake		

Day II administer 100 μCi ^{131}I p.o.

	at 24 hr
1. Rectilinear scan - 31 hole collimator	364 keV
2. 24 hr uptake	

Totals: 25 patients with all rectilinear scans and some gamma camera images, 9 patients with some rectilinear scans

Mean dose: 290 μCi

Summary

1. Total 9 rectilinear scans
2. 2 optional gamma camera images

} Complete protocol

Fig. 8. Protocol II for clinical intercomparison of ^{123}I (5% ^{124}I) with ^{131}I

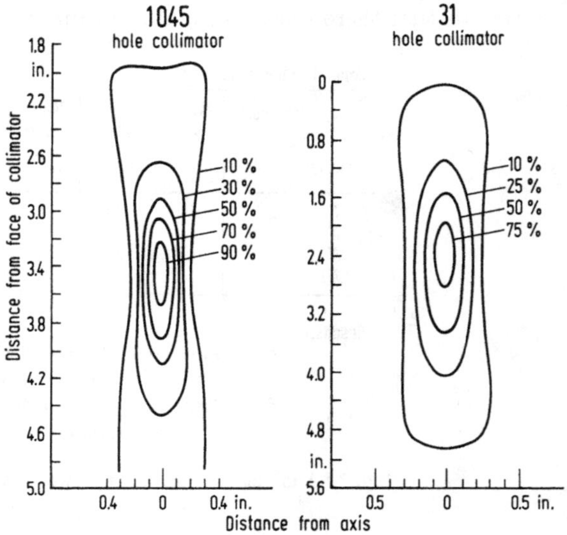

Fig. 9. Isoresponse curves ^{99m}Tc (air)

300 μCi Iodine-123 was used because such a dose in combination with the low energy collimator results in scanning conditions in most patients which require the scanner to be run at maximum speed. As seen in the protocol, both 6 and 24 hour, 28 and 159 keV scans were performed utilizing the two collimators and Iodine-123. Both collimators were used with Iodine-123 in an attempt to evaluate the possible collimator penetration of higher energy contaminants. Gamma camera images with a single pin hole collimator were optional. After completion of Iodine-123 scans on Day II, a conventional Iodine-131 thyroid scan was done with 100 μCi administered dose to each patient. A total of 34 patients were studied, with 25 patients having the 9 essential rectilinear scans used for intercomparisons.

The 31-hole collimator as used for Iodine-131 has approximately an 8 mm resolution at the 364 keV photon energy. However, the same collimator has approximately a 5.5 mm resolution at 140 keV, the approximate energy of the principle Iodine-123 gamma, as seen in Fig. 9, and compares very closely with the 5.75 mm resolution of the 1045-hole collimator [9].

The resultant nine scintiscans in each of 25 patients were all performed at the same line spacing and with the same light pipe to avoid any detectable scintiscan difference. The 225 thyroid scans plus available gamma camera images were coded and grouped by patient identification. The scans were then read independently by four experienced physicians and ranked in order of preference as well as evaluated for clinical impressions of size, shape, homogeneity, presence of nodules and adequacy for interpretations. These readings were decoded, summarized and statistically evaluated.

In Fig. 10, the ranking scores by the observers are tabulated in decreasing order of preference with the 24 hour, 159 keV, 1045-hole collimator Iodine-123 scans being first and the 24 hour, 28 keV, 1045-hole Iodine-123 scans last. A similar

Order of preference	1	2	3	4	5	6	7	8	9
Treatment / Observer	24 hr 159 keV 1.045 h.col. ^{123}I	24 hr 159 keV 31 h.col. ^{123}I	6 hr 159 keV 31 h.col. ^{123}I	6 hr 28 keV 31 h.col. ^{123}I	6 hr 159 keV 1045 h.col. ^{123}I	24 hr 364 keV 31 h.col. ^{131}I	6 hr 28 keV 1045 h.col. ^{123}I	24 hr 28 keV 31 h.col. ^{123}I	24 hr 28 keV 1045 h.col. ^{123}I
I	135	144	151	123	126	101	125	129	90
II	151	150	135	116	115	123	108	116	105
III	140	136	137	121	116	122	115	111	104
IV	144	109	114	123	119	125	122	114	133
Total rank score	570	539	537	483	476	471	470	470	432

Results:

Accepting that data are likely to have normal distribution:

A Observers $F(0.95)_{3,24} = 8.64$ $F(\text{test})_{3,24} = 1.47$

No significant difference between observers

B Treatments $F(0.95)_{8,24} = 3.12$ $F(\text{test})_{8,24} = 3.27$

Significant difference with level of risk < 0.05

Fig. 10. Analysis of variance by two-way classification for Protocol II

ranking technique was reported in the previous protocol when Iodine-131 was clearly preferred by all observers over all other treatments. In this protocol, however, Iodine-131 is not clearly superior to the Iodine-123 scans. As seen in the bottom of the figure, no significant difference was observed between observers, as was the case also in the previous protocol. The analysis of variance of the pooled data suggest a slight significant difference between the treatments that are compared only at a level of risk less than 0.05. Furthermore, no significant difference could be found when component treatments were analyzed as pairs, especially the first ranked—24 hour, 159 keV, 1045-hole collimator Iodine-123 scans versus the Iodine 131 scans. This also corresponded to the general opinion of the observers that in most cases there was very little difference between scan treatments in this study.

Most of the 28 keV scans were ranked lower than other treatments, similar to the results of the previous protocol. In most cases, three of four observers usually agreed well on their impression of size, shape, homogeneity, presence of nodules, diagnosis and adequacy for interpretation of the scans. A small number of 28 keV scans were considered inadequate for interpretation. Overall in both studies,

28 keV scans have been of little value, probably due to their high content of scattered higher energy photons.

In order to evaluate the effect of collimator septal penetration by high energy isotope contaminants, an objective study was carried out. In this study, the collimator, crystal and photomultiplier tube scanning head assembly used in all of our thyroid scans was hooked up to a multichannel analyzer and various collimators (31-hole, 73-hole, 1045-hole) were tested while viewing several isotopes. Cerium-139 was used as a mock Iodine-123 source as it has a single 166 keV gamma emission. As can be seen in Fig. 11, when using the 1045-hole collimator, a 5%

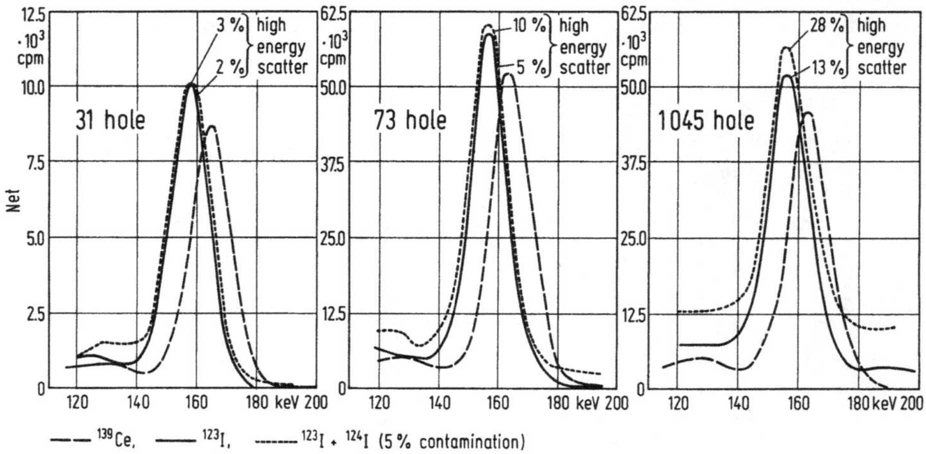

Fig. 11. Collimator system responses with 100 μCi each of ^{139}Ce, ^{123}I, and ^{123}I + ^{124}I (5% contamination)

contamination of Iodine-123 with Iodine-124 adds 28% scatter to the count rate. This scatter would degrade the resolution of scans. Therefore, Iodine-124 contamination obviates the sensitive low energy—high resolution collimator use, and the use of a high energy collimator, because of its decreased sensitivity, markedly reduces the dose advantage of Iodine-123 over Iodine-131. Iodine-124 at 5% contamination results in 3—4 times the radiation dose of Iodine-123!

The sequence of scans and gamma camera images to be used in some illustrative cases is seen in the normal patient in Fig. 12. The 28 keV and 159 keV six hours series is seen at the top with the same studies at 24 hours in the middle with the gamma camera scintiphotos and the 24 hour Iodine-131 scan at the bottom. In Fig. 13 is seen a patient with Graves' Disease. Next, Fig. 14, is the heterogeneous uptake seen in a multinodular goiter. This can be compared in Fig. 15 to a single cold nodule which is particularly well demonstrated in the 159 keV 1045-hole collimator scans. In Fig. 16, an autonomous nodule is depicted. These five examples give an opportunity to sample the intercomparison of Iodine-123 with Iodine-131. Also of interest in Fig. 17 is the case of a hypothyroid child with a congenital goiter. The low uptake made it undesirable to administer Iodine-131 and scans were only obtained utilizing Iodine-123, representing one of the nine cases not having undergone the complete protocol of scan studies.

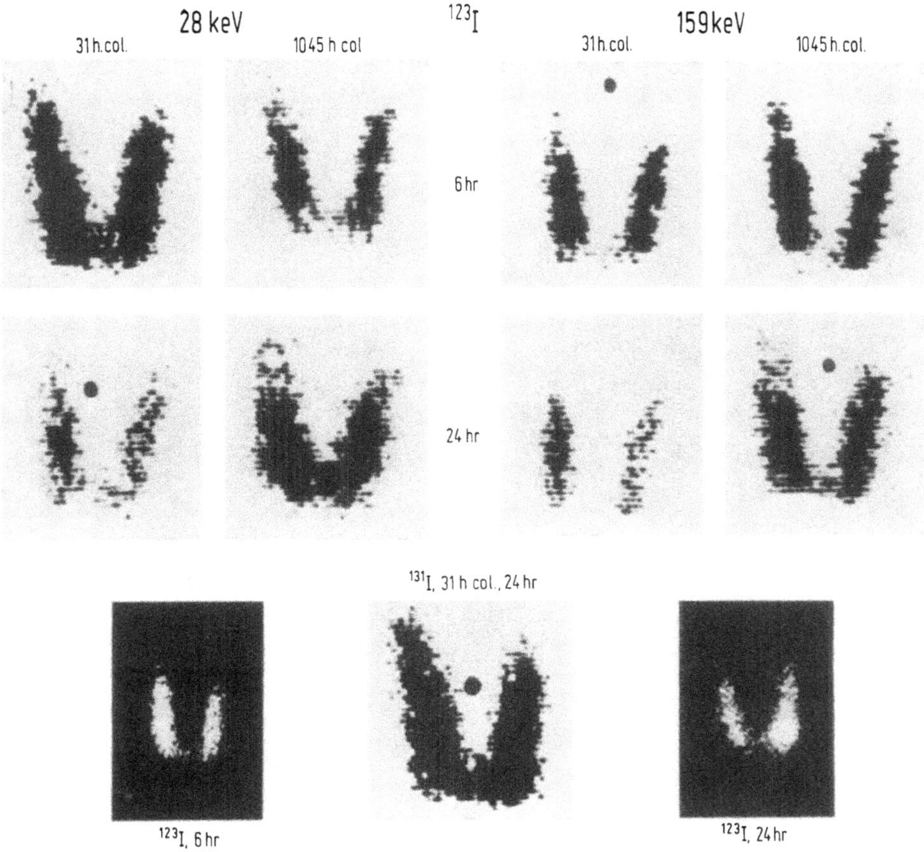

Fig. 12. Normal 24 hour uptake 38%

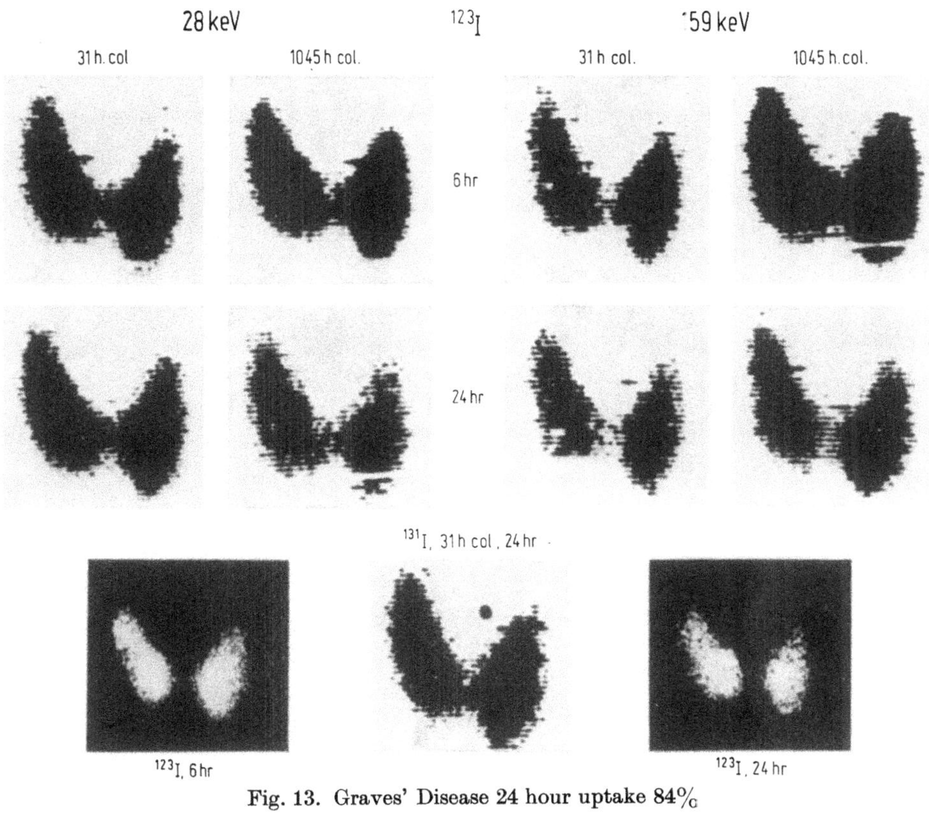

Fig. 13. Graves' Disease 24 hour uptake 84%

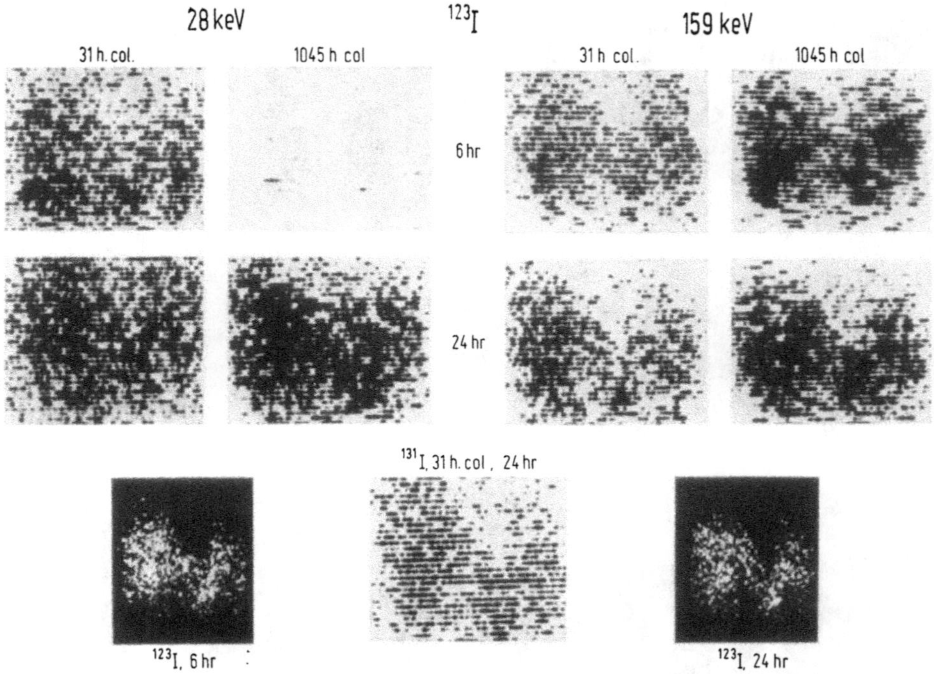

Fig. 14. Multinodular goiter 24 hour uptake 17%

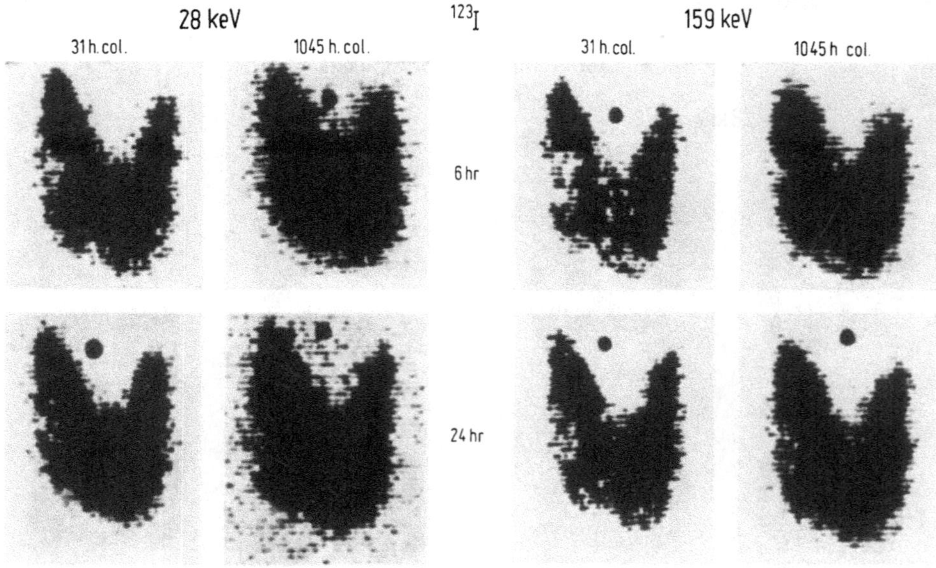

Fig. 15. Cold nodule 24 hour uptake 23%

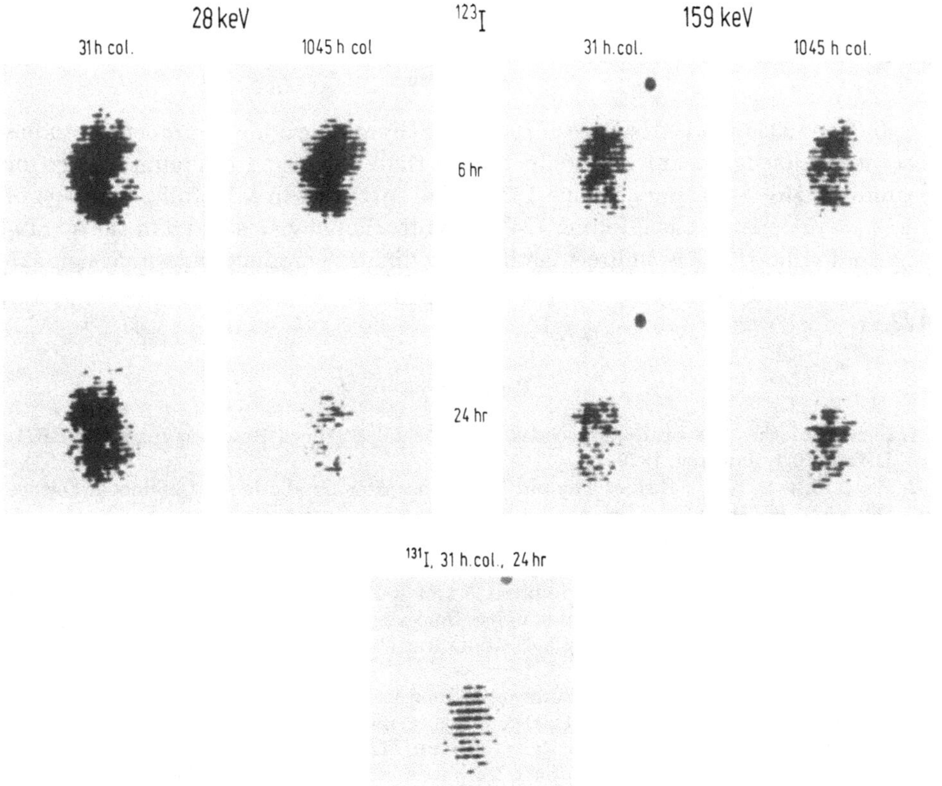

Fig. 16. Autonomous nodule 24 hour uptake 25%

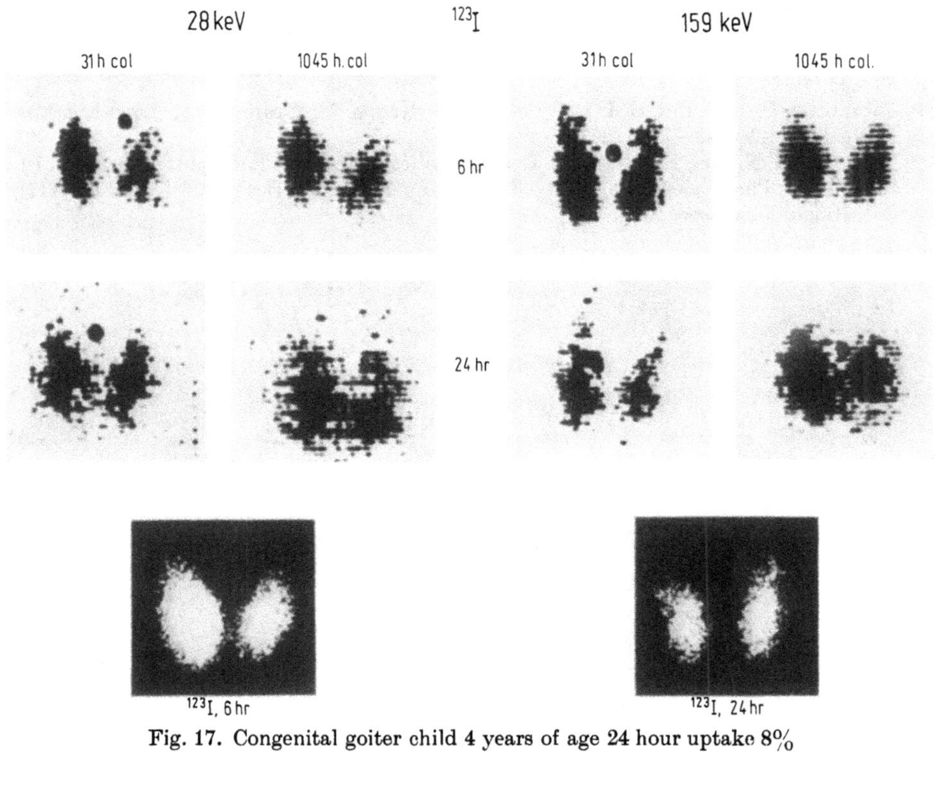

Fig. 17. Congenital goiter child 4 years of age 24 hour uptake 8%

Summary

Iodine-123 is an ideal isotope of Iodine for thyroid imaging and function studies because of its significant dose reduction to the patient and optimum energy for scanning. However, pure Iodine-123 is the only way to take full advantage of these characteristics since Iodine-124 contamination obviates many of the advantages of Iodine-123. The indirect method of Iodine-123 production from Xenon-123 appears to be the only practical production solution to obtaining "pure" Iodine-123.

References

1. Survey of the Use of Radionuclides in Medicine. Stanford Research Institute, BRH/DMRE 70-1, January, 1970.
2. Hempleman, L. H.: Risk of Thyroid Neoplasms after Irradiation in Childhood. Science **160,** 159–163 (1968).
3. Myers, W. G., Anger, H. O.: Radioiodine-123 (abstract). J. Nucl. Med. **3,** 183 (1962). Conference reported in Nucleonics **20**(10), 102 (1962).
4. —: Comparisons of Iodine-131, Iodine-125 and Iodine-123 for In Vitro Applications in Diagnosis. In: Transactions of the VIIth International Congress of Internal Medicine, Munich 1962, Vol. 2, pp. 858–863, E. Wollheim and B. Schlegel (Eds.), Stuttgart: Georg Thieme.
5. —: Radioisotopes of Iodine. In: Radioactive Pharmaceuticals, pp. 217–243, AEC Symposium Series, No. 6 (CONF-651111), Oak Ridge, Tennessee, 1965.
6. Rhodes, B. A., Wagner, H. N., Jr., Garrard, M.: Iodine-123, Development and Usefulness of a New Radiopharmaceutical. Isotopes and Radiation Technology, Vol. 4, No. 3, Spring 1967.
7. Sodd, V. J., Blue, J. W.: Cyclotron Generator of High Purity Iodine-123 (abstract). J. Nucl. Med. **9,** 349 (1968).
8. Wellman, H. N., Kereiakes, J. G., Yeager, T. B., Karches, G. J., Saenger, E. L.: A Sensitive Technique for Measuring Thyroidal Uptake of Iodine-131. J. Nucl. Med. **8,** 86–96 (1967).
9. Richards, P., Atkins, H. L.: A Collimator System for Scanning at Low Energies. J. Nucl. Med. **8,** 143–153 (1967).
10. Wagner, H. N., Jr., Emmons, H.: Characteristics of an Ideal Radiopharmaceutical. In: Radioactive Pharmaceuticals, pp. 1–32, AEC Symposium Series, No. 6 (CONF-651111), Oak Ridge, Tennessee, 1965.

A "Proportion Graph" Method for Measuring Binding Systems*

J. P. Raynaud

Unité de Recherches sur le Métabolisme Moléculaire et la Physio-Pathologie des Stéroides de l'Institut National de la Santé et de la Recherche Médicale, Laboratoire Hormones, Bicêtre (Val de Marne), France

The "specific" binding of small ligands to non-enzymatic proteins, for example transport proteins, target cell receptors and repressors, has been increasingly investigated over the last few years. This has been the case for certain hormonal steroids which not only bind these proteins with great affinity ($K \geq 10^8\,M^{-1}$), but also, to a lesser extent, many other non-specific binding proteins (of unknown significance) which are often present in the biological sample under study. During the analysis of a specific binding protein in such a mixture, non-specific binding is frequently, if not always, encountered and raises definite practical problems. For, even though the affinity constant of non-specific binding is relatively low, the number of non-specific binding sites may be large and it may therefore be impossible to detect the presence of a specific binding amid the total binding in the mixture.

A new graphical technique, called "proportion graph", has been developed which enables the number of binding sites and the intrinsic association constant of the specific binding system (S) to be determined, at equilibrium, even in the presence of a non-specific binding system[1].

This method is suitable for a computerized statistical analysis of data for either 1 or 2 specific binding systems (S or $S_I + S_{II}$), or 1 specific and 1 non-specific binding system ($S + NS$) or 2 specific and 1 non-specific binding systems ($S_I + S_{II} + NS$). The only convenient way of running the necessary number of experiments is by using a radioactive ligand of high specific activity.

General Comments and Nomenclature

This graphical technique is based on several assumptions:

1. the binding systems are independent,
2. all sites of each system are identical,
3. there is no cooperative effect,
4. each system obeys the law of mass action and
5. protein does not affect the activity of the ligand except by binding.

* Partial support has been obtained from the Centre National de la Recherche Scientifique, la Délégation Générale à la Recherche Scientifique et Technique, la Fondation pour la Recherche Médicale Française, the Ford Foundation and Roussel-Uclaf.

[1] Details may be found in references [1—3].

At equilibrium:

$$(B) = \sum_{i=1}^{n} \frac{(N_i)\,(U)}{1/K_i + (U)}$$

where N denotes the total number of sites, U the unbound ligand, B the bound ligand, i the index of each set of binding sites and K the intrinsic association constant. All concentrations (...) are in moles/litre and K is in M^{-1}. In the present discussion, neither protein concentration nor the number of sites per molecule of binding protein will be taken into consideration.

Experimental Procedure

The binding system may be either a crude or partially purified preparation. One should verify that during the experiments the ligand remains chemically unchanged and that no covalent link is formed.

Binding is measured by equilibrium dialysis using the sac technique which allows the largest surface of contact between the protein solution inside the sac and the outer solution and which enables the outer volume to be increased, as is necessary in certain competitive experiments.

In an actual determination, the ligand is placed in the outside compartment and, at equilibrium, its concentration in the sac (T) is given by:

$$(T) = (B) + (U).$$

Binding analysis does not necessitate *per se* the use of radioisotopic material but, in practice, radioisotopic ligands of high specific activity ($\geq$30 C/mmole) are almost always used, owing to the very low concentration of ligand needed in such experiments (range $10^{-12} - 10^{-6}$ M).

The radioactive ligand concentration in the outside compartment is (a) giving (α) at equilibrium in the dialysis sac. The ligand concentration may be increased by isotopic dilution, i.e. by adding to the outer solution various concentrations of non-radioactive ligand (A) becoming $(\mathscr{A})$ in the sac at equilibrium. One can thus easily determine the total concentration in the sac at equilibrium, (T), which is equal to $(\alpha) + (\mathscr{A})$; since the specific activity $\frac{(^*a)}{(a) + (A)}$ is known, and since the radioactivity in the sac (α) can be determined, $(\alpha) + (\mathscr{A})$ may be derived from:

$$\frac{(^*a)}{(a) + (A)} = \frac{(^*\alpha)}{(\alpha) + (\mathscr{A})} \quad (^* \text{ means radioactivity}).$$

The bound and the unbound ligand are expressed in terms of their proportion to any concentration of total ligand (T); b and u are respectively given by $(B)/(T)$ and $(U)/(T)$. When using radioactive material, they are obtained directly from radioactivity determinations:

$$b = \frac{(^*B)}{(^*\alpha)} \text{ and } u = \frac{(^*U)}{(^*\alpha)}\,.$$

Furthermore, for each value of (T), $b + u = 1$.

Binding Systems

1. One Specific System (*S*)

In a "proportion graph", log *b* (or log *u*) is plotted vs. log (*T*) (Fig. 1). The method is described in detail here so that its use in the more complex binding situations, in fact in precisely those cases where it is most useful, can be clearly understood.

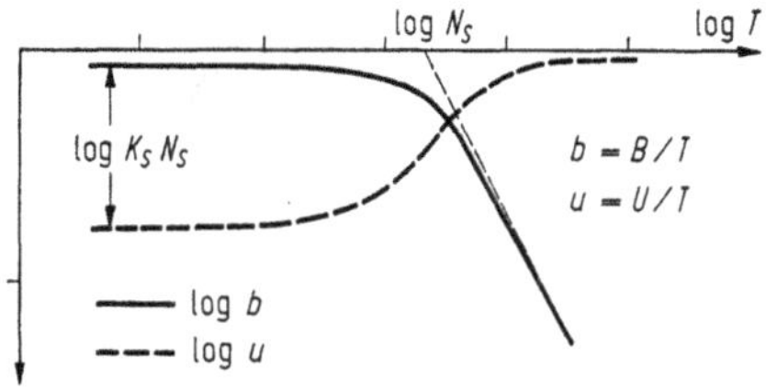

Fig. 1. Proportion graph for a single specific binding system (*S*)

The scale of the ordinate can be expanded to give a better picture of the shape of the curve. When *T* increases and all the sites are occupied, log *b* follows the linear dilution function log (T) — log (N_S) (of slope -1 if the scales of the ordinate and the abcissa are identical); by extrapolating log *b*, log (N_S) is obtained on the abcissa. It is obvious that appropriately low concentrations of ligand are or are not obtained depending on the specific activity of the tracer α. In order to obtain an initial plateau before saturation of the sites and therefore K_S close to its true value, the amount of radioactive tracer α should be as small as possible, giving $(U) < 1/K_S$.

When $(U) \to 0$, $\log b - \log u$ becomes approximately equal to $\log K_S(N_S)$. By determining log (N_S) and $\log K_S(N_S)$, K_S may be calculated.

When (T) increases, log *u* approaches zero.

The values obtained graphically are used as initial estimates for minimization and statistical analysis.

From an experimental point of view, the "proportion graph" can naturally be used for any binding determination not necessarily involving radioactive material, but only provided that *b* and *u* are determined at suitable concentrations.

Competition Studies

When increasing amounts of a competitor are added to a fixed amount of ligand, the quantity of bound ligand decreases. If the concentration of the competitor is high, the ligand should be almost completely displaced. The intrinsic affinity constant of the competitor, K_c, can be obtained by determining the constant K_S as before and then the apparent association constant of the ligand K_{app}, in the presence of a given amount of competitor (Fig. 2).

K_c is thus given by the classic equation $K_c = (K_S/K_{app} - 1)\cdot 1/C$, C being the free competitor concentration. The result is checked firstly by using different concentrations of competitor, secondly by carrying out binding experiments using the two components separately. The ligand and the supposed competitor should give the same number of sites N_S, unless the competitor binds to sites other than those to which the ligand binds.

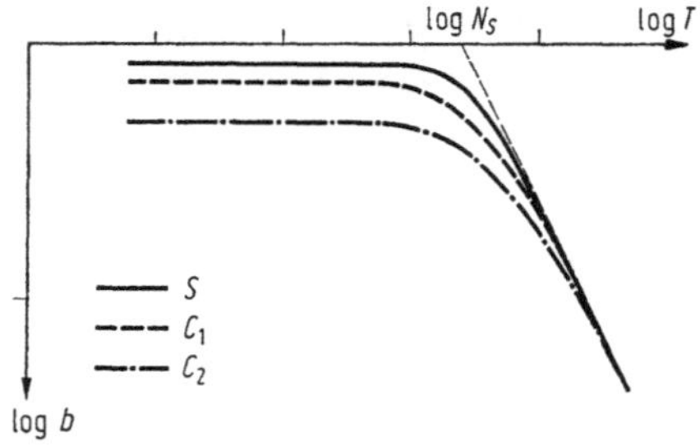

Fig. 2. Competitive binding curves for a single specific binding system as represented by the proportion graph

2. One Non-Specific System (NS)

Qualitatively, the binding of a hormone by proteins of a tissue non-responsive to this hormonal compound is considered non-specific, as opposed to the binding of the active hormone by a protein found only in the corresponding target tissue, which is said to be specific.

Quantitatively, the affinity constants of specific binding proteins are correlated to their possible efficiency at physiological concentration of the hormonal ligand. Non-specific binding proteins on the other hand only bind slightly at physiological hormone concentrations in spite of an overall larger number of sites.

In fact, a high affinity constant and a relatively small number of sites are characteristics of specific binding proteins, whereas non-specific binding proteins will be defined as those having relatively low affinity constants and usually rather large concentrations of binding sites.

It is easy to demonstrate that, when K_{NS} is small or when (N_{NS}) is large, $(B_{NS}) \cong K_{NS}(N_{NS})\,(U)$.

The binding is proportional to (U) as if there were a "partition system" of coefficient $K_{NS}(N_{NS})$. The representation on the "proportion graph" of $\log b = f(\log (T))$ is parallel to the abcissa (Fig. 3) and $\log K_{NS}(N_{NS})$, "e_1", is obtained

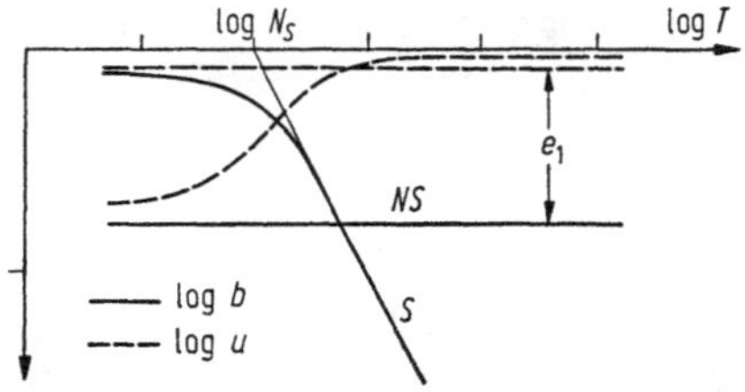

Fig. 3. Comparison of a specific binding system with a non-specific one by the proportion graph

from $\log b - \log u$; this value is independent of (T). In graphical terms, the "partition system" is in fact an approximation, since it represents the tangent at the origin of the hyperbola of the non-specific binding. It should be emphasized that the value $K_{NS}(N_{NS})$ may include several non-specific bindings of different capacity and affinity.

3. One Specific System and One Non-Specific System together $(S + NS)$

In this case,

$$(B) = (B_S) + (B_{NS}) = \frac{(N_S)(U)}{1/K_S + (U)} + K_{NS}(N_{NS})(U).$$

In the "proportion graph" (Fig. 3), the curve $\log b$ is seen to have three distinct parts. When (T) is small, b_{NS} is small and $\log b$ approaches $\log b_S$.

When (T) increases, b_S becomes small with respect to b_{NS} and $\log b$ then approaches b_{NS}. Therefore $\log b$ is nearly asymptotic to $\log b_S$ at $(T) \to 0$. In between one finds a curve which is influenced by both the values of b_S and b_{NS}. These characteristics can be easily understood from Fig. 3 which represents the two kinds of binding separately.

Fig. 4 shows the S shaped curve for $\log b$ characteristic of an $S + NS$ system. The representation of $\log b_{NS}$ is no longer a straight line: the non-specific binding is relatively small at low (T) due to the presence of the specific binding system. One should also note that $\log b$ never has a slope -1 since it equals $b_{NS} + b_S$, and that b_{NS}/b_S increases with (T).

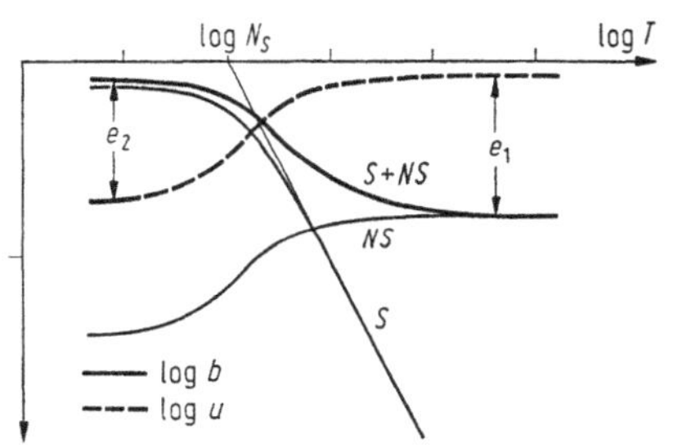

Fig. 4. Proportion graph representation of a specific binding system in presence of a non-specific one

The binding parameters of this system can be evaluated by a graphical method using the characteristics of the curve as demonstrated below.

Consider the limits of $\log b - \log u$ when $T \to \infty$ and $T \to 0$.

$$\lim_{T \to \infty} \log b - \log u = \log K_{NS}\,(N_{NS}) \text{ ("}e_1\text{" on Fig. 4)},$$

$$\lim_{T \to 0} \log b - \log u = \log\,(K_S(N_S) + K_{NS}(N_{NS})) \text{ ("}e_2\text{" on Fig. 4)}.$$

From the limits of $\log b - \log u$ one can therefore determine $K_S(N_S)$ and $K_{NS}(N_{NS})$. K_S is then obtained by a graphical construction.

The parallel to the abcissa at point $b = h_1$ or h_2 where

$$\log h_1 = \log \frac{K_S(N_S)/2 + K_{NS}N_{NS}}{1 + K_S(N_S)/2 + K_{NS}N_{NS}},$$

$$\log h_2 = \frac{1}{1 + K_S N_S/2 + K_{NS}(N_{NS})}$$

will intersect $\log b = f(\log(T))$ at a point the abcissa of which is

$$\log 1/K_S(K_S(N_S)/2 + K_{NS}(N_{NS}) + 1)$$

from whence K_S may be evaluated.

With these approximations, the best solution and its limits may be calculated by means of regression and statistical analysis using a computer.

The possibility of detecting a specific binding in the presence of a non-specific one depends on the relative sizes of $K_S(N_S)$ and $K_{NS}(N_{NS})$ (see Fig. 5).

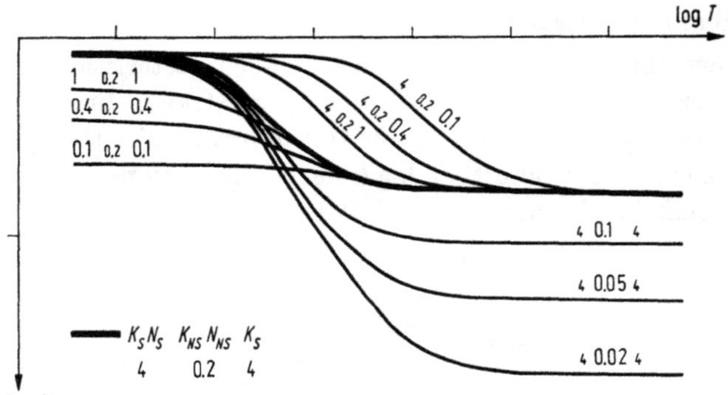

Fig. 5. Influence of the binding parameters on the shape of the curve in an $S + NS$ system

Competition Studies

Competition for the non-specific binding cannot be observed since it is unsaturable. The competition in the specific system may however give rise to a new apparent association constant $K_{\text{app}S}$ from which the competitor affinity may be calculated.

4. Two Specific Systems together ($S_I + S_{II}$)

In Fig. 6 are drawn theoretical curves with arbitrary parameters on the basis of which it is possible to analyse two different sets of specific binding sites. As can be seen, it is only in certain well-defined situations that the two sets of binding can be differentiated. The graphical evaluation takes into account the fact that $\lim_{(T) \to 0} \log b - \log u = \log\left(K_{SI}(N_{SI}) + K_{SII}(N_{SII})\right)$ and that when $(T) \to \infty$,

$(B) \rightarrow (N_{SI}) + (N_{SII})$. On the basis of these limits, the 4 binding parameters may be determined by carrying out a regression analysis on the middle part of the curve.

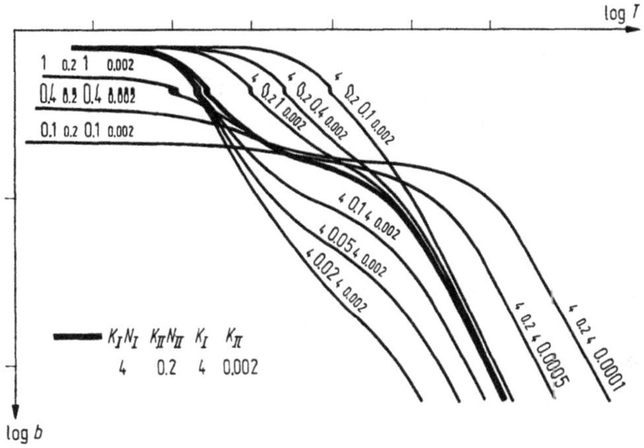

Fig. 6. Influence of the binding parameters on the shape of the curve in an $S_I + S_{II}$ system

In this as well as in the following case, competition studies are difficult to carry out, but may be informative. Competition obviously depends on the relative sizes of the two binding systems.

5. Two Specific Systems and One Non-Specific System together ($S_I + S_{II} + NS$)

In Fig. 7 are drawn theoretical curves with the same parameters as used for $S_I + S_{II}$. The features of the simpler system are still present, but one can observe

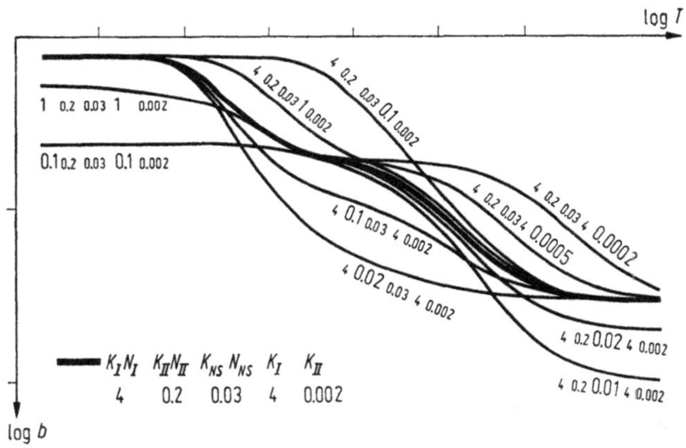

Fig. 7. Influence of the binding parameters on the shape of the curve in an $S_I + S_{II} + NS$ system

degeneration of the general system into various combinations resembling $S_I + S_{II}$ or $S + NS$ or even simply S. There are situations, however, where the triple binding may be clearly represented graphically.

Mathematical and Statistical Analysis

The rough estimates obtained for the parameters by the graphical method enable these non-linear functions to be linearized by means of a Taylor expansion. The best fit of the experimental data to an assumed model is then obtained by means of a least squares method. Following this, the standard errors of the parameters are computed by a matrix method and the adequacy of fit is checked by statistical analysis. This is in fact done by an IBM 360 computer using a Fortran program.

Results

This method has proved to be extremely useful in differentiating specific and non-specific binding in quantitative terms when studying plasma and tissue steroid binding proteins.

For example the $S + NS$ system has been successfully applied to the binding of estrogenic compounds to a 105 000 · 90 g · min supernatant of prepuberal sow endometrium or rat uterus homogenate and the S system to that of Progesterone or Corticosterone to a purified protein (corticosteroid binding protein CBG). The results when using Testosterone as a ligand will be recorded here as an illustration of the possible applications of the method (see Table 1).

Table 1. *Binding of Testosterone*

Material:	SBP	SBP + CBG	Plasma (dilution 1/40) SBP + CBG + albumin		
Model:	S	$S_I + S_{II}$	$S + NS$	$S_I + S_{II}$	$S_I + S_{II} + NS$
K_{SI} (10^9 M^{-1})	1.2 ± 0.1	1.4 ± 0.2	0.71 ± 0.14	0.88 ± 0.15	1.9 ± 0.5
N_{SI} (10^9 M)	4.5 ± 0.5	9.0 ± 2.0	2.1 ± 0.4	1.7 ± 0.3	0.6 ± 0.2
K_{SII} (10^6 M^{-1})		30 ± 10		0.1 ± 0.05	33 ± 18
N_{SII} (10^6 M)		22 ± 5		6700 ± 3300	14 ± 6
$K_{NS} N_{NS}$			0.59 ± 0.03		0.50 ± 0.03
Residual variance			0.010	0.009	0.0025

The binding of Testosterone to a pure protein extracted from human plasma Cohn fraction IV (sex binding protein SBP) fits the model S. In the case of a less well purified fraction (containing CBG), the $S_I + S_{II}$ model is necessary and for the plasma itself the data will only fit the model $S_I + S_{II} + NS$ (Fig. 8).

A comparison of the values obtained for K_{SI} and K_{SII} in these experiments shows that an appropriate model is necessary to fit the data. Furthermore, the method is found to be valid in different experimental situations.

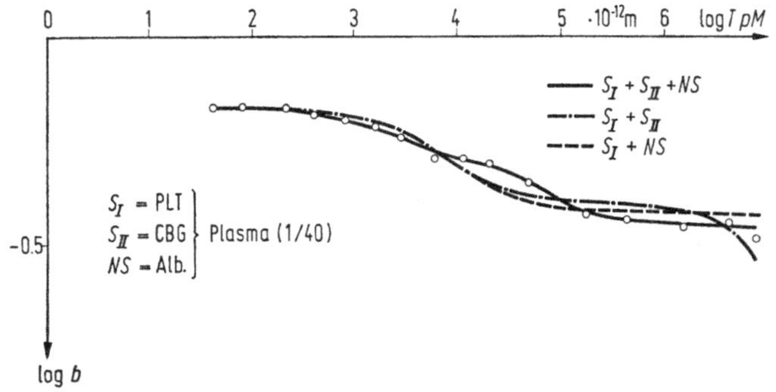

Fig. 8. Binding of Testosterone to 1/40 diluted plasma at 4 °C. The three models are represented in order to deduce the only one which fits the data fairly well

Discussion

Although it would seem that a check by varying the number of sites would validate the method, it can be seen from Table 1 that this test is not rigorous enough to distinguish the appropriate model from amongst the unsuitable ones without carrying out a statistical analysis.

To analyse binding, several orders of ligand concentration must be used. Transformation of binding data in order to obtain linearity results in a loss in information. Besides, with the traditional methods of plotting (LINEWEAVER and BURK, EADIE or SCATCHARD) linearity is only achieved with a single binding system and even then the weight of the points is much distorted by the transformation (cf. the reciprocal plot of LINEWEAVER and BURK).

The analysis of several binding systems is sometimes possible with the Eadie and Scatchard plot, but the method is tedious and imprecise. The nonlinearity of this plot can however be used to verify the presence of two or more binding systems.

When S and NS systems are present together, the above graphical methods neither adequately represent the data nor bring out significant differences.

In the case of the Lineweaver-Burk plot, when there is a second binding system with a large number of sites present, the deviations from the straight line are difficult to observe, especially in the region $1/U \simeq 0$ and are therefore impossible to calculate by statistical analysis. In the case of the Scatchard plot, if non-specific binding is present, B becomes too large for an accurate representation.

The proposed "proportion graph" on the other hand has the advantage that it enables all the characteristics of the binding to be detected over a large range of ligand concentrations and also the binding of the ligand to particular sets of binding sites to be predicted for a given total ligand concentration.

It is rather fortunate that, considering all the assumptions made in section I, one can obtain identical results when comparing protein affinity as studied on the pure, purified and crude material. This indicates that the principle of the analysis might be correct and that its successful application depends only on the relative amounts of the various binding systems present and on the relative strength of their affinities.

The method can also be used to follow the stages of purification of a protein. Furthermore, the detection of 2 different sets of sites on the same protein molecule is possible since, if 2 sets are present, the ratio N_{SI}/N_{SII} remains constant at each purification step.

References

1. BAULIEU, E. E., RAYNAUD, J. P.: Eur. J. Biochem. **13**, 293 (1970).
2. — —: Symposium on the Synthesis and Use of Labelled Lipids and Sterols, Milan 1968. Progr. Biochem. Pharmacol. **5**, 46 (1969).
3. LEBEAU, M. C., MERCIER-BODARD, C., OLDS, J., BOURQUIN, D., BRÉCY, T., RAYNAUD, J. P., BAULIEU, E. E.: Ann. Endocrinol. **30**, 183 (1969).

C^{11}-Aktivierung mit dem Betatron oder Linearbeschleuniger und die mögliche biologisch-medizinische Verwendung als Tracer und Scansubstanz

Activation of C^{11} by Betatron or Linear Accelerator and its Potential Biological and Medical Uses as Tracer and Scanning Substance

D. MÖRSDORF, S. WUNDER, H. SCHWALL, H. BEHSCHAD und P. SCHENK

Czerny-Krankenhaus, Heidelberg*, und Institut für Chemie der Universität Saarbrücken**

Summary

A potential new biophysical tracer is described: ^{11}C, which has a half-life of 20.5 min with 0.5 MeV gamma radiation and can be employed with existing equipment and drugs in common use.

Substances containing ^{12}C are activated by means of gamma rays of a threshold energy of at least 19 MeV, produced by betatron or linear accelerator. There is difficulty in incorporating sufficient tracer into a non-toxic quantity of a physiological substance. The work was supported by animal experiments and chemical analyses.

Einleitung

Kurzlebigere Isotope (z. B. C^{11}, $T_{1/2} = 20{,}5$ min, 0,5 MeV γ-Strahlung) ermöglichen mittels Funktionsdiagnostik und Scanbild zusätzliche Untersuchungen am Menschen bei geringerer Dosisbelastung.

Dosis der (γ, β)-Isotopen

In der biochemischen Forschung werden unter anderem C^{14}-Verbindungen (β-Strahler) als Tracer (PASCHKE) im Stoffwechsel benutzt. Daneben ist die Nuklearmedizin, um Isotopenaussagen von Organen zu erhalten (z.B. J^{131} für die Schilddrüsen-Funktionsprüfung oder Schilddrüsen-Scandarstellung, Au^{198} für das Leberscanning usw.) auf die γ-Strahlung von diesen Isotopen angewiesen (BROWNELL-SWEET, KNIPPING et al.). Da diese Isotope eine recht lange Halbwertszeit haben, ist die verabreichte Dosis im Organismus durch die Isotope leider recht hoch (Säuglings- und Kinderscanning sowie Placentascanning sind wegen des Strahlenschadens daher nur bedingt möglich).

Wie die genauere Analyse der Dosis zeigt, ist die Strahlendosis der Halbwertszeit $T_{1/2}$ der Isotopen (s. (1)) proportional. Eine möglichst geringe Halbwertszeit hat das Isotop C^{11}, dessen Erzeugung unten genauer beschrieben wird. Den Unterschied der Strahlenbelastung erkennt man klar durch das folgende Beispiel von

* D. MÖRSDORF (z. Z. beurlaubt), S. WUNDER, H. BEHSCHAD, P. SCHENK.

** H. SCHWALL.

C^{11} und Au^{198} bei einer Leberdarstellung. Nach GLOCKER-MACHERAUCH gilt in 1. Näherung für die Dosis

$$D^{\infty}_{\gamma+\beta}(\text{rad}) = D^{\infty}_{\beta} + D^{\infty}_{\gamma} = \frac{73{,}9}{24}\bar{E}_{\beta}\, Q_0^{*}\, T_{\text{eff}} + 1{,}41\cdot 10^{-3} Q_0^{*}\, T_{\text{eff}}\, \Gamma \varrho_0 g_p. \qquad (1)$$

E in MeV, T_{eff} in h, Q_0^{*} in µCi/g, Γ in Rcm²/hmCi, ϱ_0 in g/cm³, g_p (Parameter). Die Vergleichswerte sind in der Tab. 1 für Au^{198} und C^{11} zusammengestellt (100 µCi in 1 kg H_2O ≙ Leberdarstellung).

Tabelle 1

Isotop	Q_0^*	$T_{1/2}$	T_{eff}	$\bar{E}_\beta$	E_γ	Γ	ϱ_0	g_p	D_β^∞	D_γ^∞	$D_\beta^\infty + D_\gamma^\infty$
Au^{198}	10^{-1}	2,7 d	2,7 d	0,3	0,4	2,35	1	~80	6,5	1,8	8,3 rad
C^{11}	10^{-1}	20,5 min = 0,014 d	0,014 d	0,38	0,5	~3	1	~80	0,04	2·0,01[1]	0,06 rad

Das Verhältnis 8,3 : 0,06 ≈ 140 beweist die enorme Strahlendosisreduzierung des C^{11} gegenüber Au^{198} (für J^{131} usw. verfahre man ebenso!).

C^{11}-Erzeugung

Wie können wir nun das radioaktive Isotop C^{11} in Molekülverbindung herstellen? Ein Weg, der über ca. 20 Jahre bekannt ist, möge ausführlich erwähnt werden (BASIL et al., FROST et al., FINCK et al., HORSLEY et al., MATTHEWS et al.). In welcher Verbindung Kohlenstoff auch vorliegt, als Graphit oder Molekülverbindung in Festsubstanzen oder Lösungen (z.B. 1 Ampulle Na-saures Benzoat), ist für den Vorgang der Aktivierung primär ohne Bedeutung. Der C^{12}-Gegenstand wird einer Bestrahlung durch γ-Quanten (mindestens 19 MeV-Schwellenenergie eines Betatrons oder Linearbeschleunigers) ausgesetzt. Nach einer bestimmten Zeit enthält die Probe (gewünschte Substanz) nach folgendem Schema die gesuchte Aktivität Q (Ci). Das Schema lautet:

$$C_6^{12}(\gamma, n)\, C_6^{11} \xrightarrow[-e^+]{T_{1/2} = 20{,}5\ \text{min}} B_5^{11} + 2\cdot 0{,}5\ \text{MeV}\ \gamma\text{-Strahlung}. \qquad (2)$$

Die Aktivität Q des C^{11} ergibt sich rechnerisch bei einem monoenergetischen Fluß Φ nach der Gleichung

$$\frac{dN}{dt} = n\Phi\sigma - \lambda N \quad (\text{GLOCKER-MACHERAUCH}). \qquad (3)$$

Nach τ min Bestrahlungszeit ist die spezifische Aktivität

$$Q^* = Q/m = 1{,}6\cdot 10^{-11}\frac{\Phi\sigma}{A}\,(1 - e^{-0{,}7\tau/T_{1/2}})\ \text{Ci/g} \qquad (4)$$

bzw.

$$Q^* = \alpha(\tau)\, B(\tau) \qquad (5)$$
$$(\alpha(\tau) \approx 16 - 18\ \mu\text{Ci/g})$$

bei unserer Anordnung ± 20%[2].

[1] Durch 2γ-Quanten bedingt. [2] 15 mCi ± 50% bei Linearbeschleuniger.

Wie man aus der Formel (5) ersieht, wird die spezifische Aktivität durch zwei Faktoren $\alpha(\tau)$ und $B(\tau)$ bestimmt. $\alpha(\tau)$ ist eine geräteabhängige Variable, die während 10—30 min Bestrahlungszeit bei max. Dosisleistung 1500—1800 r/min (42 MeV-Betatron) ungefähr auf 20% konstant bleibt. $B(\tau)$ ist in der Tab. 2 aufgeführt. Daraus ergibt sich, daß eine Bestrahlungszeit von 20,5 min die Hälfte der maximal erreichbaren Aktivität ergibt. Auch die Gerätebelastung ist während dieser Zeit nicht zu groß.

Tabelle 2

(min)	$T_{1/2}(C^{11})$ (min)	$\tau/T_{1/2}$	$e^{-0,7\tau/T_{1/2}}$	$B(\tau)\cdot 100\%$
10,25	20,5	$\frac{1}{2}$	0,7	30
20,5	20,5	1	0,5	50
41,0	20,5	2	0,25	75
61,5	20,5	3	0,12	87

Wenn wir also 1 g C^{12} 20,5 min bestrahlen, erhalten wir bei dieser maximalen Dosisleistung (rad/min) 8—9 μCi/g (Abb. 1). Um mit diesen niedrigen spezifischen Aktivitäten Scanbilder herzustellen, benötigt man bei den gewöhnlich gewünschten Aktivitäten von 20—500 μCi mindestens ca. 10 g Substanz ($\triangleq$40 μCi). Nur wenige physiologische Substanzen sind indes bekannt, um in dieser hohen Dosierung nicht toxisch zu wirken.

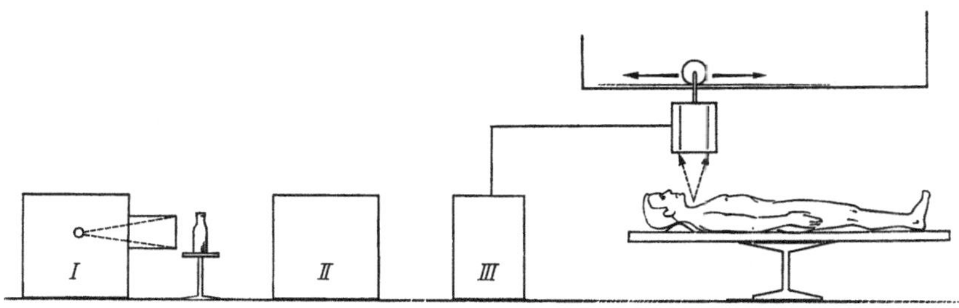

Abb. 1. Anwendung der Isotope C^{11}

In mehreren Selbstversuchen benutzten wir Methionin (Fa. Vitis, Fa. Merck) in einer Menge von 10 g ($\leqq$50 μCi). Durch die enorme Plasmakonzentration entsteht ein unnötig hoher „Untergrund", so daß das Scanbild im Versuch nicht deutlich ausfiel.

Im Tierversuch benutzten wir weiterhin bei Kaninchen 2 g Benzoesäurederivate ($\triangleq$10 μCi, Fa. Schering), wobei sich die Blase im Scanbild positiv darstellte. Die physiologische Kleinheit der Nieren ($\approx$ 1 cm $\varnothing$) als auch der hohe Untergrund verschlechtern hier die Ergebnisse.

Anschließend sind Ganzkörperprofile von Menschen mit Methionin 5 Minuten nach Injektion (ca. 30 μCi), Glukose 5 Minuten nach Injektion (ca. 20 μCi) und Glukose nach weiteren 40 Minuten (ca. 5 μCi) wiedergegeben (Abb. 2). Aus diesen Kurven ergibt sich eine größere spezifische Anreicherung von Methionin in der Lebergegend gegenüber Glukose.

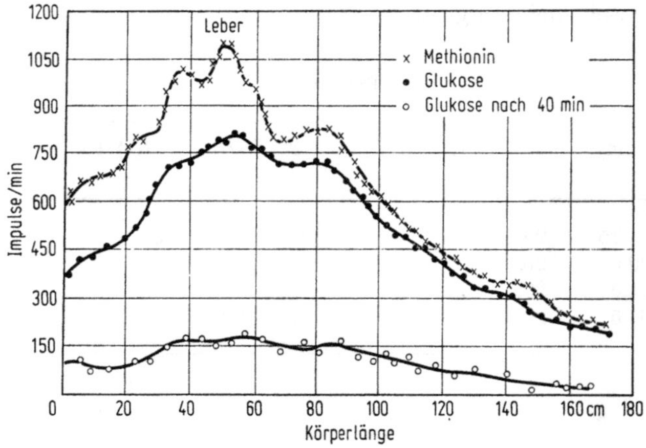

Abb. 2. Ganzkörperprofile von Menschen

Aufgrund dieser ersten Untersuchungen sind wir der Meinung, daß die höhere Substanzaktivierung mit einem Linearbeschleuniger deutlichere Scanbilder und bessere Ganzkörperprofile bei 10—100 mg Substanz ergibt (10^6 rad-Dosis bei einer Dosisleistung von etwa $0{,}5 \cdot 10^6$ rad verspricht mit 30 MeV γ-Energie in 1 g Kohlenstoff eine Aktivität von 500 μCi $\pm$ 30% zu erzeugen). Wir haben daher geplant, Untersuchungen mittels Linearbeschleuniger-Aktivierung im nuklearmedizinisch-biologischen Bereich auszuführen.

Nebenwirkungen

Bei diesen Experimenten sind uns folgende Nachteile bekannt:

a) chemische Zweitsubstanzen (Cadena, Rowlands),

b) Toxizität der Zweitsubstanzen,

c) Szilard-Chalmers-Effekt,

d) Be — 2·Ausbeute ($T_{1/2} = 53$ Tage) (Artus).

Obwohl bis jetzt die Linearbeschleunigerbenutzung (LA) noch eine Domaine der physikalischen Institute ist, werden in Frankreich, Japan und Deutschland usw. schon LA für therapeutische Anwendungen in Strahlenkliniken vereinzelt aufgestellt. Bei der Installierung möge man nicht auf den geringen Umbau der LA-Geräte für C^{11}-Erzeugung verzichten. Gerade diese neue bio-physikalische C^{11}-Tracermethode (O, N, Cl usw. können auch erzeugt werden) kann in der Nuklearmedizin dazu beitragen, die Dosis zu reduzieren und in der Pharmakologie schnell bestimmte Angriffsorte von Medikamenten ermitteln helfen.

Weiterhin benutzen wir im Vorversuch: Macrodex (Fa. Knoll) für Placenta und Niere, Nephrotest (Fa. Cassella) für Niere, Urografin (Fa. Schering) für Niere, Methionin (Fa. Vitis, Fa. Merck) für Pankreas und Leber.

Zusammenfassung

Es wird eine neue biophysikalische Tracer (Isotopen)-Methode beschrieben, die sich der Anwendung herkömmlicher Geräte in Kombination mit gängigen Medikamenten bedient. Neben den ersten Selbstversuchen sind wir mit der chemischen Analyse beschäftigt.

An dieser Stelle danke ich ganz besonders Herrn Prof. Dr. KÄRCHER, daß er für mich bei dem BfwF eine Assistentenstelle ermöglicht hat (D. MÖRSDORF).

Herrn Prof. Dr. EISTERT danke ich für die Assistentenstelle am Chemischen Institut der Saarbrücker Universität (H. SCHWALL).

Ebenfalls danken wir Herrn Prof. Dr. KUTTIG für die Überlassung des Betatrons zu den Versuchen.

Literatur

ARTUS, H., et al.: Zeitschrift für Physik **178,** 109 (1964).

—: Zeitschrift für Physik **189, 355** (1966).

BASIL, R., et al.: Possibilities for the analysis of O^{16} (γ, n). Compt. rend. **239,** 422 (1954).

BATE, L. C.: Nuclear methods of oxygen analysis ($O^{16}(t, n)$ F^{18}). Nucleonics **21,** No. 7 (July), S. 72—75 (1963).

BÖTTGER, MÜLLER: Zyklotron für medizinische Forschung. Atompraxis **13,** 313—315 (1967).

BREUER, H., POHLIT, W.: Der $(\gamma, 2n)$ - Prozeß in O^{16}. Nuclear Physics **30,** 417—423 (1962).

BROWNELL, G., SWEET, W.: Localization of brain tumours. Nucleonics **11,** No. 11 (Nov.), S. 40—45 (1953).

CADENA, D. G., ROWLAND, J. R.: J. Chem. Soc. London B **4,** 488—491 (1968).

FINCKH, E., KOSIEK, R., LINDENBERGER, K., MAIER, K., MEYER-BERKHOUT, U., SCHLECHTER, U., ZIMMERER, J.: Fotoprotonen aus C, N, O und Ne. Zeitschrift für Physik **174,** 337—350 (1963).

FROST, D., MICHEL, L.: Zusatzdosis bei Gewebeaktivierung. Strahlentherapie **132,** Heft 2, S. 228—245 (1967).

— —: Zusätzliche Dosis durch n bei schnellen $\bar{e}$. Strahlentherapie **124,** Heft 3, S. 321—350 (1964).

— — SCHWARZ, E.: Radiomarkierung $C^{12} \rightarrow C^{11}$. Strahlentherapie **127,** Heft 4, S. 510—515 (1965).

GLOCKER-MACHERAUCH: Röntgen-Kernphysik für Biophysiker. Stuttgart: G. Thieme 1965.

HORSLEY, R., JOHNS, H., HASLAM, R.: Energy absorption in human tissue. Nucleonics **11,** No. 2 (Feb.), S. 28—31 (1953).

KNIPPING, W., BOLT, W., VALENTIN, H., VENRATH, H., ENDLER, P.: Regionale Funktionsanalyse. Münchener Med. Wochenschrift **99.** Jahrg., S. 46 (1957).

KOSIEK, R.: Energiezerteilung der Fotoprotonen aus N^{15}. Zeitschrift für Physik **179,** 544—555 (1964).

MATTHEWS, C., DOLLERY, C. T., CLARK, J., WEST, J. B.: Hammersmith Hospital, London/England. Radioactive Gases in Radiopharmaceuticals. Printed USAEC, Oakridge, Tennessee, April 1966, S. 568—592.

MYRS, HUNTER, U.: C^{11} in bone. In: Sympos. of Medical Scintigraphy, 1968, S. 108—155.

PASCHKE, K.: C^{14}-Autoradiographie. Münchener Med. Wochenschrift **36,** 2043—2053 (1968).

Axial Skeletal Scintigraphy in Mammary Carcinoma*

C. S. B. GALASKO

Department of Surgery, Royal Postgraduate Medical School and Hammersmith Hospital, London, U. K.

Although the treatment of "early" mammary cancer is still a matter of controversy, local surgery with or without radiotherapy offers the only hope of cure. Once the cancer has disseminated radical local therapy is irrational. It cannot stem the course of the disease. It is important, therefore, to try to establish the extent of the disease accurately before planning treatment.

Because of the lack of reliable information as to the optimum treatment for the "early" case attention has been directed to establishing controlled, prospective, randomized clinical trials in the hope of determining the treatment which offers the best chance of a "cure". However, in any group of women with an apparently localised lesion there will be many with occult distant metastases. The outcome in these patients cannot be affected by any form of treatment which is aimed at "curing" the disease. This will introduce inaccuracies in the assessment of the therapeutic regimes being compared.

The axial skeleton is the commonest site for distant metastases (GARLAND, 1954; HAAGENSEN, 1956; COLE, 1964; FISHER et al., 1968) and where a careful autopsy has been carried out skeletal lesions have been found in 85 per cent of patients dying from the disease (JAFFE, 1958). Many surgeons, therefore, carry out a radiographic skeletal survey to exclude metastatic disease before submitting a patient to radical treatment.

Unfortunately, extensive destruction of bone must be present before metastatic lesions become apparent radiologically (BORAK, 1942; SHACKMAN and HARRISON, 1948; JAFFE, 1958; EDELSTYN et al., 1967). In about 50 per cent of patients metastases demonstrated at autopsy were not seen on X-rays (BACHMAN and SPROUL, 1955). Because of this insensitivity of X-rays other methods have been sought for the detection of skeletal metastases. The most promising technique is scanning for the radioactive isotopes of those elements which are selectively concentrated by bone (COREY, 1962; KOFMAN et al., 1963; ERJAVEC, 1965; CHARKES, SKLAROFF and YOUNG, 1966; FRENCH and MCCREADY, 1967; SPENCER et al., 1967; BESSLER, 1968).

Such examinations, however, have been limited previously to bones which were suspected clinically or radiologically of containing metastases. The methods used were too laborious and a scan of the axial skeleton would have taken 6 to 8 hours.

---

* Based on some of the material submitted in a thesis to the University of the Witwatersrand for the degree of Master of Surgery.

Previous attempts to detect symptomless lesions have been limited to the spine (GYNNING et al., 1961; WALDESKOG, 1962) or to the pelvis and lumbar spine (SKLAROFF and CHARKES, 1968), since the methods used by these workers took too long for a full survey.

We turned, therefore, to the gamma camera as a rapid method for the determination of the distribution of radioactive isotopes. Although no work had been reported on its use in skeletal scintigraphy it had been shown to be a useful instrument for the examination of the brain, lung, liver, thyroid, kidney, spleen and bone marrow (GOTTSCHALK and ANGER, 1964; LOKEN et al., 1965; LOKEN et al., 1966; BRUBAKER, 1967; MEALE and LANDIS, 1967; HARBERT, ASHBURN and DAVIDSON, 1968; LOKEN and WESTGATE, 1968).

These reports indicated that the camera was at least as accurate as the scanner and some even suggested that it may have been more accurate.

Many isotopes have been suggested for skeletal scintigraphy. Calcium-47 was the first, but is unsuitable for most laboratories as it has a very high gamma emission energy of 1.31 MeV and this requires special equipment. Strontium-85 is probably the isotope in commonest use for skeletal scintigraphy but it has a long half-life of 64 days, delivers a high dose of irradiation to the patient even if very small amounts are given, is excreted in the faeces and the colon must, therefore, be washed out and the count rates obtained are very low. Gallium-68 or 72 have been used (DUDLEY et al., 1950; EDWARDS et al., 1966) but may be toxic (BRUCER et al., 1953). These isotopes were, therefore, not investigated.

The other isotopes used in skeletal scintigraphy are Fluorine-18 and Strontium-87m, and therefore a pilot study was set up to compare them. Both isotopes have short half-lives, that of ^{18}F is 1.8 hours and that of ^{87m}Sr is 2.8 hours, and the irradiation dose received by the patient is small.

Recently Carbon-11 in the form of $^{11}CO_2$ has been suggested for skeletal scintigraphy (MYERS and HUNTER, 1967 and 1969) and it was also investigated. There has been no previous report on its use in man.

Methods

1. Pilot Study Comparing Fluorine-18 and Strontium-87m

Eight patients with osseous metastases from cancer of the breast were examined with a gamma camera (Pho-Gamma Mk II — Nuclear Chicago). Two separate studies were made on each patient so that the two isotopes could be compared. The dose of ^{18}F was 1.5 mCi and that of ^{87m}Sr was 2.5 mCi. The isotopes were supplied by the M. R. C. Cyclotron Unit, Hammersmith Hospital, and were given by intravenous injection. Three patients were injected first with ^{87m}Sr and 5 with ^{18}F. The interval between the two examinations was 3 to 5 days.

Regions known to contain metastases were examined at regular intervals after injection until the count-rate became too low for scanning to be practicable. When these metastases became visible on the scintigram, the rest of the axial skeleton was examined. With the patient sitting upright, pictures were taken of the left lateral, right lateral, and posterior aspects of the skull, the cervical, dorsal and lumbar spines, the left and right shoulder regions including the apical ribs, the left and right ribcages and the posterior aspect of the pelvis. Fluorine and strontium are excreted in the urine so that a full bladder masks the pubis. The patient, therefore, emptied her bladder and pictures were then taken of the left, right and central aspects of the pelvis, the upper ends of the femora and sternum, whilst she lay supine. Seventeen pictures were taken and these constituted the "axial skeletal scintigram". The times required for each picture and for the positioning of the patients, between pictures, were noted.

A radiographic survey of the axial skeleton was made in each patient. This consisted of a lateral view of skull, antero-posterior and lateral views of the cervical, dorsal, and lumbar spines, a postero-anterior view of the chest, and an antero-posterior view of the pelvis. The two scintigrams and the radiographic examination were made with a period of 5 days.

2. Carbon-11 Study

A second pilot study was set up to compare ^{18}F and ^{11}C as carbon dioxide for skeletal scintigraphy.

Permission was obtained from the Medical Research Council to examine 8 patients, with skeletal metastases from mammary carcinoma, on two separate occasions, so that the two isotopes could be compared.

One and a half millicuries of ^{18}F was injected intravenously and the axial skeletal scintigram started 1 to $1\frac{1}{2}$ hours later. The $^{11}CO_2$ was inhaled. The patient breathed in air from a cylinder via a rubber tube to which an expiratory valve and a rebreathing bag was attached. There was no soda-lime absorber included in the circuit. After breathing normally for 1 to 2 minutes the patient then took 4 to 6 forced breaths with the expiratory valve open. The valve was then closed, 30 mCi of $^{11}CO_2$ was injected into the tubing and 5 to 6 breaths taken in by the patient.

The $^{11}CO_2$ was administered close to an air extractor. After these 5 to 6 breaths, the face mask was removed, the air extractor switched on and the patient walked to an adjacent room where the gamma camera was situated. Regions shown on the X-ray to contain metastases were then scintigrammed at regular intervals until the count rate became too low for scintigraphy to be practical. This usually occurred 2 to $2\frac{1}{2}$ hours after the inhalation of the $^{11}CO_2$.

3. Patients Studies

One hundred and seventeen women were examined. Seventeen patients who clinically were thought to have a mammary carcinoma, but in whom histology of the resected mass showed a benign lesion, had an axial skeletal scintigram. These patients were considered as "controls". Fifty patients had an apparently "early" mammary cancer both on clinical and radiological grounds. Each patient, on admission, had a careful clinical examination and a radiographic skeletal survey, and each had an axial skeletal scintigram. Fifty women presented with an advanced mammary cancer, either due to extensive local disease or because of distant metastases. Each patient had a careful clinical examination, a radiographic skeletal survey and an axial skeletal scintigram.

All patients have been followed up carefully for 18 months and those with advanced mammary cancer have had their radiographic skeletal surveys repeated every 3 months. The patients with apparently "early" mammary cancer had their X-rays repeated if there was an indication such as pain.

One and a quarter hours after the intravenous administration of 1.5 mCi of ^{18}F each patient had an axial skeletal scintigram.

In all cases the camera was used with a 4½ inch collimator, which was obtained by bolting the 3 inch and the 1½ inch multi-channel collimators together.

Results

1. Pilot Study Comparing Fluorine-18 and Strontium-87m

All scintigrams were completed in 90 minutes. Each picture took approximately 4 to 5 minutes and 17 were required for the complete survey.

The optimum time for examination (Table 1) was 1¼ to 3 hours after injection of ^{18}F (Fig. 1) and 3 to 4½ hours after injection of ^{87m}Sr.

Table 1. *Optimum time for scanning*

Patient	Fluorine-18 (hours)	Strontium-87m (hours)
1	1½–2	3 –4
2	1¼–2	3 –4½
3	2 –2½	3½
4	2 –2½	3½–4
5	2 –2½	3 –3½
6	1½–2¾	3 –4
7	3	Lesions not seen
8	1½–3	3 –4½

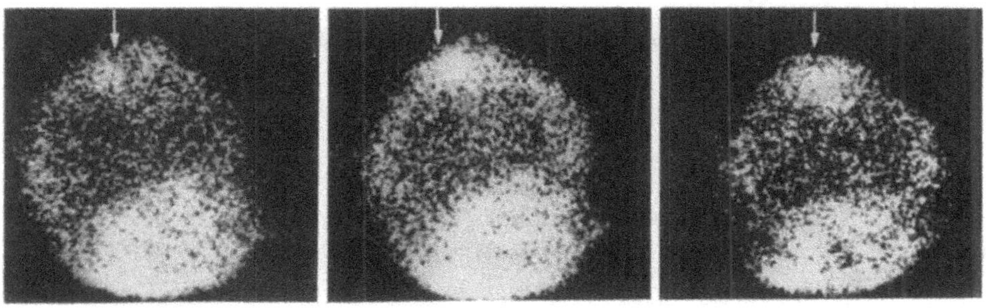

Fig. 1. Fluorine-18 scintigram of skull. The lesion (arrowed) is just visible after 2 minutes (left), becomes clearer after 30 minutes (middle), and is well defined after 100 minutes (right)

In all patients better results were obtained with ^{18}F than with ^{87m}Sr (Fig. 2). With ^{18}F all lesions visible on radiographs were apparent on the scintigram. With ^{87m}Sr some lesions were not demonstrated in 4 of the 8 patients. In 3 of these most of the lesions were shown, but in one, with metastases in the 2nd, 3rd and 4th lumbar vertebrae, no lesions were demonstrated.

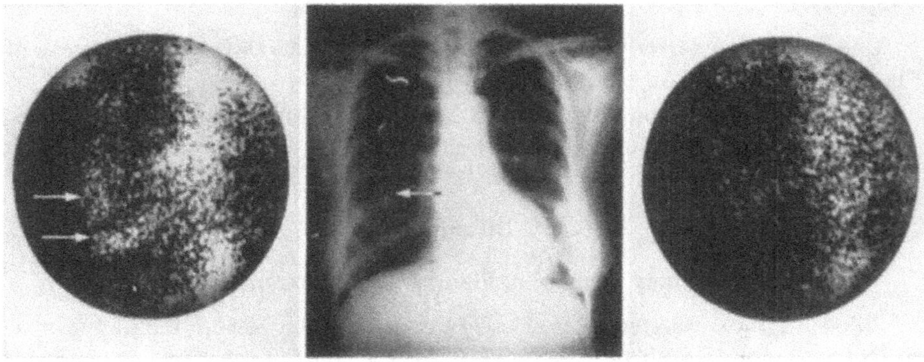

Fig. 2. Comparison of scintigrams in a patient with metastases in the right 7th and 8th ribs. The metastases are not seen clearly on the radiograph or the ^{87m}Sr scintigram (right) but are well defined on the ^{18}F scintigram (left)

In 6 of the 8 patients additional lesions, which were not visible on radiographs, were found with ^{18}F. In all, radiographs taken 3 to 9 months later have confirmed the presence of metastases at the sites indicated on the scintigram. Four of the 8 showed additional lesions with ^{87m}Sr.

2. Carbon-11 Study

After 6 patients had been studied the investigation was abandoned as the $^{11}CO_2$ did not localise either in the normal bone or in the metastases. The regions examined included the skull, cervical, dorsal and lumbar spines, and pelvis. The extremities were not examined.

The $^{11}CO_2$ was adequately absorbed as high count rates were obtained over sites distant from the lungs. The scintigram obtained did show the body outline suggesting that the $^{11}CO_2$ was taken up by the tissues generally and not by any specific organ or tissue.

3. Patients Studies

Control Patients. All the scintigrams in the 17 patients with benign lesions of the breast, were normal and no false positives were obtained.

"Early" Mammary Cancer. No patient had clinical or radiological evidence of dissemination. In 12 (24%) of the 50 patients lesions were seen on the scintigram and in 9 of these patients the lesions have since been confirmed, either on sub-

sequent radiographs or at autopsy (Figs. 3 and 4). The time after mastectomy, when the lesions demonstrated on the scintigrams, became apparent on X-ray or were demonstrated at autopsy is shown in Table 2.

Advanced Mammary Cancer. Of the 50 patients with advanced mammary cancer lesions were apparent on X-ray in 25 (50%) and in the scintigram in 42 (84%). These 42 patients included the 25 with positive X-rays. Six other patients developed radiological evidence of metastases on subsequent radiographs, so that 31 (62%) patients at some time had radiological evidence of skeletal metastases.

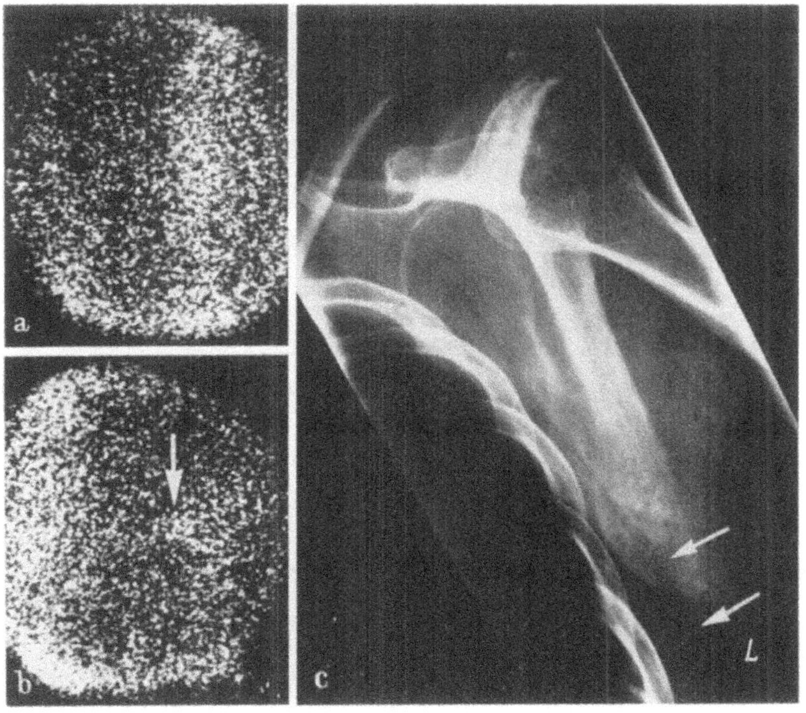

Fig. 3. Scintigram of the scapular region. Patient with apparently "early" cancer. a The right side is normal. b On the left a lesion (arrowed) can be seen in the inferior angle of the scapula. The X-ray was normal. c Repeat X-ray taken 3 months later. A metastasis (arrowed) is apparent in the inferior angle of the scapula

Thirty-six (72%) of the 50 patients, including 17 patients who had no metastases on X-ray, had additional lesions present on the scintigram. Five of these patients died at home, before the repeat X-rays could be carried out, and had no autopsy. It was thus impossible to confirm the lesions in these patients. All the remaining 31 patients have had the additional lesions subsequently confirmed, either on X-ray or at autopsy (Figs. 5 and 6). The time when these lesions were confirmed is shown in Table 3.

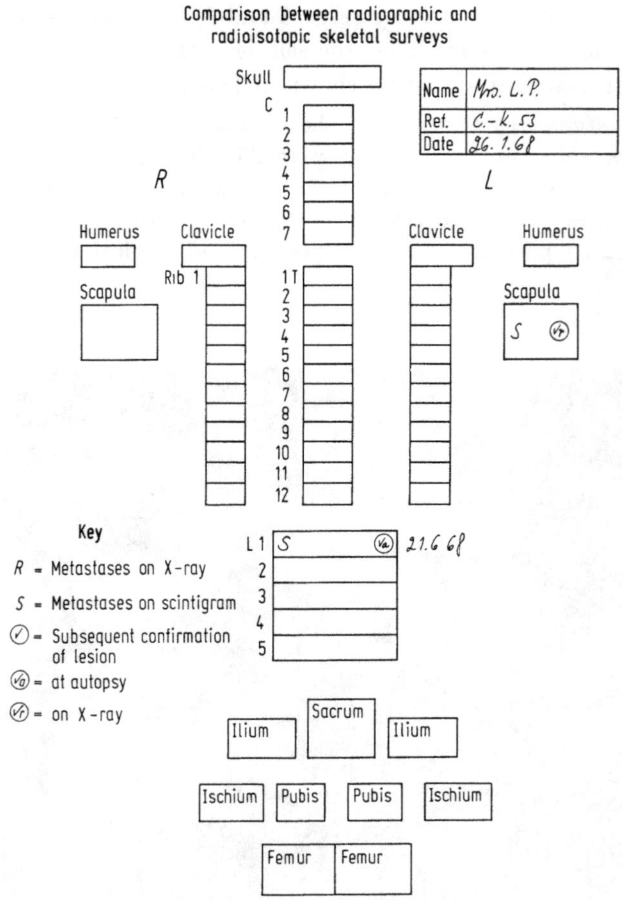

Fig. 4. Mrs. L. P. Apparently "early" mammary cancer. At the time of mastectomy no metastases were seen on the radiographic skeletal survey. Lesions were apparent on the scintigram in the left scapula and the first lumbar vertebra. The lesion in the scapula became apparent on X-ray 3 months later (Fig. 3). The metastasis in the lumbar vertebra was never evident on X-ray but was seen at autopsy 5 months after the mastectomy

Table 2. "*Early*" *mammary cancer*

Time taken for lesions, which were demonstrated on the scintigram but were not apparent on X-ray, to be confirmed either radiologically or at autopsy. All patients had an apparently "early" mammary cancer.

No. of patients		50
No. with lesions on scintigram		12 (24%)
Percentage confirmation of lesions	at 3 months	33%
	at 6 months	40%
	at 12 months	67%

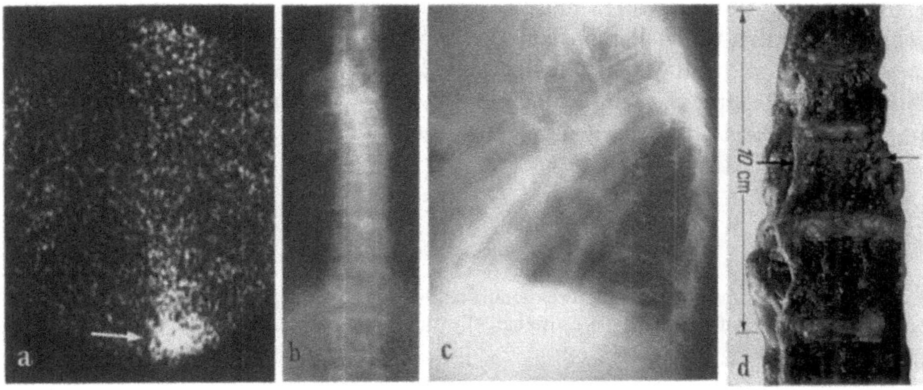

Fig. 5. Scintigram of the dorsal spine. Patient with advanced cancer.

a On the scintigram a lesion (arrowed) is evident in the 11th dorsal vertebra. b and c Antero-posterior and lateral X-rays of the dorsal spine taken at the same time as the scintigram. They are normal. d The lesion was demonstrated 6 weeks later at autopsy (arrowed)

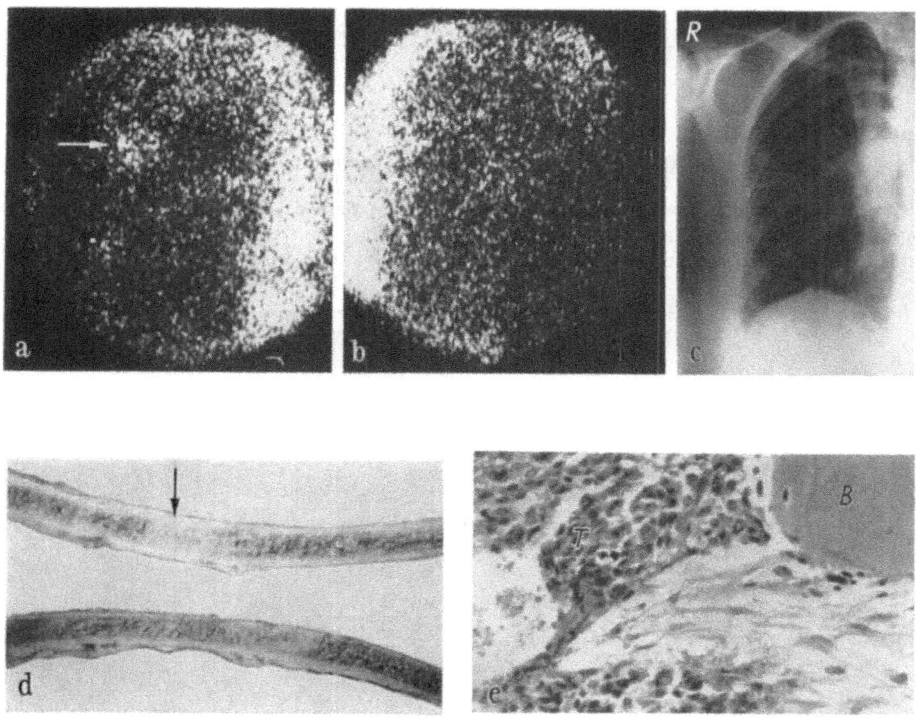

Fig. 6. Scintigram of the rib cage. Patient with advanced carcinoma.

a A lesion (arrowed) is seen in the right rib cage. b Scintigram of the left side. This is normal. c X-ray of the right ribs. No lesion is seen. d The lesion was demonstrated at autopsy (arrowed). e Histological section of the metastasis (B = Bone spicule; T = Tumour)

It took anything up to 18 months for the lesions, which were evident on the scintigram only, to be demonstrated either on subsequent radiographs or at autopsy (Tables 4 and 5). In some patients lesions appeared radiologically at different intervals after the scintigram and, therefore, the number of regions and not number of patients have been shown in these tables.

Table 3. *Advanced mammary cancer*

Time taken for lesions, which were demonstrated on the scintigram but were not apparent on X-ray, to be confirmed either radiologically or at autopsy. All patients had advanced mammary cancer.

No. of patients		50
No. with additional lesions which were not present on X-ray		36 (72%)
Percentage confirmation of lesions	at 3 months	55%
	at 6 months	72%
	at 12 months	96%
	at 18 months	100%

Table 4. *Patients with advanced mammary cancer*

Time (in months) taken for lesions which were demonstrated on the scintigram to appear radiologically. Any patient could have had more than one region (e.g. skull, dorsal spine, pelvis) involved and may, therefore, appear more than once.

Time (months)	No. of regions
2	12
3	12
4	8
5	8
6	2
7	2
8	2
9	5
10	4
11	3
12	3
13	2
14	—
15	—
16	—
17	—
18	2

Table 5. *Patients with advanced mammary cancer*

Time (in months) after lesions had been demonstrated on the scintigram, when they were seen at autopsy. In each case the X-rays had remained negative until death. Any patient could have had more than one region involved, and may appear more than once.

Time (months)	No. of regions
1	4
2	8
3	—
4	—
5	—
6	4
7	—
8	1
9	3

Discussion

The method was chosen because conventional scanning techniques are too lengthy for an examination of the entire axial skeleton. BAKER and his colleagues (1962) and ANGER (1966) have described whole body scanners which can examine the body within a short period of time. SIMPSON, ROTENBERG and BAKER (1962) found that with their scanner results of skeletal scintigraphy were very difficult to interpret and RONAI, WINCHELL and ANGER (1968) have used the whole body scanner of ANGER on 5 patients with known skeletal metastases, and found that they had to use it in conjunction with a positron camera.

Fluorine-18 was the isotope chosen for the main study as it was shown to be superior to ^{87m}Sr in the pilot study. RONAI and his colleagues (1968) showed that it was better than ^{85}Sr. Other isotopes, which concentrate in bone, are unsuitable as they either require special equipment (^{47}Ca) or may be toxic (^{68}Ga and ^{72}Ga).

The results show that it is possible to survey the skeleton, excluding the long bones of the limbs, in only 90 minutes. It is not surprising that 24 per cent of patients with apparently "early" mammary cancer on clinical and radiological grounds have occult skeletal metastases at the time of mastectomy. Despite modifications in surgery and radiotherapeutic techniques, earlier diagnosis and prophylactic ovarian ablation and chemotherapy, there has been virtually no change in the 3 or 5 year survival rates over the past few decades (COLLINS and ADAMS, 1967; BULBROOK, 1968; SEGALOFF, 1968). In addition the majority of patients who develop skeletal metastases do so within the first 2 years after mastectomy (HAAGENSEN, 1956; RENNAES, 1960), and our results indicate that it may take at least 18 months for lesions, which have been demonstrated on the scintigram, to become apparent on the radiographs. To date the patients with apparently "early" mammary cancer have only been followed up for 18 months, and it is probable that with a longer period of observation further lesions will be confirmed.

WALDESKOG, GYNNING and LANGELAND (1962) found lesions which were not apparent on X-ray in 10 of the 90 (11%) women they examined, but their survey was limited to the spine. SKLAROFF and CHARKES (1968) found on scintigraphy radiologically negative lesions in 10 of the 64 (16%) patients they had examined, but their method was limited to the spine and pelvis. Since the entire axial skeleton was examined in our series it was not surprising that a higher proportion of patients with an apparently "early" cancer (24%) were found to have lesions which had not been apparent radiologically. In this study, the scintigram was not used to plan therapy, but as a result of the study routine scintigraphy is to be carried out at the Hammersmith Hospital before submitting patients with an apparently "early" carcinoma to mastectomy.

Skeletal lesions were found in 84 per cent of patients with advanced mammary cancer, and this figure is similar to that obtained in post-mortem studies (JAFFE, 1958). Metastases were seen on X-rays, obtained at the same time as the scintigram, in 50 per cent of patients. When the X-rays were repeated every three months metastases were seen in only another 6 (12%) patients. In all those patients who have been followed up for 18 months or who have come to autopsy lesions have been confirmed. It has not been possible to confirm every lesion in every patient as some regions of the skeleton are not examined at routine autopsy, but no false positives were obtained in any region which had been examined at autopsy. As lesions may appear on X-ray at varying intervals after the scintigram an occasional lesion still awaits confirmation in a few patients.

The disease is more extensive in patients with advanced than in those with apparently "early" mammary cancer, and therefore it is not surprising that lesions were confirmed more rapidly in patients in the former category.

The recognition of lesions on the scintigram depends on the difference in uptake of the isotope between the lesions and the normal bone. The majority of metastases invoke an osteoid reaction in the invaded bone (PERKINSON et al., 1955; COREY, 1962; POSNER et al., 1963) and it is this immature osteoid tissue which concentrates the isotope and is responsible for the difference in uptake between the lesion and the normal bone.

Acknowledgements

I wish to thank Mr. D. D. VONBERG and his staff of the Medical Research Council Cyclotron Unit, London; Professor J. F. FOWLER, Department of Medical Physics; Professor R. E. STEINER and Dr. F. DOYLE, Department of Diagnostic Radiology, and especially Professor R. B. WELBOURN and Mr. J. I. BURN, Department of Surgery, Royal Postgraduate Medical School, London, for the facilities they provided and for their help and encouragement.

References

1. ANGER, H. O.: Whole body scanner Mark II. J. nucl. Med. **7**, **331** (**1966**).
2. BACHMAN, A. L., SPROUL, E. E.: Correlation of radiographic and autopsy findings in suspected metastases in the spine. Bull. N. Y. Acad. Med. **31**, **146** (**1955**).
3. BAKER, R. G., ROTENBERG, A. D., CEDERLUND, J. F., JOHNS, H. E.: Whole body scanning in Medicine I. Physical aspects. Canad. Med. Assoc. J. **87**, **367** (**1962**).
4. BESSLER, W. T.: Skeletal scintigraphy as an aid in practical roentgenographic diagnosis. Amer. J. Roentgenol. **102**, 899 (1968).
5. BORAK, J.: Relationship between the clinical and roentgenological findings in bone metastases. Surg. Gynec. and Obstet. **75**, 599 (1942).

6. Brubaker, D. F.: Clinical advantages of the scintillation camera. J. nucl. Med. **8,** 361 (1967).
7. Brucer, M., Andrews, G. A., Bruner, H. D.: A study of gallium-72. Radiology **61,** 534 (1953).
8. Bulbrook, R. D.: Studies of endocrinology of breast cancer. In Jarrett, A. S. (ed.): The treatment of cancer of the breast. Maidenhead: Syntex Ltd. 1968, p. 40.
9. Charkes, N. D., Sklaroff, D. M., Young, I.: A critical analysis of strontium bone scanning for detection of metastatic carcinoma. Amer. J. Roentgenol. **96,** 647 (1966).
10. Cole, Mary P.: The place of radiotherapy in the management of early breast cancer. A report of two clinical trials. Brit. J. Surg. **51,** 216 (1964).
11. Collins, V. P., Adams, R. M.: Paradox of breast cancer. Amer. J. Roentgenol. **99,** 965 (1967).
12. Corey, K. R.: The use of Ca-47 and Sr-85 in the diagnosis of bone tumours. In: Medical uses of Ca-47. Vienna: Int. Atomic Energy Agency 1962, p. 95.
13. Dudley, H. C., Imirie, G. W., Jr., Istock, J. T.: Deposition of radiogallium (Ga-72) in proliferating tissues. Radiology **55,** 571—578 (1950).
14. Edelstyn, G. A., Gillespie, P. J., Grebbell, F. S.: The radiological demonstration of osseous metastases. Experimental observations. Clin. Radiol. **18,** 158 (1967).
15. Edwards, C. L., Hayes, R., Ahumada, J., Kniseley, R. M.: Gallium-68 citrate; a clinically useful skeletal scanning agent. J. nucl. Med. **7,** 363—364 (1966).
16. Erjavec, M.: Scintillation scanning of bone tumours and metastases by the use of Sr-85. Radiobiol. Radiother. **6,** 119 (1965).
17. Fisher, B., Ravdin, R. G., Ausman, R. K., Slack, N. H., Moore, G. E., Noer, R. J.: Surgical adjuvant chemotherapy in cancer of the breast; results of a decade of co-operative investigation. Ann. Surg. **168,** 337 (1968).
18. French, R. J., McCready, V. R.: The use of 18-F for bone scanning. Brit. J. Radiol. **40,** 655 (1967).
19. Garland, L. H.: The rationale and results of simple mastectomy plus radiotherapy in primary cancer of the breast. Amer. J. Roentgenol. **72,** 923 (1954).
20. Gottschalk, A., Anger, H. O.: Diagnostic uses of the scintillation camera. J. nucl. Med. **5,** 371 (1964).
21. Gynning, I., Langeland, P., Lindberg, S., Waldeskog, B.: Localisation with Sr-85 of spinal metastases in mammary cancer and changes in uptake after hormone and roentgen therapy. A preliminary report. Acta radiol. **55,** 119 (1961).
22. Haagensen, C. D.: Disease of the breast. Philadelphia: W. B. Saunders Co. 1956.
23. Harbert, J. C., Ashburn, W. L., Davidson, J. D.: An improved method of renography using the split crystal scintillation camera. J. Urol. (Baltimore) **99,** 681—687 (1968).
24. Jaffe, H. L.: Tumours and tumorous conditions of the bones and joints. Philadelphia: Lee and Febiger 1958.
25. Kofman, S., Sky-Peck, H. H., Thibaudeau, Y., Ray, R. D., Taylor, S. G.: The use of strontium-85 in evaluation of bone metastases. A preliminary report. J. nucl. Med. **4,** 9 (1963).
26. Loken, M. K., Westgate, H. D.: Using Xe-133 and a scintillation camera to evaluate pulmonary function. J. nucl. Med. **9,** 45—50 (1968).
27. — Wigdahl, L. O., Gilson, J. M., Staab, E. V.: Mercury-197 and Mercury-203 chlormerodrin for evaluation of brain lesions using a rectilinear scanner and scintillation camera. J. nucl. Med. **7,** 209—218 (1966).
28. — — Staab, E. V., Gilson, J. M.: Evaluation of conventional scanning and Anger camera for brain studies. J. nucl. Med. **6,** 335 (1965).
29. Meale, A., Jr., Landis, G. A.: Clinical use of the scintillation camera. J. nucl. Med. **8,** 371 (1967).
30. Myers, W. G., Hunter, W. W., Jr.: Radiocarbon-11 for scanning. J. nucl. Med. **8,** 305 (1967).
31. — —: C-11 in bone and lung scanning. Medical Radioisotope Scintigraphy. I. A. E. A., Symposium, Salzburg, 1969, vol. II, p. 43—53.
32. Perkinson, J. D., Whitney, I. B., Monroe, R. A., Lotz, W. E., Comer, C. L.: Metabolism of F-18 in domestic animals. Amer. J. Physiol. **182,** 383 (1955).

33. Posner, A. S., Eanes, E. D., Harper, R. A., Zipkin, I.: X-ray diffraction analysis of the effect of fluoride in human bone apatite. Arch. Oral Biol. **8**, 549 (1963).
34. Rennaes, S.: Cancer of the breast in women. A clinical and statistical investigation of the material at the Norwegian Radium Hospital 1932—51. Acta, chir. Scand. Suppl. **1**, 266 (1960).
35. Ronai, P., Winchell, H. S., Anger, H. O.: Skeletal survey for metastatic tumours of bone using 18-F and 85-Sr with scintillation camera and whole body scanner. J. nucl. Med. **9**, 517 (1968).
36. Segaloff, A.: Progress in the treatment of breast cancer. In Jarrett, A. S. (ed.): The treatment of cancer of the breast. Maidenhead: Syntex Ltd. 1968, p. 8.
37. Shackman, R., Harrison, C. V.: Occult bone metastases. Brit. J. Surg. **35**, 385 (1948).
38. Simpson, W. J. K., Rotenberg, A. D., Baker, R. G.: Whole body scanning in Medicine II. Clinical aspects. Canad. Med. Assoc. J. **87**, 371 (1962).
39. Sklaroff, D. M., Charkes, N. D.: Bone metastases from breast cancer at the time of radical mastectomy. Surg. Gynec. and Obstet. **124**, 763 (1968).
40. Spencer, R., Herbert, R., Rish, M. W., Little, W. A.: Bone scanning with 85-Sr, 87m-Sr and 18-F. Brit. J. Radiol. **40**, 641 (1967).
41. Waldeskog, B., Gynning, I., Langeland, P.: Localisation with Sr-85 of spinal metastase in mammary cancer and changes in uptake after hormone and roentgen therapy. In: Medical Uses of Ca-47. Vienna: Int. Atomic Energy Agency 1962, p. 104.

Die renale Angioscintigraphie

Ein Beitrag zur Differentialdiagnose raumfordernder, intrarenal gelegener Prozesse

Renal Angioscintigraphy

A Contribution to the Differential Diagnosis of Expansive Intrarenal Processes

J. SEIFERT

Strahlenklinik des Klinikums Essen der Ruhr-Universität Bochum

Summary

Angioscintigraphy can improve the accuracy of diagnosis of renal tumors at little discomfort to the patient. It supplements the angiographic findings by indicating the distribution and nature of precapillaries and capillaries, and the site and extension of the tumor. A radioactive material (^{131}I or ^{113m}In) is introduced through the catheter previously used for angiography or aortography. Results obtained with 26 angioscintigraphs are briefly described; ^{113m}In is preferred because, for the same biological half-life, it enables more activity or a larger volume to be administered without increasing the radiation dose.

Der Nachweis von Tumoren der Nieren, der Nierenbeckenkelchsysteme und der Ureteren obliegt im wesentlichen der Röntgendiagnostik, und im Rahmen der vielfältigen Untersuchungsmethoden mit röntgenologischer Dokumentation gebührt der renalen Angiographie ohne Zweifel eine vorrangige Stellung. Insbesondere nach Vervollkommnung [1, 2] der Kathetermethoden kann vor allem die gezielte Angiographie einer Niere durch überlagerungsfreie Darstellung in einem Serienangiogramm in vielen Fällen einen infiltrativ wachsenden, hypernephroiden Tumor (Grawitz-Tumor) gegenüber anderen, expansiven Nierenprozessen differenzieren. Hypernephroide Tumoren bieten in der Regel ein gefäßtypisches, angiographisches Bild mit primären und sekundären Zeichen eines gefäßreichen, infiltrativ wachsenden, zur Nekrose neigenden Tumors. Im Gegensatz dazu können jedoch totalnekrotische Tumoren als gegen das normale Nierenparenchym abgesetzte, gefäßfreie Bezirke imponieren, die bei scharfer Abgrenzung mit einer Zyste verwechselt werden können. So haben LÖHR und Mitarbeiter [3] in einer Zusammenstellung von Renovasographien bei parenchymatösen Nierentumoren fast 30% gefäßfreie bzw. gefäßarme, hypernephroide Tumoren gesehen. Darüber hinaus sind die infiltrierend wachsenden Nierencarcinome angiographisch gefäßarm und in einem großen Prozentsatz nicht sicher gegen einen chronisch schrumpfenden, narbigen, postentzündlichen Prozeß abzugrenzen. Weiterhin gibt die renale Angiographie bei Tumoren des Nierenbeckenkelchsystems in seltenen Fällen Aufschluß über ein malignes Wachstum.

Unter Berücksichtigung der oben angeführten Tatsachen und Erfahrungen kann man auch unter Zuhilfenahme differenzierter, röntgendiagnostischer Untersuchungsverfahren mit etwa knapp einem Drittel nicht sicher abgeklärter oder fehlbeurteilter Nierentumoren rechnen. Zu diesem Schluß sind auch einige, in der letzten Zeit erschienene Arbeiten kritischer Autoren über die Renovasographie gekommen.

Es ist nun ein Anliegen, diese „offenen" Fälle durch weitere, möglichst wenig eingreifende Untersuchungsverfahren noch zu differenzieren. In einem Teil dieser Fälle konnten wir eine Abklärung durch die Anwendung des Substraktionsverfahrens und der Harmonisierung gleichartiger Röntgenaufnahmen erreichen. Auf Grund theoretischer Überlegungen und Ergebnisse eigener Beobachtungen bei angioscintigraphisch nachgewiesenen Tumoren im kleinen Becken [4] haben wir die Angioscintigraphie bei Nierentumoren geprüft.

Voraussetzungen

Wir gehen davon aus, daß schnell wachsende Tumoren relativ gefäßreich sind. Bei Durchmusterung histologischer Präparate sowohl hypernephroider Tumoren wie auch infiltrierend wachsender Nierencarcinome erstaunt die hohe Zahl von versorgenden Tumorgefäßen, die aus verschiedenen Gründen sich eines angiographischen Nachweises entziehen. So ist es im Röntgenkontrastbild, eingeengt durch das Auflösungsvermögen, bei besten, physikalischen wie photographischen Bedingungen und unter Zuhilfenahme des Substraktionsverfahrens möglich, Gefäße im Bereich der Nierenrinde mit einem Kaliber von 200 μ an aufwärts zu differenzieren [5]. In zentraleren Nierenabschnitten steigt diese Grenze auf etwa 400 μ an. Darüber hinaus liegen im Tumorgebiet andere hämodynamische Verhältnisse vor als in tumorfreien Abschnitten. Dies ist einmal bedingt durch die morphologische Eigenart der Tumorgefäße, die in der Größenordnung der Capillaren nur aus Endothelschläuchen ohne muskulären oder bindegewebig-kollagenen Anteil bestehen. Wir haben dies bei durch Oestrogene hervorgerufenen, hypernephroiden Tumoren bei Goldhamstern beobachtet. Zum anderen lassen vorgeschaltete arterio-venöse „shunts" das in einer definierten Zeiteinheit angebotene Kontrastmittel nach dem Prinzip der intravasalen Turbulenzströmung weitgehend abfließen, so daß Tumorbezirke im zentralen Anteil selbst mit größeren Lumenquerschnitten für den Kontrastmitteldurchfluß im Nebenschluß liegen können. Wir sehen hierin eine wesentliche Teilerklärung für den angiographischen Begriff des sogenannten „gefäßarmen" bzw. „gefäßfreien" Nierentumors, der histologisch noch gefäßanteilig gut versorgt und histiochemisch zellaktiv erscheint.

Methodik

Wir führen die renale Angioscintigraphie unmittelbar im Anschluß an die selektive Nierenangiographie oder in besonders gelagerten Fällen im Anschluß an die Etagenaortographie durch. Die Indikation zur renalen Angioscintigraphie stellen wir nach Kenntnis des angiographischen Befundes, wenn noch differentialdiagnostische Schwierigkeiten bestehen. Durch den liegenden, gebogenen Oedman-Katheter, der über die Punktion einer Femoralarterie in die zu untersuchende A. renalis eingebracht worden ist, instillieren wir ein definiertes Volumen von Molekülkolloiden mit 131-Jod-Markierung als 131-Jod-Albumin-Makropartikeln oder von Aggregationskolloiden als 113m-Indium-Eisen-Makrokolloide. Die durchschnittliche Partikelgröße (anteilig zu mindestens 75%) beträgt bei den 131-Jod-Albumin-Makropartikeln etwa 60 μ und die der 113m-Indium-Eisen-Makrokolloide etwa 40 μ, sie wird jeweils vor der Instillation unter dem Hämocytometer kontrolliert.

Tabelle 1. *Halbwertszeiten*

	Physikalische Halbwertszeit	Biologische Halbwertszeit
131-Jod-Albumin-Makropartikeln	8,1 d	18 h
113m-Indium-Eisen-Makrokolloide	1,7 h	12 h

Die zu verabfolgende Dosis richtet sich nach dem darzustellenden Organ bzw. Organabschnitt. Für die Angioscintigraphie der Niere geben wir in der Regel 50 μCi 131-Jod-Albumin-Makropartikeln einseitig oder 100 μCi im Anschluß an eine Etagenaortographie mit Darstellung beider Nieren oder aber 300 μCi 113m-Indium-Eisen-Makrokolloide.

Tabelle 2. *Strahlenbelastung* [6]

Isotop	Niere	Ganzkörper	Blut	Geschlechtsorgane
131-Jod	0,9 rad	0,03 rad	0,05 rad	0,03 rad
113m-Indium	0,6 rad	0,002 rad	0,005 rad	0,003 rad

Die Durchführung der renalen Angioscintigraphie setzt voraus, daß im Anschluß an die röntgendiagnostische Untersuchung ein Nachweis der intraarteriell instillierten Isotopen mit einer Scintillationskamera oder mit einem Scanner möglich und auch räumlich durchführbar ist. Im allgemeinen beginnen wir das Scan etwa 20 bis 30 Minuten nach der Inkorporation, besondere Eile ist bei dem kurzlebigen Indium geboten. Aber nach unseren Erfahrungen kann noch 3 Stunden nach Verabfolgung der 131-Jod-Albumin-Makropartikeln und $1\frac{1}{2}$ Stunden nach Verabfolgung der Indium-Eisen-Makrokolloide ein verwertbares Nierenangioscan geschrieben werden. Unsere gewonnenen Scans wurden mit dem Picker-Magna-III-Scanner oder mit der Scintillationskamera nach Anger hergestellt.

Ergebnisse

Unsere Erfahrungen stützen sich auf insgesamt 26 Angioscintigraphien, die sich auf 13 Beckenangioscintigraphien und neben 2 Coeliacoangioscintigraphien auf insgesamt 9 renale Angioscans verteilen.

Tabelle 3. *Renale Angioscan-Aufstellung*

Angiographischer Befund		Angioscan	Operationsergebnis
Zysten:	6 Fälle	4 Fälle	4 Fälle bestätigt
	1 Fall unsicher		
Tumoren:	1 Fall	5 Fälle	5 Fälle bestätigt
	2 Fälle unsicher		

In 2 Fällen der durch das Angioscan nachgewiesenen Zysten haben wir ein Nierenscintigramm mit 200 μCi 203-Hg-Salyrgan 48 Stunden nach der renalen Angioscintigraphie mit 113m-In-Fe-Makrokolloiden geschrieben und das Ergebnis verglichen bzw. die Scans subtrahiert. Der umschriebene Fixationsdefekt (zystischer Bezirk) konnte dann sozusagen „aktivitätsfrei" zur Deckung gebracht werden.

Die Deutung der renalen Angioscans erfolgt nach den Kriterien

a) Aktivitätsverteilung,

b) Lokalisation,

c) Ausdehnung und Grad der Fixationsänderung der instillierten Isotopen.

Beispiele anhand der Abb. 1—7.

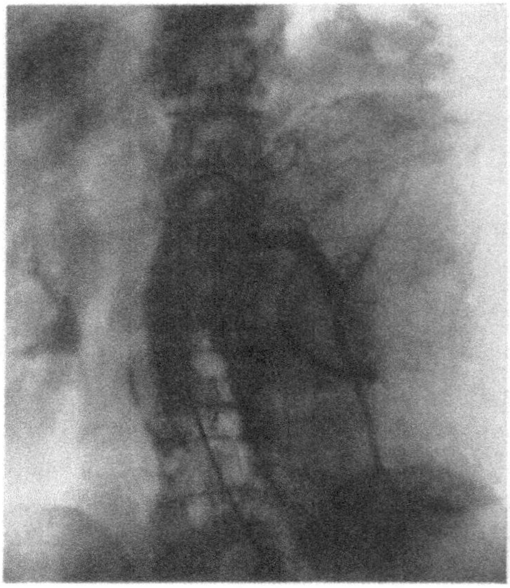

Abb. 1

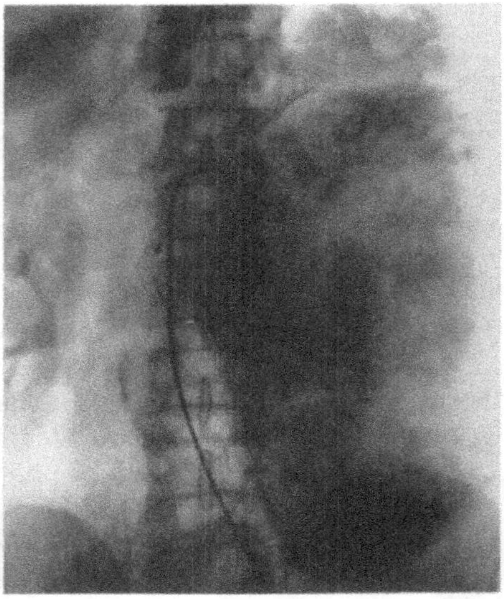

Abb. 2

Abb. 1 u. 2. Arterielle und arterio-venöse Phase eines zerfallenden Hypernephroms

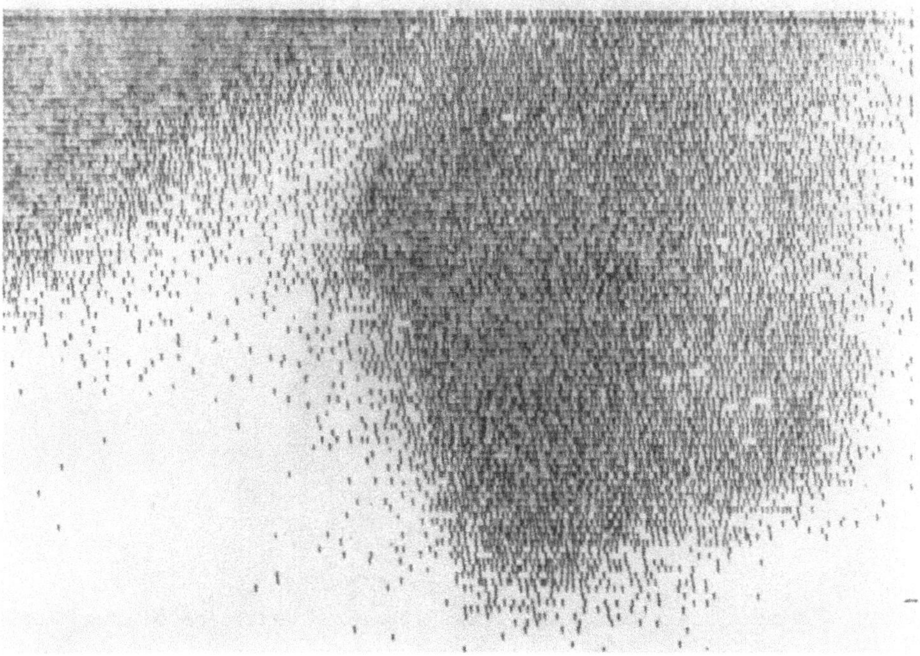

Abb. 3. Typisches Angioscintigramm der linken Niere, gleicher Fall wie Abb. 1 u. 2. 300 μCi 113m-In-Fe-Makro-Komplex. Durch die arteriovenösen „shunts“ Anreicherung auch in der Leber

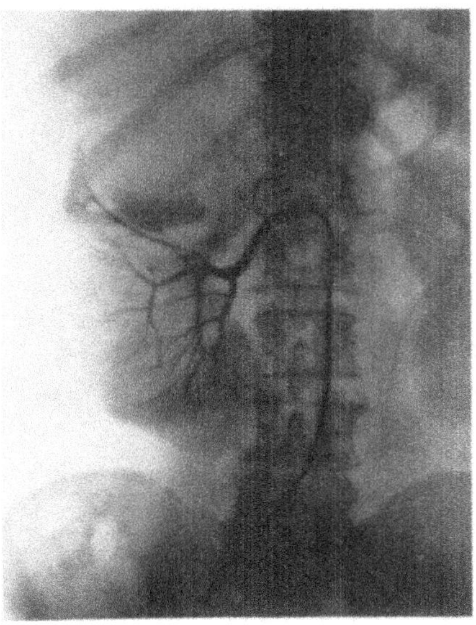

Abb. 4

Abb. 5

Abb. 4 u. 5. Selektives Renovasogramm und entsprechendes Angioscan nach Instillation von 50 μCi MAA-J-131. Typischer Befund für eine Nierenzyste

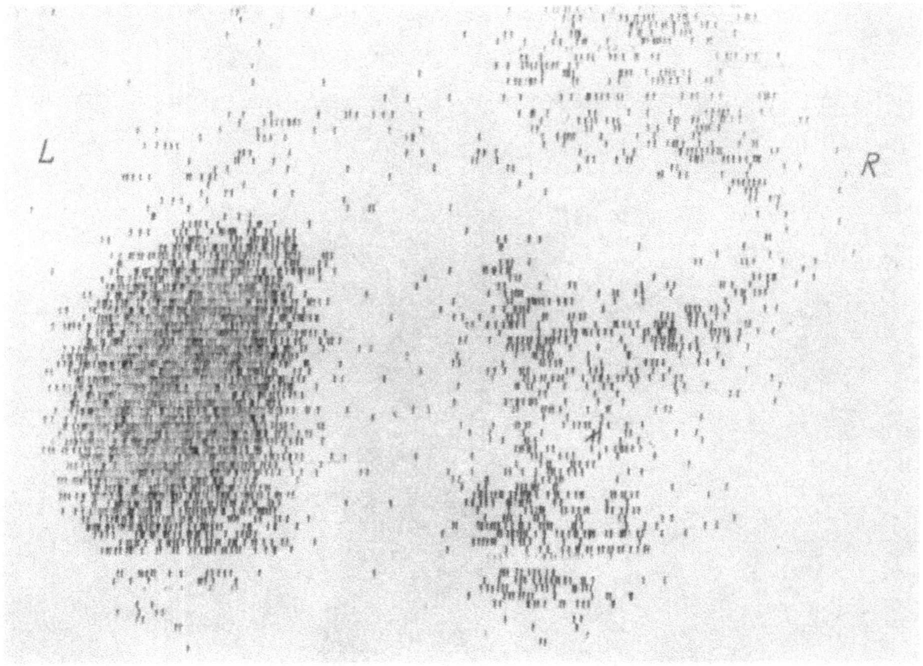

Abb. 6. Nierenscintigramm mit 203-Hg-Salyrgan in einer Dosis von 200 μCi. Gleicher Fall wie Abb. 4 u. 5

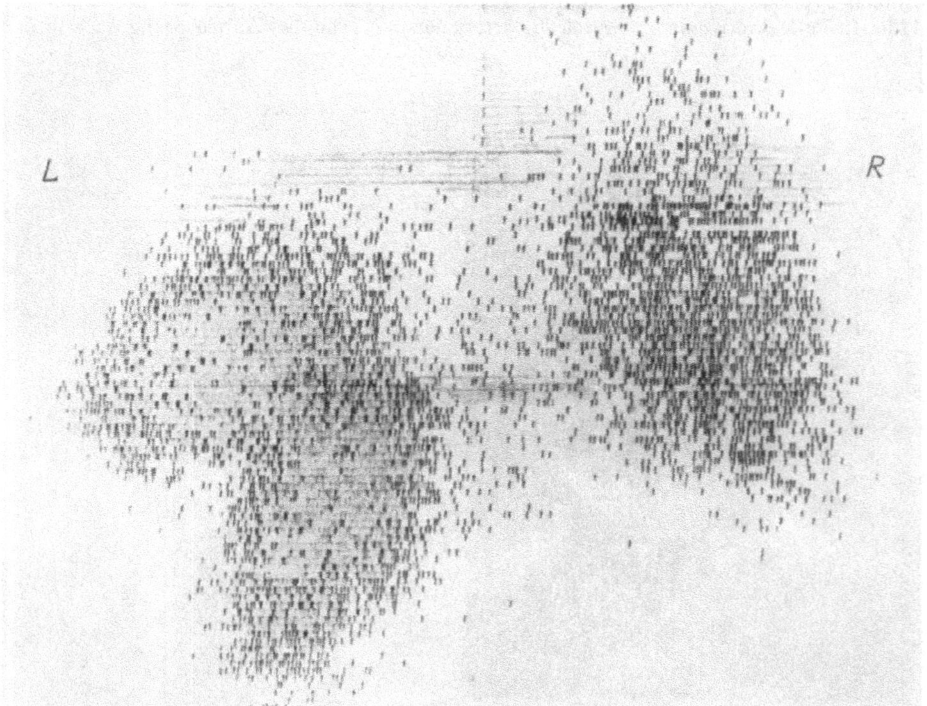

Abb. 7. Angioscan nach Etagenaortographie, Instillation von 1 mCi 113m-In-Fe-Komplexverbindung. Pyelonephritische Schrumpfniere rechts bei Zyste links, operativ bestätigt

Diskussion

Die Angioscintigraphie beruht auf dem Nachweis von Mikroembolien in einem selektiv sondierten Organ oder Organabschnitt, die durch in ihrer Größe definierte und mit 131-Jod oder 113m-Indium markierte Molekülkolloide hervorgerufen werden. Bei der renalen Angioscintigraphie haben wir unsere theoretischen Überlegungen hinsichtlich der morpho-histologischen Eigenart der Tumorgefäße und der hämodynamischen Besonderheiten dergestalt bestätigen können, daß wir in den bis jetzt untersuchten, angiographisch unsicher oder fehlbeurteilten Fällen die Diagnose durch das Angioscan abklären konnten. Da das differenzierte Angioscintigramm oder Angioscintiphoto in erster Linie vom Verteilungsmuster abhängt, das durch den Gefäßquerschnitt des Tumorgebietes bzw. des Nierenparenchyms vorgegeben ist, sind wir dazu übergegangen, unsere Angioscans ausschließlich mit 113m-Indium-Eisen-Makrokolloiden durchzuführen. Bei gleicher biologischer Halbwertszeit und etwa gleicher Kolloidgröße sind wir in der Lage, eine höhere Aktivitätsmenge, eventuell auf ein größeres Volumen verteilt, zu verabfolgen, ohne die Organ- und Ganzkörperstrahlenbelastung wesentlich zu erhöhen.

Wir glauben, daß die renale Angioscintigraphie eine wertvolle Ergänzungsuntersuchung darstellt, die einen wesentlichen diagnostischen Beitrag bei intrarenal gelegenen, angiographisch „gefäßarmen" oder „gefäßfreien" Tumoren leistet. Die Angiographie gibt Auskunft über die Lage und die arterielle Zugehörigkeit des zu untersuchenden Gefäßabschnittes, die Angioscintigraphie über die Verteilung und Natur der Praecapillaren und Capillaren sowie über den Sitz und die Ausdehnung des zu beurteilenden Prozesses. Auch kann indirekt eine Aussage über die Durchblutungsverhältnisse ungleicher Gewebsanteile gewonnen werden.

Literatur

1. SELDINGER, S. I.: Acta radiol. **39**, 368—376 (1953).
2. MAYOR, G., CAMPONOVO, F., PUPATO, F., ZINGG, E., WELLAUER, J., RÜTTIMANN, A., DEL BUONO, M., SCHNAUDER, A.: Urologe **2**, 76—96 (1963).
3. LÖHR, E.: Fortschr. Röntgenstr. **109**, H. 6, 695—701 (1968).
4. SEIFERT, J.: Vortrag a. d. Deutschen Röntgenkongreß, Stuttgart 1969.
5. ECOIFFIER, J.: Ann. d. Radiol. **11**, 81—87 (1968).
6. IBRAHIM, S.: J. nucl. Med. **9**, 325 (1968).

Zellkinetische Untersuchungen zur Klärung der Ursache periodischer Schwankungen des Neutrophilenspiegels im Blut bei einem Fall von zyklischer Neutropenie

Examination of the Cellular Kinetics to Explain the Periodic Fluctuations in the Neutrophil Count in a Case of Cyclic Neutropenia

G. Meuret

Abteilung für Nuklearmedizin der Medizinischen Universitätsklinik Freiburg i. Br. und Abteilung für klinische Physiologie, Zentrum für klinische Grundlagenforschung der Universität Ulm

Summary

The kinetics of granulopoiesis and neutrophils was studied in a case of cyclic neutropenia using labeling techniques (^{3}HTDR, ^{3}H-DFP) combined with serial quantitative evaluations of the granulopoietic cells. It was demonstrated that the oscillation of the neutrophil count was caused by rhythmic fluctuations of the neutrophil production rate due to periodic changes in the proliferation activity of granulopoiesis. The results could be interpreted as indicating pathological regulation of granulopoiesis.

Das Krankheitsbild der zyklischen Neutropenie ist gekennzeichnet durch ein autonomes, periodisches, synchrones Auftreten und Wiederverschwinden von Beschwerden zusammen mit Fieber, Neutropenie und Monozytose. Durch Beobachtungen und Untersuchungen an etwa 50 Fällen konnte Stellung genommen werden zu Fragen wie Geschlechtsverteilung [26, 27], Symptomatik und Krankheitsverlauf [4, 5, 6, 7, 18, 19, 22, 26, 27, 29, 31, 32], Hormonhaushalt [17, 26, 27, 28, 33], Vererbung [11, 16, 20, 23, 27], z. T. auch zu hämatologischen und pathologisch-anatomischen Problemen [15, 21, 24, 25, 26, 27, 28, 30, 34]. Die Frage nach der Ursache der Krankheitssymptome jedoch blieb nahezu ungeklärt.

Es ist Aufgabe dieser Arbeit, das granulozytäre Zellerneuerungssystem bei einer Patientin mit zyklischer Neutropenie quantitativ, morphologisch und durch Anwendung verschiedener Zellmarkierungsmethoden mit Radioisotopen zellkinetisch zu beschreiben, um zur Klärung der Pathophysiologie dieser Erkrankung beizutragen.

Material und Methoden

Kasuistik: Untersucht wurde eine 53-jährige Patientin (H. K., Journal Nr. 13/9267 der Med. Univ.-Klinik Freiburg), die seit dem 50. Lebensjahr unter Beschwerden litt, welche einförmig und regelmäßig in 3-wöchigen Abständen rezidivierten.

Während des Klinikaufenthaltes wurde das Vorliegen einer zyklischen Neutropenie festgestellt. — 126 Tage lang wurde nahezu täglich der Zellstatus des Blutes kontrolliert und in bestimmten Abschnitten des Krankheitszyklus zellkinetische Untersuchungen durchgeführt.

Untersuchungen der Granulopoese mit Tritium-markiertem Thymidin (^3HTDR): Da die angewandte Methode an anderer Stelle ausführlich besprochen wurde [9, 10, 12, 13, 14], soll hier lediglich auf das Prinzip eingegangen werden. — Der Patientin wurde 0,1 μC/g Körpergewicht ^{3}HTDR* intravenös verabfolgt. Bis zum 12. Tag nach der Injektion wurden häufig Knochenmarkpräparate hergestellt, bis zum 15. Tag Leukozytenkonzentrat-Ausstriche aus der Mundspülflüssigkeit. Die luftgetrockneten und in absolutem Methylalkohol fixierten Ausstriche wurden mit dem Kodak AR10 stripping-film autoradiographiert, nach einer Expositionszeit von 69 Tagen entwickelt, fixiert und nach GIEMSA gefärbt. — Pro Zeitpunkt wurden mindestens 500 myeloische Zellen differenziert und die Silberkornzahlen in der Filmschicht über den Zellkernen bestimmt. Zellkerne, die mehr als 3 Silberkörner aufwiesen, wurden als markiert gewertet. Der prozentuale Anteil markierter Zellen einer Reifungsklasse ergab den „Markierungsindex".

Nach den üblichen morphologischen Kriterien wurden 8 granulopoetische Zelltypen als Kompartments unterschieden: M_1 = Myeloblast, M_2 = Promyelozyt, M_3 = halbreifer Myelozyt, M_4 = reifer Myelozyt, M_5 = Metamyelozyt, M_6 = Jugendlicher, M_7 = stabkerniger Neutrophiler, M_8 = segmentkerniger Neutrophiler. — Die in Abb. 2 angegebenen Normalwerte für die verschiedenen Kompartments sind das Ergebnis zytologischer Differenzierung von fast 30 000 granulopoetischen Zellen in 14 Knochenmarkpräparaten einer Normalperson [12].

Bestimmung der Neutrophilenspeicher und des Neutrophilenumsatzes im Blut: Etwa 500 ml Blut wurden in einem 100 ml ACD enthaltenden, sterilen und pyrogenfreien Plastikbeutel abgenommen, mit 1 μC/ml Blut Tritium-markiertem Diisopropylfluorophosphat (^{3}H-DFP)** versetzt und nach einstündiger Inkubation bei Zimmertemperatur re-infundiert. Während der Autotransfusion, die etwa 10 Minuten dauerte, wurden zu Anfang, in der Mitte und am Ende aus dem Infusionsschlauch Blutproben entnommen für Leukozyten-Zählungen und -Differenzierungen und zur Herstellung von Leukozytenkonzentraten.

Zur Bestimmung der autotransfundierten Blutmenge wurde das gesamte Transfusionsbesteck vor, nach der Blutentnahme und nach der Re-Infusion gewogen.

Unter Berücksichtigung der transfundierten Blutmenge und der Zahl markierter Neutrophiler pro μl Infusionsblut ließ sich die Gesamtzahl der re-infundierten markierten Neutrophilen berechnen. (In den Autoradiogrammen der Leukozytenkonzentrat-Ausstriche aus dem Plastikbeutel betrug der Markierungsindex der Neutrophilen 100%.)

Nach der Re-Infusion wurden anfangs stündlich, später in größeren Zeitintervallen bis zum 3. Tage regelmäßig Blutproben entnommen und Leukozytenkonzentrate hergestellt. 3—5mal täglich erfolgten Leukozyten-Zählungen und -Differenzierungen im peripheren Blut. Die Präparate wurden mit der stripping-film-Methode autoradiographiert (Expositionszeit 80 Tage) und der Anteil markierter Neutrophiler durch mikroskopische Auswertung von je 3 000 Neutrophilen pro Zeitpunkt festgestellt.

Analog dem Vorgehen von ATHENS und Mitarb. [2] wurden für Neutrophile 3 verschiedene Speichergrößen ermittelt:

Die Werte des gesamten Blutneutrophilenspeichers *(GBNS)* ergaben sich durch Multiplikation der Zahl der autotransfundierten markierten Neutrophilen (N_N) mit deren Verdünnungsfaktor im Blut am Ende der Re-Infusion.

Der zirkulierende Neutrophilenspeicher stellt die auf das Blutvolumen (Körperoberfläche × 2 680 [3]) bezogene Neutrophilenzahl dar.

Die Differenz zwischen *GBNS* und *ZNS* ergab den marginalen Neutrophilenspeicher *(MNS)*.

Die intravasale Halbwertszeit der autotransfundierten Neutrophilen wurde graphisch im halblogarithmischen Koordinatensystem bestimmt (Abb. 4). Bei Kurven mit 2 Abschnitten wurde hierzu der 2. Abschnitt verwendet, d. h. es wurden lediglich die später als 2 Stunden nach der Autotransfusion ermittelten Werte berücksichtigt.

Ein Näherungswert für die stündliche Umsatzrate der Neutrophilen (*NUR*) ergab sich durch Multiplikation von *GBNS* mit ln 2, dividiert durch die intravasale Halbwertszeit der Zellen.

---

* Fa. Radiochemical Centre Amersham, spez. Akt. 5 Ci/mM.

** Fa. Radiochemical Centre Amersham, spez. Akt. 4820 mCi/mM.

Untersuchungen, die mit dieser Methode an Normalpersonen durchgeführt wurden, ergaben ähnliche Werte wie sie von Athens und Mitarb. durch Markierung mit DF^{32}P und Messung der Radioaktivität der gesamten Leukozyten gefunden wurden. Zur Beurteilung der bei der Patientin gewonnenen Ergebnisse (Tab. 1) wurden deshalb die von diesen Autoren angegebenen Normalbereiche [2] herangezogen.

Ergebnisse

Begleitet von bestimmten klinischen Symptomen der Patientin schwankte der Neutrophilenspiegel im Blut periodisch zwischen leicht neutropenischen Werten und einem nahezu völligen Fehlen der Neutrophilen. Aufgrund der registrierten Neutrophilenkurve (Abb. 1) und der beobachteten Symptomatik wurde

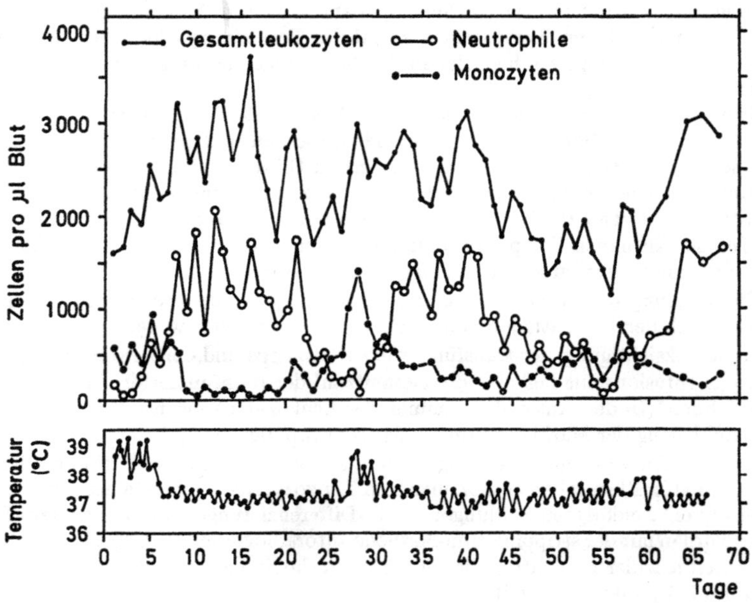

Abb. 1. Verhalten des Leukozytenspiegels, der Neutrophilen- und Monozytenzahlen im peripheren Blut und der Körpertemperatur während zweier Krankheitsphasen einer Patientin mit zyklischer Neutropenie

der etwa 26 Tage dauernde Krankheitszyklus in folgende 4 Abschnitte unterteilt (s. auch Abb. 8):

1. Ein Abschnitt der „Remission" (Dauer ~12 Tage), mit Neutrophilenwerten zwischen 1 500 und 2 500 µl Blut.

2. In dem folgenden „Initialabschnitt der Neutropenie" (~9 Tage) fiel der Neutrophilenspiegel ab, die Monozytenzahl stieg an, und es traten Beschwerden auf, die mit dem Grad der Neutropenie zunahmen.

3. Im „Abschnitt der maximalen Neutropenie" (Neutrophilenzahl meist geringer als 100 pro µl Blut) erreichten auch die Beschwerden und die Monozytose ihren Höhepunkt.

4. Nach diesem etwa 1—2 Tage dauernden Zustand setzte mit dem 4. Krankheitsabschnitt ein Anstieg der Neutrophilen im Blut und ein Abfall der Monozyten ein, verbunden mit einem schließlich vollständigen Verschwinden der Krankheitssymptome („Reparation", ~5 Tage).

Die Zahlen der übrigen Blutzellen waren unauffällig und folgten nicht den zyklischen Veränderungen der Neutrophilen und Monozyten.

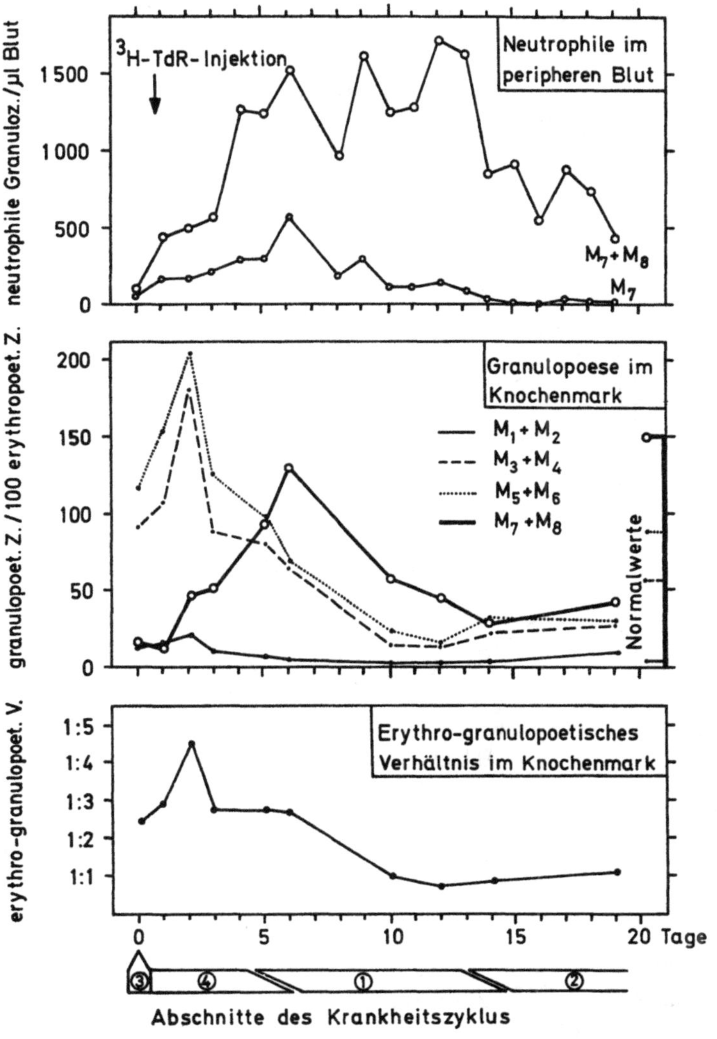

Abb. 2. Neutrophilenspiegel im Blut, Granulopoese und erythro-granulopoetisches Verhältnis im Knochenmark während einer Phase einer zyklischen Neutropenie (Normalwerte nach FLIEDNER [12]).

M_1 = Myeloblast; M_2 = Promyelozyt; M_3 = halbreifer Myelozyt; M_4 = reifer Myelozyt; M_5 = Metamyelozyt; M_6 = Jugendlicher; M_7 = stabkerniger neutrophiler Granulozyt; M_8 = segmentkerniger neutrophiler Granulozyt

Krankheitsabschnitte: 1 Remission, 2 Initialabschnitt der Neutropenie, 3 Zeitpunkt der maximalen Neutropenie, 4 Reparation

Zytologische Befunde im Knochenmark

Aufgrund der relativ konstanten Erythrozyten- und Retikulozytenwerte im peripheren Blut wurde eine während des Krankheitsgeschehens gleichbleibende Erythropoese angenommen, zu der die Befunde der Granulopoese in Beziehung gesetzt wurden. Die derart umgerechneten Ergebnisse von Knochenmarkdifferentialzählungen (1 500—3 000 kernhaltige Zellen pro Zeitpunkt) sind in Abb. 2 dargestellt. Die Kurven lassen erkennen, daß synchron mit dem Ansteigen und Absin-

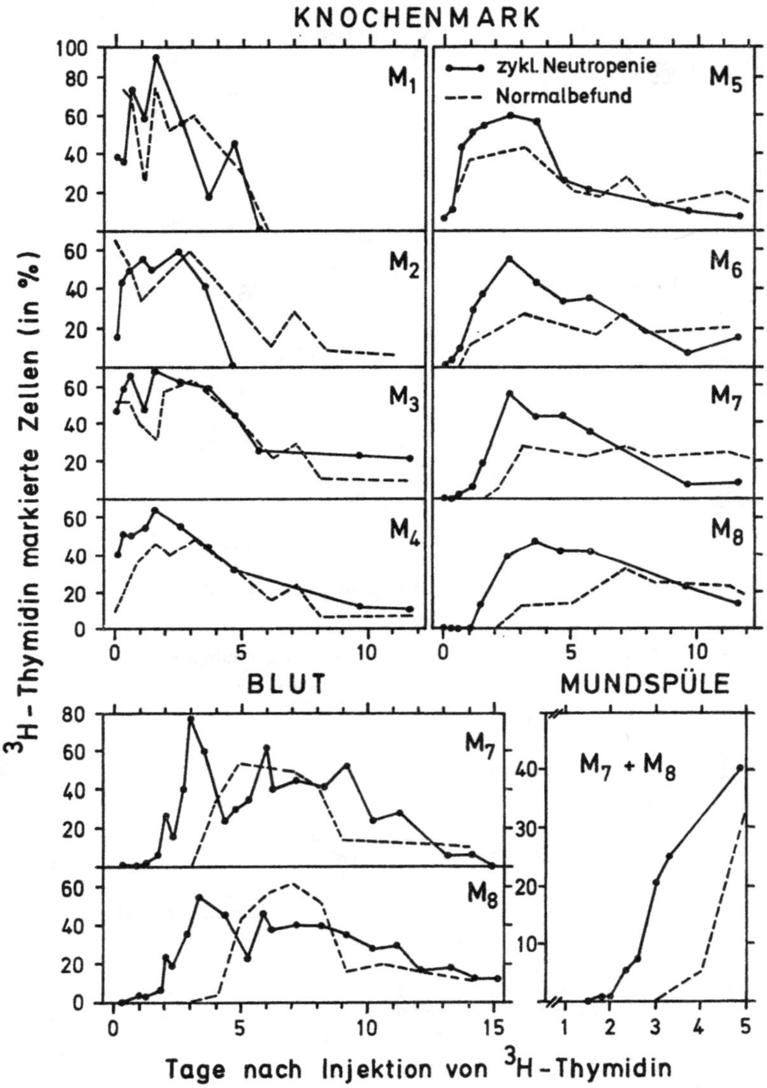

Abb. 3. Verhalten der Markierungsindizes nach intravenöser Injektion von ^{3}H-Thymidin bei granulopoetischen Zellen im Knochenmark, bei neutrophilen Granulozyten im peripheren Blut und im Sediment der Mundspülflüssigkeit (M_1—M_8 siehe Legende von Abb. 2)

ken des Neutrophilenspiegels im Blut auch bei der Granulopoese im Knochenmark quantitative Veränderungen abliefen. Die übrigen Zellsysteme des Knochenmarks verhielten sich unauffällig.

Zellkinetik der Granulopoese während der Reparation

Etwa 1 Tag nach dem Zeitpunkt der minimalen Neutrophilenzahl im Blut wurde der Patientin ^{3}HTDR injiziert (s. Abb. 2). Die Veränderungen der Markierungsindices granulopoetischer Vorstufen während der ersten 12 Tage nach der Injektion von ^{3}HTDR sind in Abb. 3 dargestellt. Zum Vergleich wurden in die Diagramme Befunde einer ^{3}HTDR-Untersuchung bei einem Patienten mit normaler Hämatopoese eingezeichnet [12].

Die Ergebnisse im Proliferationsspeicher der Granulopoese (Zellformen $M_1 - M_4$) wichen nicht signifikant von den Befunden eines hämatologisch Gesunden ab. Eindeutige Unterschiede ergaben sich jedoch bei den reiferen, initial nicht markierten Zellkompartments der Granulopoese (Zellformen $M_5 - M_8$). Der Einstrom markierter Zellen aus dem Proliferationsspeicher in die Speicher M_5 und nachfolgende erfolgte wesentlich früher als normalerweise. Die ersten markierten jugendlichen Neutrophilen (M_6) wurden schon 8,5 Stunden nach der ^{3}HTDR-Injektion beobachtet. Markierte stabkernige Neutrophile (M_7) fanden sich in den Knochenmarkpräparaten 15 Stunden, segmentkernige Neutrophile (M_8) 38 Stunden nach der Injektion (Normalwerte [12]: 12, 18, 60 Stunden). Der höchste Markierungsindex wurde bei M_6 und M_7 38 Stunden, bei M_8 61 Stunden nach der ^{3}HTDR-Applikation erreicht. Diese Werte waren mehr als doppelt so hoch wie die zu gleichen Zeitpunkten gemessenen Normalwerte.

Im späteren Verlauf der Untersuchung (etwa nach dem 8. Tag) fanden sich keine eindeutigen Abweichungen der Markierungsindices von der Norm mehr.

Im venösen Blut konnten 24 Stunden nach der ^{3}HTDR-Injektion die ersten markierten stab- und segmentkernigen Granulozyten (M_7 und M_8) nachgewiesen werden. 43 Stunden nach der Injektion wurden auch in den Autoradiogrammen der Präparate aus dem Sediment der Mundspülflüssigkeit markierte neutrophile Granulozyten gefunden.

Die Reifungszeit der Neutrophilen, definiert als Zeitintervall zwischen dem ersten Auftreten markierter Metamyelozyten und dem Erscheinen markierter Neutrophiler im Blut, lag zum Zeitpunkt der Untersuchung bei 13 Stunden. Sie war damit gegenüber der Norm (96—120 Stunden) [12] auf etwa $^1/_7$ verkürzt.

Intravasale Halbwertszeit, Speichergrößen und Umsatzrate der Neutrophilen im Blut

Während einer Remission (2 400 Neutrophile pro µl Blut) und eines Initialabschnittes der Neutropenie (240 Neutrophile pro µl Blut) wurden zellkinetische Daten der Blutneutrophilen durch Autotransfusion von ^{3}H-DFP-markierten Neutrophilen ermittelt.

Zum Zeitpunkt der Untersuchung in der Remission lag die intravasale Halbwertszeit der Neutrophilen bei 5 Stunden (Abb. 4, Tab. 1); im neutropenischen Krankheitsabschnitt dagegen betrug sie nur 2,3 Stunden.

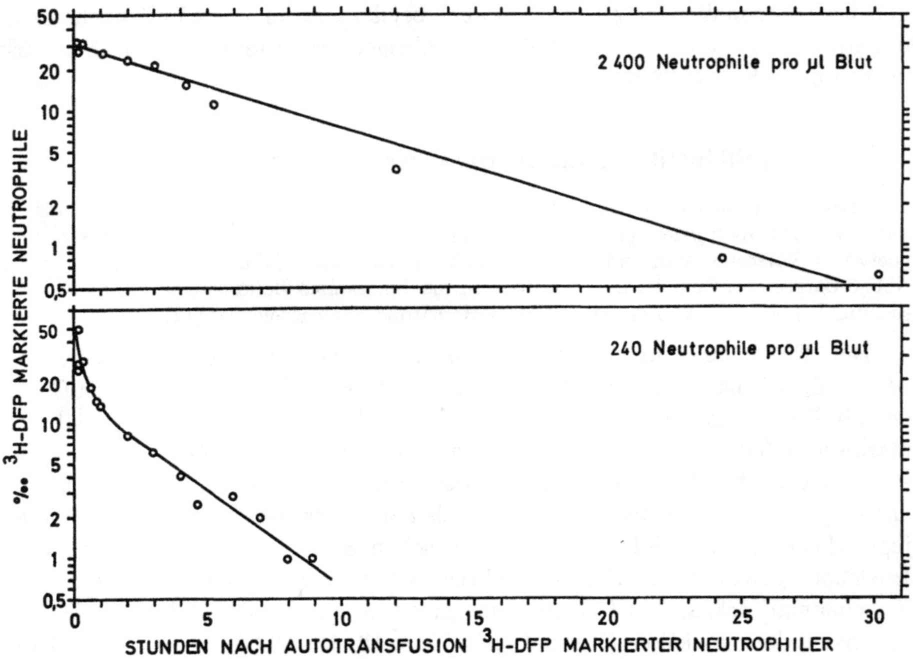

Abb. 4. Markierungsindex neutrophiler Granulozyten im Blut nach Autotransfusion in vitro mit ^{3}H-DFP-markierten Granulozyten in zwei Krankheitsabschnitten

Tabelle 1. *Speichergrößen, intravasale Halbwertszeit und Umsatzrate der Neutrophilen in zwei verschiedenen Krankheitsabschnitten bei einer Patientin mit zyklischer Neutropenie* (Normalbereiche nach Athens und Mitarb. [2] und Albritton [1])

	Neutroph.[a]	*GBNS*[b]	*ZNS*[c]	*MNS*[d]	$T_{1/2, N}$[e]	*NUR*[f]
Remission	2400	59	16	43	5	8,2
Neutropenie	240	10	1,6	8,4	2,3	3
Normal	4500 (3—7·10³)	70 (14—160)	31 (11—46)	39 (0—85)	6,7 (4—10)	6,8 (2,1—14)

[a] Neutrophile pro µl Venenblut.
[b] Gesamter Blutneutrophilenspeicher (·10⁷/kg Körpergewicht).
[c] Zirkulierender Neutrophilenspeicher (·10⁷/kg Körpergewicht).
[d] Marginaler Neutrophilenspeicher (·10⁷/kg Körpergewicht).
[e] Intravasale Halbwertszeit der Neutrophilen (Stunden).
[f] Neutrophilenumsatzrate (·10⁷/kg Körpergewicht/Stunde).

Während der Remission lagen die Größen der drei verschiedenen Neutrophilenspeicher *(GBNS, ZNS, MNS)* in dem von Athens und Mitarb. angegebenen Normalbereich [2]. — Zum Zeitpunkt der 2. Untersuchung war im zirkulierenden Blut nur noch $^{1}/_{10}$ der Neutrophilenmenge der Voruntersuchung vorhanden. Die Größe des *GBNS* war auf $^{1}/_{6}$, des *MNS* auf $^{1}/_{5}$ abgefallen, die Neutrophilenumsatzrate dagegen nur um $^{1}/_{3}$, so daß dieser Wert noch im Normalbereich lag.

Diskussion

Ursache der rhythmischen Veränderungen des Neutrophilenspiegels im Blut der untersuchten Patienten

Schwankungen der Neutrophilenzahl im Blut können Veränderungen folgender Komponenten zugrunde liegen:

1. der Zellverteilung zwischen den Neutrophilenspeichern,
2. der Neutrophilenumsatzrate,
3. der Neutrophilenproduktionsrate.

Der Erörterung dieser Punkte liegen Daten zugrunde, die durch ^{3}H-DFP markierte Neutrophile zum Zeitpunkt einer fast maximal ausgeprägten Neutropenie und während einer Remission gewonnen wurden. Da kein Hinweis auf eine sprunghafte Änderung der Neutrophilenkinetik vorliegt, bedingt z. B. durch einen Infekt, durch Leukozytenantikörper oder eine Veränderung der Milzfunktion, kann angenommen werden, daß bei diesen Untersuchungen während des Zyklus vorkommende Extremwerte bestimmt wurden, zwischen denen die Umsatzraten und Speichergrößen während der übrigen Abschnitte der Krankheit lagen.

Zu 1. Analysiert man unter der o. g. Fragestellung die ermittelten Größen der Neutrophilenspeicher, so findet man während der Neutropenie einen im Vergleich zur Remission um 90% verminderten Zellgehalt im *ZNS*, dem ein Abfall von nur 83% im *GBNS* gegenübersteht (Tab. 2). Das Verhältnis von *GBNS* zu *ZNS* hatte sich während der Neutropenie etwa um 40% von *ZNS* zugunsten von *GBNS* verschoben.

Tabelle 2. *Verhalten der Neutrophilenspeicher während einer Remission und einer fast maximal ausgeprägten Neutropenie*
(Größen *GBNS* und *ZNS*: Zellen·10^7/kg Körpergewicht)

	GBNS	*ZNS*	*GBNS*:*ZNS*	[*ZNS*]*
A. Remission	59	16	1:0,27	16
B. Neutropenie	10	1,6	1:0,16	2,24
Abfall (A—B)	83%	90%	40%**	86%

* [*ZNS*] = *ZNS*, wenn *GBNS*:*ZNS* = 1:0,27.
** Abfall von *ZNS* als Anteil von *GBNS*.

Wird diese Zellverschiebung zwischen den Speichern rechnerisch eliminiert, so würde sich für *ZNS* ein Abfall von 86% ergeben, der ausschließlich Folge des Zellverlusts von *GBNS* war.

Aufgrund dieser Analyse kann gesagt werden, daß die Neutropenie fast ausschließlich Folge einer Verkleinerung des Gesamtspeichers der Neutrophilen und nur zu 4% durch eine Änderung der Zellverteilung in den Speichern bedingt war.

Zu 2. Die Ursache des zyklisch auftretenden Neutrophilenabfalls war mit Sicherheit nicht ein erhöhter Neutrophilenuntergang, da die Neutrophilenumsatzrate während der neutropenischen Krankheitsabschnitte nicht anstieg, sondern abfiel.

Zu 3. Eine Vorstellung der Neutrophilenproduktionsrate während des Krankheitszyklus soll mit Hilfe des in Abb. 5 dargestellten Modells der Granulopoese gewonnen werden. Es besteht aus hintereinandergeschalteten Kompartments, in denen sich die verschiedenen Zellformen der Granulopoese während der Proliferation, Reifung und Speicherung eine Zeitlang aufhalten.

Aus der Größe des Zelleinstroms in das Kompartment und des Zellabstroms resultiert entweder ein Fließgleichgewicht oder eine bestimmte Änderung der Zellzahl in diesem. Vorausgesetzt, daß innerhalb des Speichers keine Zellen zugrunde gehen, gilt demnach folgende Beziehung:

Änderung der Speichergröße = Zelleinstrom − Zellabstrom oder Zelleinstrom = Änderung der Speichergröße + Zellabstrom (Daten in Zellen pro Zeiteinheit).

Durch die Neutrophilenumsatzrate *(NUR)* und die Änderung des Gesamt-Blutneutrophilenspeichers *(GBNS)* läßt sich mit dieser Formel zunächst die Rate des Neutrophileneinstroms ins Blut bestimmen. Diese Größe, eingesetzt als Neutrophilenabstrom, ergibt zusammen mit der Änderung im Neutrophilenspeicher des Knochenmarks die Produktionsrate von Neutrophilen aus den Vorläuferkompartments.

Mit den zur Verfügung stehenden Daten wurde diese Analyse wie folgt durchgeführt:

Das in Abb. 5 dargestellte Verhalten der Neutrophilenumsatzrate während des Krankheitszyklus ergab sich unter den oben näher ausgeführten Voraussetzungen.

Das in Tab. 2 dargestellte Verhalten der Speichergrößen von *GBNS* und *ZNS* ließ es als wahrscheinlich erscheinen, daß sich diese während der Krankheitsphasen annähernd parallel änderten. Zur Ermittlung der Neutrophileneinstromrate ins Blut könnte dann auch die Änderung von *ZNS* in die Formel eingesetzt werden.

Das Verhalten von *ZNS* während des Zyklus ergab sich aus einer Kurve, die an den Neutrophilenzahlen im Blut orientiert war. Hierbei wurde, gestützt auf die relativ konstanten Werte der Erythrozyten und Retikulozyten, vorausgesetzt, daß sich das Blutvolumen während des Zyklus nicht wesentlich veränderte. Mit Hilfe dieser Kurve (s. Abb. 5, „Neutrophile im Blut") konnte für jeden Zeitpunkt (t) des Zyklus ein Maß für die Speichergröße (y) und für die tägliche ($\Delta t = \text{const}$) Änderung des Speichers (Δy) abgelesen werden.

Die graphisch durchgeführte Addition von Δy zu den Werten der Neutrophilenumsatzrate in den entsprechenden Zeitpunkten ergab eine Kurve für die Einstromrate der Neutrophilen ins Blut (Abb. 5, „Neutrophilen-Einstrom ins Blut").

Das Verhalten der Neutrophilenproduktion ergab sich schließlich durch graphische Addition der Änderungsraten des Neutrophilenspeichers im Knochenmark zu der Kurve des Neutrophileneinstroms ins Blut (Abb. 5, „Neutrophilenbildungsrate im Knochenmark").

Diese zellkinetische Analyse ergab, daß die Neutrophilenproduktion rhythmisch verlief mit derselben Phasendauer wie die zyklischen Veränderungen des Neutrophilenspiegels im Blut. Den Schwankungen der Neutrophilenproduktion folgten jeweils gleichgerichtete Veränderungen des Neutrophilenspiegels im Blut nach. Die bei der untersuchten Patientin beobachteten zyklischen Veränderungen des Neutrophilenspiegels waren somit Folge rhythmischer Schwankungen der Neutrophilenproduktion im Knochenmark.

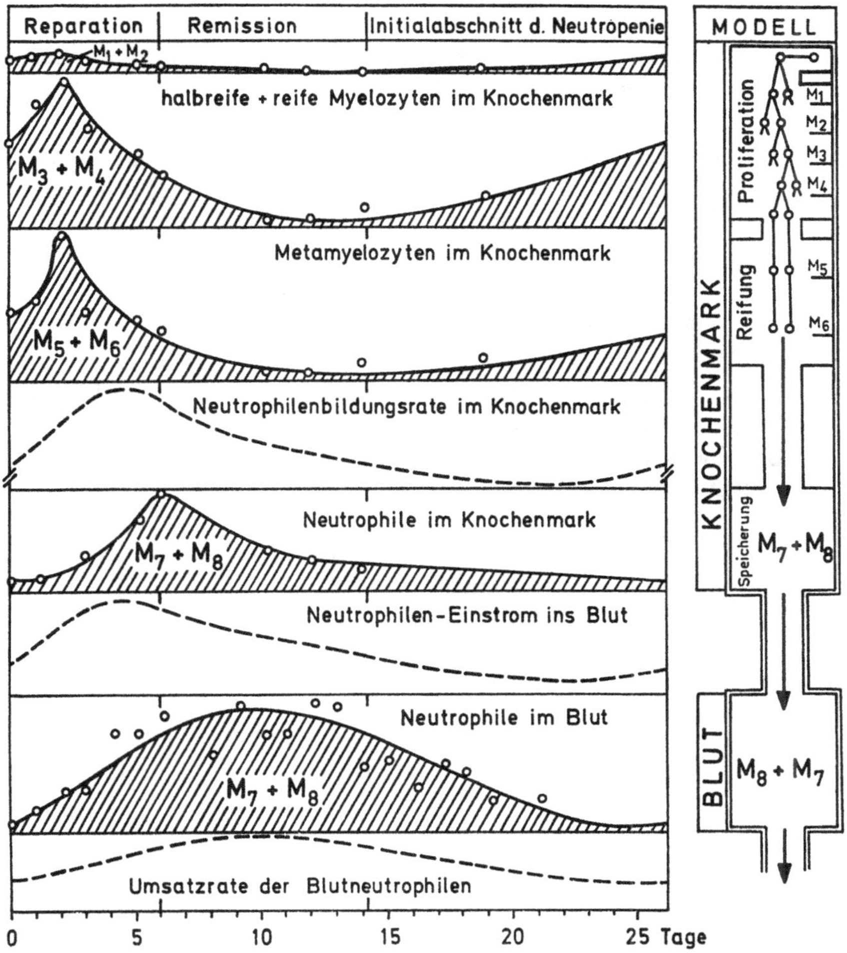

Abb. 5. Halbschematische Darstellung des Verhaltens der Speichergrößen und der Zellkinetik der verschiedenen Zellkompartments der Granulopoese während eines Krankheitszyklus (M_1—M_8 siehe Legende von Abb. 2)

Zellkinetik und funktionelles Verhalten der Granulopoese während des Krankheitszyklus

Einen Überblick über die Befunde der verschiedenen Zellkompartments der Granulopoese im Verlauf einer Krankheitsphase vermittelt Abb. 5 (die Kurven wurden im letzten Teil des Initialabschnittes der Neutropenie bis zu den Werten eines maximal neutropenischen Zustandes ergänzt).

Während des Zyklus konnten folgende funktionell verschiedene Abschnitte beobachtet werden:

Reparation (Anfang): Bei minimalen Neutrophilenwerten im Blut nahmen zunächst im Proliferations- und Reifungsspeicher die Zellzahlen rasch zu. Durch eine, wie die Befunde der ^{3}HTDR-Untersuchung zeigten (Abb. 3), ungestörte

Proliferation und stark beschleunigte Reifung setzte dann rasch eine Zunahme der Neutrophilenbildungsrate mit vermehrtem Neutrophileneinstrom ins Blut ein; die entleerten Neutrophilenspeicher im Knochenmark und im Blut wurden wieder aufgefüllt.

Die vorliegende Situation der Granulopoese entspricht etwa einem experimentell herbeigeführten Zustand nach Entspeicherung der Neutrophilenpools durch Leukopherese. Dieser Eingriff wird von einer normalen Granulopoese ebenfalls mit einer Mehrproliferation und Mehrproduktion von Neutrophilen beantwortet [8].

Reparation (Mitte): Nachdem die Neutrophilenwerte im Blut der Patientin etwa 500 pro µl erreicht hatten, nahmen die Zellzahlen im Proliferationsspeicher bereits wieder ab.

Reparation (Ende): Die zu Beginn der Reparation im Proliferationsspeicher vermehrt neugebildeten Zellen strömten nach der Ausreifung in den Neutrophilenspeicher des Knochenmarks ein, der am Ende der Reparation seine maximale Zellzahl erreichte.

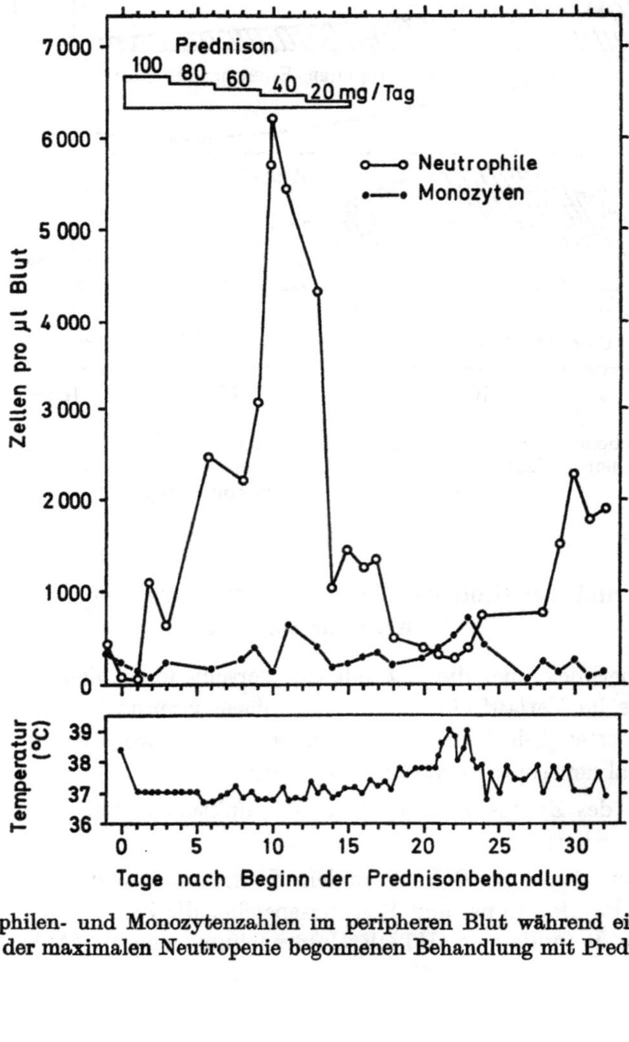

Abb. 6. Neutrophilen- und Monozytenzahlen im peripheren Blut während einer zum Zeitpunkt der maximalen Neutropenie begonnenen Behandlung mit Prednison

Remission: Durch den Abfall der Zellzahlen im Proliferations- und Reifungsspeicher ging die Neutrophilenbildungsrate während der Remission stetig zurück. Etwa 3 Tage nach dem Maximum im Knochenmark wurden auch im Blut die höchsten Neutrophilenzahlen gezählt, die jedoch niemals Normalwerte erreichten.

Die während dieser Krankheitsabschnitte erhobenen Befunde zeigten zunächst ein funktionell normales Verhalten der Granulopoese zu Beginn der Reparation. Die Zunahme der Proliferation als Antwort auf die Neutropenie war jedoch „unterdosiert“ (Ende der Reparation), so daß keine Normalisierung des Neutrophilenspiegels erreicht wurde (Beginn der Remission). Im weiteren Verlauf des Krankheitszyklus zeigte sich ein zunehmendes Mißverhältnis zwischen Neutrophilenbildung und Neutrophilenspiegel im Blut.

Initialabschnitt der Neutropenie: In diesem Abschnitt fielen die Neutrophilenbildungsrate und dadurch auch die Zellzahlen in den Neutrophilenspeichern auf Minimalwerte ab. Gleichzeitig aber nahm die Zahl der Neutrophilen-Präkursoren $M_1 - M_6$ wieder zu.

Möglicherweise lag dieser Beobachtung eine Reifungsstörung beim Übergang der jugendlichen zu den stabkernigen Neutrophilen zugrunde. Würde diese Annahme zutreffen, so könnte der Neutrophilenanstieg bis auf Werte von 6 000 pro µl (Abb. 6), der auf eine Prednisonmedikation in dieser Situation erfolgte, als eine durch das Medikament ermöglichte Ausreifung von im Knochenmark „aufgestauten“ Präkursoren interpretiert werden. Diese Deutung würde auch den Neutrophilenabfall erklären, der trotz Weiterführung der Therapie rasch nach dem Maximum einsetzte.

Zum *Zeitpunkt der maximalen Neutropenie*, bei Neutrophilenwerten unter 100 pro µl Blut, sprang plötzlich die Proliferation der Granulopoese für einige Tage wieder voll an; Zeichen einer gestörten Reifung waren dann nicht mehr nachweisbar.

Ein Modell zur Simulation der beobachteten Granulopoese

In Abb. 7 ist ein Modell dargestellt, mit dem das Verhalten der Zellzahlen in den verschiedenen Speichern der Granulopoese während des Krankheitszyklus nachgeahmt wird. In diesem Schema ist das granulozytäre Zellerneuerungssystem als ein technisches Modell dargestellt, in dem die einzelnen Anteile des Systems (Kompartments) durch kaskadenartig hintereinandergeschaltete und durch Flüssigkeitsheber verbundene Gefäße simuliert sind. Der vom Stammzellspeicher ausgehende Zellstrom muß zwischen den Kompartments M_6 und M_7 ein besonders hohes Niveau überwinden. Ein Regulationssystem sondiert den Neutrophilenspiegel im Blut und steuert durch Ventile den Zelleinstrom aus dem Stammzellspeicher in das System sowie den Zelldurchfluß zum Reifungsspeicher. Durch eine zwischengeschaltete Spiralfeder spricht diese Steuerung jedoch erst bei extrem niedrigen Neutrophilenzahlen im Blutspeicher an und gibt dann für kurze Zeit den Zellfluß frei. Schon nach einem geringen Neutrophilenanstieg im Blut „kippt“ der Regulationsmechanismus zurück und bremst erneut die Zellproliferation.

Zellen, die während der auf einige Tage befristeten Freigabe des Zellflusses neu gebildet wurden, durchfließen wellenartig hintereinander die verschiedenen Speicher. Gegen Ende der Remission haben die während der Reparation vermehrt neu

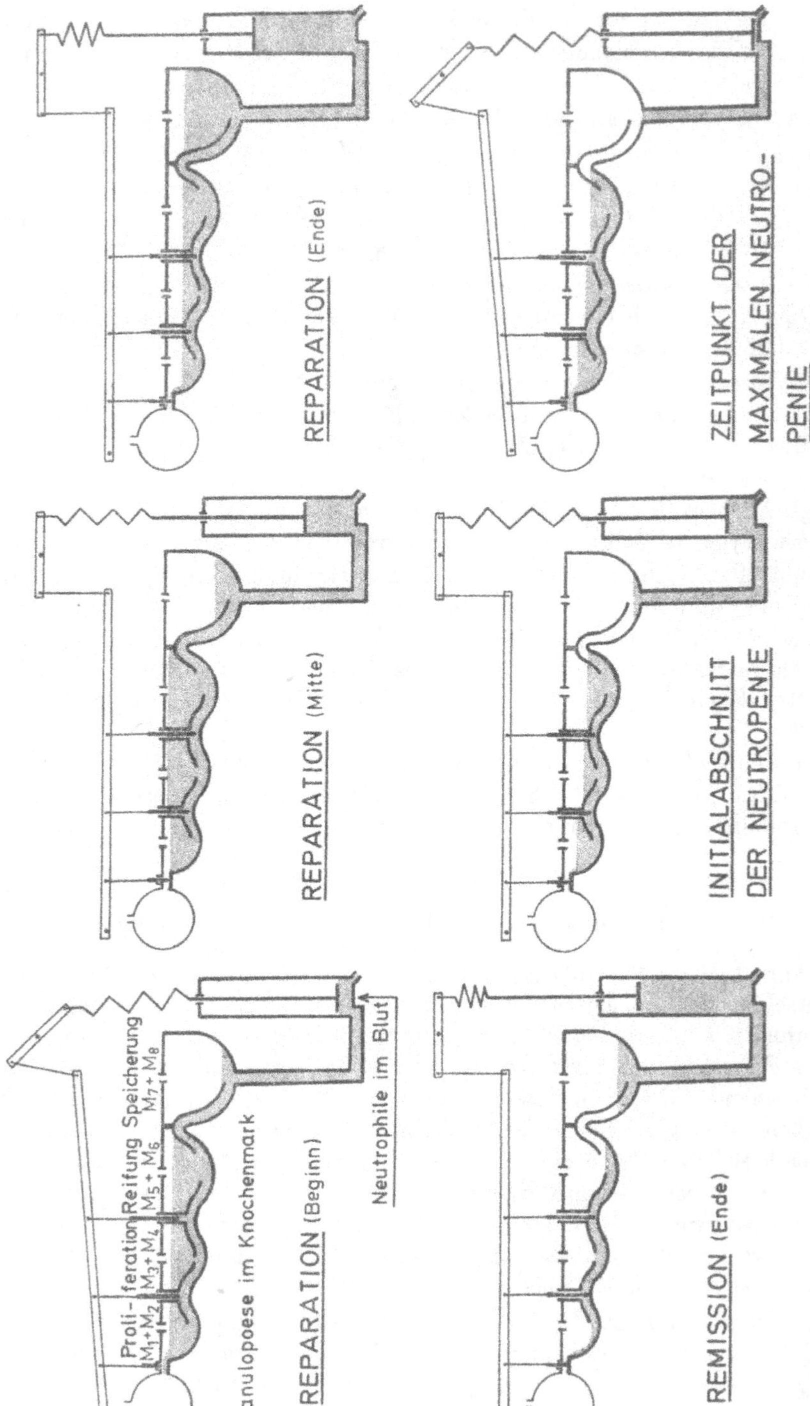

Abb. 7. Ein Modell zur Simulation der Granulopoese bei der Patientin mit zyklischer Neutropenie (Erklärung siehe Text)

gebildeten Zellen die Kompartments $M_1 - M_6$ der Granulopoese verlassen. In diesen liegt dann ein extrem niedriger Zellspiegel vor, so daß der Zellfluß zwischen den Kompartments M_6 und M_7 abreißt. Während des Initialabschnittes der Neutropenie stauen sich von den Präkursionsspeichern nachfließende Zellen vor dem Hindernis zwischen M_6 und M_7, bis zu Beginn der Remission der Speicher M_6 wieder ein Niveau erreicht, das den Zelleinstrom in das Kompartment M_7 ermöglicht.

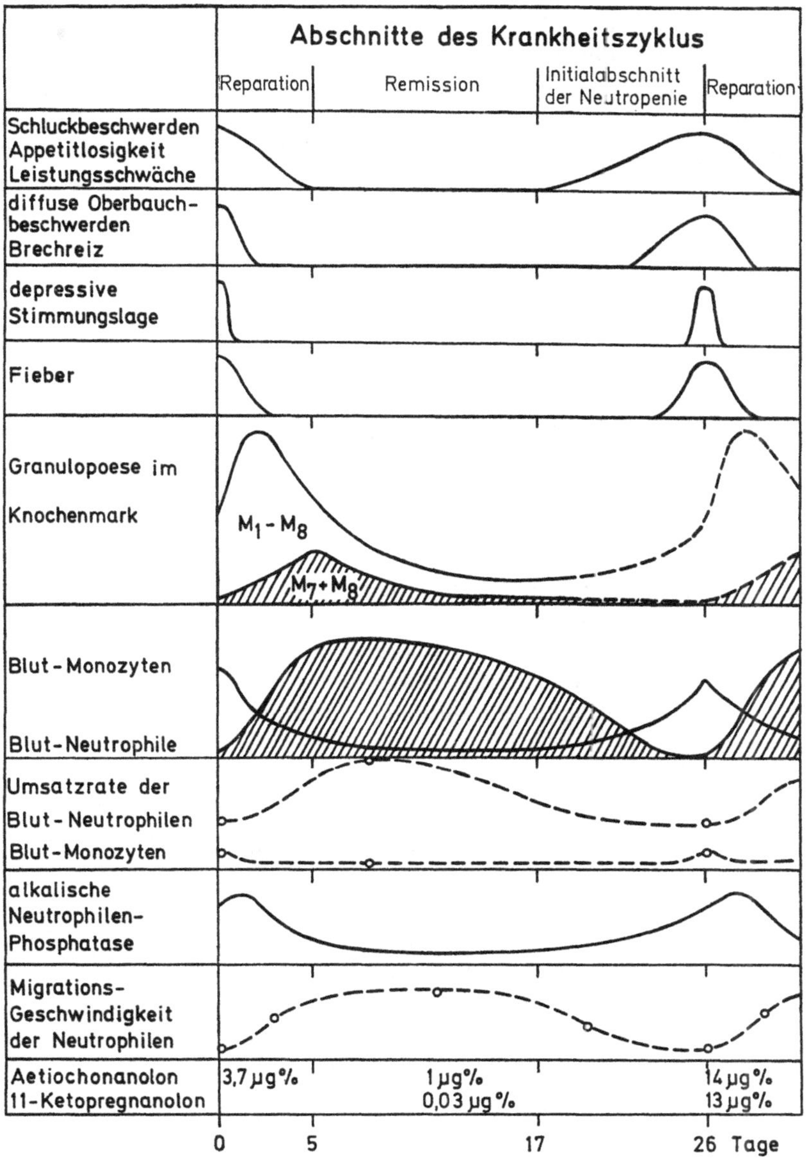

Abb. 8. Schematische Darstellung des beobachteten Krankheitssyndroms der zyklischen Neutropenie. (Die ausgezogenen Kurven beruhen auf täglichen Beobachtungen bzw. auf Messungen in 1- bis maximal 5-tägigen Intervallen; bei größeren Zeitabständen zwischen den Messungen (○) wurden die angenommenen Kurvenverläufe punktiert)

Obwohl nicht beurteilt werden kann, ob oder inwieweit ein solches Modell zur Deutung der Untersuchungsergebnisse herangezogen werden darf, so scheint doch der Vorstellung einer gestörten Regulation eine gewisse Wahrscheinlichkeit zuzukommen. Diese Vorstellung könnte auch andere, rhythmisch auftretende Krankheitssymptome der Patientin erklären.

Die zyklische Neutropenie als Krankheitssyndrom

Ähnlich wie bei anderen Fällen trat auch bei dieser Patientin ohne erkennbare Ursache ein Krankheitssyndrom auf, das jahrelang mit nahezu konstanter Periodik rezidivierte.

Würde es gelingen, die Ursache der vielfältigen Symptome (Abb. 8) in den Organen selbst festzustellen, von denen sie jeweils ausgehen, so würde man wahrscheinlich eine große Zahl völlig verschiedenartiger und voneinander unabhängiger Störungen nachweisen können, die eigenartigerweise synchron auftreten und wieder verschwinden.

Als Erklärungsmöglichkeit für die Ursache der Periodik derart verschiedener Krankheitssymptome scheint der Annahme einer gestörten zentralen Regulation eine gewisse Bedeutung zuzukommen. Wie bei dem in Abb. 7 dargestellten hypothetischen Modell der Granulopoese könnte ein vermindert ansprechbares zentrales Steuerungssystem vorliegen, das die Funktion normaler Organe nur noch in weiten Grenzen zu regeln vermag.

Zusammenfassung

Bei einer Patientin mit zyklischer Neutropenie wurde mit Hilfe von Zellzählungs- und -differenzierungsmethoden, zusammen mit zum Teil für die Fragestellung entwickelten Zellmarkierungstechniken durch Radioisotope, die Zellkinetik der Granulopoese und des Neutrophilensystems analysiert. Hierdurch wurden einige zur Klärung der Pathophysiologie dieser Erkrankung wichtige Ergebnisse gewonnen:

1. Die bei der Patientin mit einer Phasendauer von etwa 26 Tagen zyklisch verlaufende Zunahme und Abnahme der Neutrophilenzahlen im Blut war eine Folge von rhythmischen Schwankungen der Neutrophilenproduktion im Knochenmark.

2. Funktionell normal war die Granulopoese während maximal neutropenischen Krankheitsabschnitten. Bei Neutrophilenwerten unter 100 pro µl Blut setzte eine überschießende Proliferationsaktivität ein mit beschleunigter Zellreifung und erhöhter Neutrophilenproduktion, die zu einer Zunahme des Neutrophilenspiegels im Blut führte. Nachdem im Blut Neutrophilenwerte um 500 pro µl erreicht waren, ging die Proliferationsaktivität der Granulopoese wieder zurück und erreichte schließlich Minimalwerte. Während des hierdurch ausgelösten Abfalls der Neutrophilenbildungsrate und des Neutrophilenspiegels lagen Zeichen einer Reifungsstörung beim Übergang der jugendlichen in die stabkernigen Neutrophilen vor. Nachdem wieder der Zeitpunkt der maximalen Neutropenie erreicht war, sprang die Proliferation der Granulopoese plötzlich wieder überschießend an, bei ungestörter Ausreifung.

3. Die Funktionsweise des granulopoetischen Zellerneuerungssystems bei der Patientin wurde durch ein technisches Modell simuliert, bei dem die Proliferationsaktivität der Granulopoese durch ein unempfindliches Steuerungssystem nach Art eines „Kippmechanismus" geregelt wird.

4. Als Ursache des rhythmisch auftretenden Krankheitssyndroms, das mit Neutropenie, Monozytose, Fieber, Schwäche, gastrointestinalen Beschwerden und depressiver Verstimmung einherging, wird eine gestörte zentrale Steuerung zur Diskussion gestellt, die durch eine verminderte Ansprechbarkeit die Funktion normaler Organe nur noch in weiten Grenzen zu regeln vermochte.

Literatur

1. Albritton, E. C. (Hrsg.): Standard values of blood. Philadelphia 1952, S. 72.
2. Athens, J. W., Raab, S. O., Haab, O. P., Mauer, A. M., Ashenbrucker, H., Cartwright, G. E., Wintrobe, M. M.: Leukokinetic studies III. The distribution of granulocytes in the blood of normal subjects. J. clin. Invest. **40**, 159 (1961).
3. Baker, R. J., Rozoll, D. D., Meyer, K.: The use of surface area as a basis for establishing normal blood volume. Surg. Gynec. Obstet. **104**, 183 (1957).
4. Barling, B.: Chronic cyclic granulopenia. Proc. Roy. Soc. Med. **61**, 653 (1948).
5. Barsby, B. E., Close, H. G.: Recurrent neutrophil agranulocytosis. Lancet **1**, 99 (1942).
6. Becker, F. T., Coventry, W. D., Tuura, J. L.: Recurrent oral and cutaneous infections associated with cyclic neutropenia. A. M. A. Arch. Derm. **80**, 731 (1959).
7. Borne, S.: Cyclic neutropenia in an infant. Pediatrics **3**, 70 (1949).
8. Craddock, C. G., Perry, S., Lawrence, J. S.: The dynamics of leukopoiesis and leukocytosis as studied by leukophoresis and isotopic techniques. J. clin. Invest. **35**, 286 (1956).
9. Cronkite, E. P., Fliedner, T. M.: Granulopoiesis. New England J. Med. **270**, 1347 und 1403 (1964).
10. — Bond, V. P., Fliedner, T. M., Killmann, S. A.: The use of tritiated Thymidine in the study of hemopoietic cell proliferation. Ciba Foundation Symposium on Hemopoiesis. London: Churchill 1960, S. 70.
11. Dörken, H.: Lymphosarkom-Leukämie, zyklische Agranulozytose und latente Neutropenie in einer Familie. Klin. Wschr. **32**, 173 (1954).
12. Fliedner, T. M.: Der Lebenszyklus neutrophiler Granulozyten. Habilitationsschrift Freiburg 1964.
13. — Cronkite, E. P., Robertson, J. S.: Granulopoiesis I. Senescence and random loss of neutrophilic granulocytes in human beings. Blood **24**, 402 (1964).
14. — — Killmann, S. A., Bond, V. P.: Granulocytopoiesis II. Emergence of pattern of labeling of neutrophilic granulocytes in human beings. Blood **24**, 683 (1964).
15. Fullerton, H. W., Duguid, H. L. D.: A case of cyclical agranulocytosis with marked improvement following splenectomy. Blood **4**, 269 (1949).
16. Hahnemann, B. M., Alt, H. L.: Cyclic neutropenia in a father and daughter. J. A. M. A. **168**, 270 (1958).
17. Jackson, H., Merrill, D.: Agranulocytic angina associated with the menstrual cycle. New England J. Med. **210**, 175 (1934).
18. Laski, B., Sass-Kortsak, A., Hillman, D. A.: Cyclic neutropenia and agammaglobulinemia. Amer. J. Dis. Child. **88**, 820 (1954).
19. Leale, M.: Recurrent furunculosis in an infant showing an unusual blood picture. J. A. M. A. **54**, 1845 (1910).
20. Levine, S.: Chronic familial neutropenia with marked periodontal lesions. Oral Surg. **12**, 310 (1959).
21. Löffler, W., Maier, C.: Über einen Fall von Feltyschem Syndrom mit zyklischer Agranulozytose. Cardiologica **12**, 195 (1947).
22. Monterieff, A.: Recurrent neutropenia. Arch. Dis. Child. **26**, 438 (1951).

23. MORLEY, A. A., CAREW, J. P., BAIKIE, A. G.: Familial cyclical neutropenia. Brit. J. Haemat. **13,** 719 (1967).
24. NATELSON, R. R.: Cyclic neutropenia with giant follicular lymphoblastoma. Report of a case and splenectomy. Blood **8,** 923 (1953).
25. OWREN, P. A.: Cyclic agranulocytosis. Acta Med. Scand. **134,** 87 (1949).
26. PAGE, A. R., GOOD, R. A.: Studies on cyclic neutropenia; a clinical and experimental investigation. A. M. A. J. Dis. Child. **94,** 623 (1957).
27. REIMANN, H. A., DE BERARDINIS, C. T.: Periodic (cyclic) neutropenia, and entity. A collection of sixteen cases. Blood **4,** 1109 (1949).
28. REZNIKOFF, P.: Cyclical neutropenia: A case study with bone marrow findings before and after splenectomy. Trans. Ass. Amer. Physicians **59,** 276 (1946).
29. RUTLEDGE, B. H., HANSEN-BRÜSS, O. C., THAYER, W. S.: Recurrent agranulocytosis. Bull. Johns Hopkins Hosp. **46,** 369 (1930).
30. SANDELLA, J. F.: Cyclic acute agranulocytosis. Report of a case with improvement after splenectomy. Ann. Intern. Med. **35,** 1365 (1951).
31. SIGUIER, F., SEBAOUN, J.: Maladie périodique avec Crises de Sialhorrhée et Neutropénie cycliques. Bull. Soc. Med. Hôp. Paris **70,** 27 (1954).
32. SIMONSEN, E.: Cyklisk agranulocytose. Nordisk Medicin **76,** 1061 (1966).
33. THOMPSON, W. P.: Observations on the possible relationship between agranulocytosis and menstruation with further studies on a case of cyclic neutropenia. New England J. Med. **210,** 176 (1934).
34. VAHLQUIST, B.: Cyclic agranulocytosis. Report of a case with a short survey of the disease. Acta Med. Scand. Suppl. **170,** 531 (1946).

Resorption von Eisen und chemisch verwandten Metallen in vitro und in vivo; die Spezifität eines eisenbindenden Systems in der Mucosa des Jejunums von Ratten*

Absorption of Iron and Chemically Related Metals in vitro and in vivo; the Specificity of an Iron-binding System in the Jejunal Mucosa of Rats

W. FORTH

Institut für Pharmakologie und Toxikologie der Universität des Saarlandes, Homburg/Saar

Summary

Iron-deficient rats absorb from tied jejunal loops in vivo not only more iron than normal rats but also the chemically related metals cobalt, manganese and even zinc, but not copper. The retention of these metals, the absorption of which is increased, is equal in both normal and anemic animals.

Everted sacs in vitro, prepared from jejunal segments of iron-deficient rats, transfer iron, the chemically related metals already mentioned, and in addition nickel, to a greater extent than those of normal rats; under the experimental conditions in vitro this is not the case for copper. The uptake of these metals by the mucosal tissue was also greater with segments of iron-deficient rats than with those of normal ones.

Finally, mutual inhibition could be shown during absorption of iron and cobalt as well as of iron and manganese. The latter experiments were done in situ on tied loops of jejunum of rats in vivo.

These results can be explained by the assumption of an iron-binding system in the mucosa responsible for absorption: this system shows the highest degree of specificity for iron, followed by cobalt, nickel, manganese and zinc, but apparently does not react with copper.

As acceptors for iron and the chemically related metals, the proteins of the cell membrane may play a role comparable with that of the reticulocyte membrane responsible for the uptake and incorporation of iron in the reticulocytes. The coincidence of these processes leads us to believe that the mucosal epithelium and reticulocytes possess similar mechanisms which are responsible for uptake and/or transcellular transport of iron.

Einleitung

Eisen zeichnet sich vor allen anderen Metallen dadurch aus, daß es, einmal inkorporiert, nur sehr langsam wieder ausgeschieden wird. Der Eisenverlust bei der Mauserung der Erythrozyten, die beim Menschen fast 70% des gesamten Eisenbestandes enthalten, beträgt nach v. HEVESY [43][1] nur 2%. Die Halbzeit für den Umsatz des Hämoglobineisens wurde mit Tracern für den erwachsenen Mann mit mehr als 8 Jahren ermittelt [18]. Zum Vergleich sei angeführt, daß man

* Die englische Fassung dieses Beitrages erschien in: Trace Element Metabolism in Animals (C. F. MILLS, Ed.). Edinburgh-London: E. & S. Livingstone 1970. — Abdruck in deutscher Sprache erfolgt mit freundlicher Erlaubnis des Verlages E. & S. Livingstone.

[1] Die Zahlen beziehen sich auf das Literaturverzeichnis am Schluß des Beitrages.

bei einem täglichen Calcium-Abbau im Skelett von 0,5 g [50] die Halbzeit für den Calcium-Umsatz auf weniger als $^{3}/_{4}$ Jahr berechnen kann. Beim Eisen kann demnach die *Ausscheidung* für den Ausgleich der Bilanz keine Bedeutung haben. Hierfür ist die *Resorption* allein maßgebend [56]. Man muß sich dabei vergegenwärtigen, daß der Eisenbedarf physiologischerweise in weiten Grenzen schwankt: der mittlere Bedarf eines erwachsenen Mannes wird mit ca. 1 mg/die angegeben; bei der Frau steigt er während der Menstruation auf das Doppelte und wird in der Schwangerschaft sogar vervierfacht [39, 43]. Hier stellt sich nun die Frage, ob der Darm selbst die Resorption an den Bedarf anpaßt oder ob andere übergeordnete Faktoren regulativ eingreifen; diese Frage galt bis vor kurzem als unentschieden.

Heute kann als sicher betrachtet werden, daß der Darm selbst insoweit regulatorische Fähigkeiten besitzt, als er in der Lage ist, das Nahrungseisen dem Bedarf angepaßt auszunutzen. Hinweise dafür ergaben bereits die Untersuchungen HAHNS [37, 38, 39]; seine Vermutungen wurden in der Folgezeit an nichtdurchbluteten Darmsegmenten in vitro experimentell bestätigt [16, 19, 22, 24, 25, 41, 55, 64]. Die unter diesen Bedingungen gewonnenen Resultate — hier können sich Faktoren wie der Sättigungsgrad des Transferrins mit Eisen [10, 51], der Füllungszustand der Eisendepots [7] oder die Neubildungsrate der Erythrozyten [7, 45, 51], denen ein direkter Einfluß auf die Regulation der Eisenresorption zugeschrieben wurde, eben nicht auswirken — lassen sich mit der Annahme eines *eisenbindenden Systems* in der Mucosa erklären [Lit. s. bei 31].

Hier wird im Gegensatz zur allgemeinen Konvention einfach nur von einem *eisenbindenden System* gesprochen, womit die Tatsache beschrieben wird, daß an und in der Mucosazelle Eisen-Akzeptoren vorhanden sind, die die Durchschleusungsrate bestimmen, eine Vorstellung, mit der auch CROSBY und Mitarb. arbeitete [Lit. s. 13]. Wahrscheinlich handelt es sich bei den eisenbindenden Systemen gar nicht um einen einzigen Akzeptor — z. B. Apoferritin, dem GRANICK [35, 36] aufgrund des Ferritingehaltes der Mucosa die Rolle des Akzeptors zuschrieb —, sondern um ein System von hoch- und niedermolekularen Liganden [3, 4, 5, 8, 9, 11]. Das eisenbindende System ist wahrscheinlich vornehmlich in den oberen Abschnitten des Dünndarmes (Duodenum und Jejunum) lokalisiert, die sich vor allem im Eisenmangel durch eine besondere Aktivität bei der Resorption von Eisen auszeichnen [16, 17, 19, 24, 25, 52, 54].

Nimmt man an, daß in der Mucosa des Darmes ein System mit der Eigenschaft, Eisen zu binden, für die Resorption verantwortlich ist, dann erhebt sich sofort die Frage nach der *Spezifität* dieses Systems; der Beantwortung dieser Frage ist die vorliegende Arbeit gewidmet. Zu diesem Zweck wurde untersucht, inwieweit sich die im Eisenmangel gesteigerte Fähigkeit, Eisen zu resorbieren, auch auf andere, dem Eisen chemisch nahestehende Schwermetalle erstreckt. Die Auswahl der Metalle erfolgte aufgrund ihrer Stellung zum Eisen innerhalb des periodischen Systems der Elemente (vgl. Abb. 1): *Kobalt* und *Nickel*, die als nächste Verwandte mit dem Eisen zusammen die Gruppe VIII bilden, *Mangan* aus der Gruppe VII A und *Kupfer* bzw. *Zink* aus den Gruppen I B und II B; hier wurden die von PHILIPPS und WILLIAMS [60] benutzten Gruppenbezeichnungen zugrunde gelegt.

Zur Messung der Resorption wurden mit Radioisotopen markierte Metalle verwendet. Die Untersuchungen wurden am Dünndarm normaler und anämischer

Ratten durchgeführt, und zwar in vitro, an nichtdurchbluteten Darmsegmenten, an abgebundenen durchbluteten Darmschlingen in situ und nach Verabfolgung der Metalle mit der Schlundsonde (vgl. Anhang: Methoden).

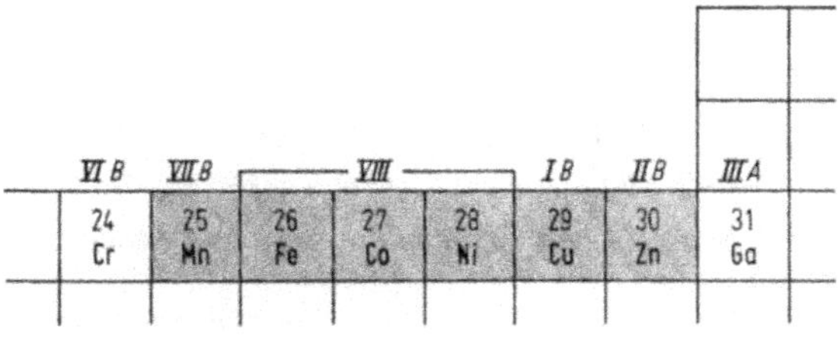

Abb. 1. Die Stellung der untersuchten Metalle im periodischen System der Elemente (Die römischen Ziffern bezeichnen die Gruppen (nach PHILLIPS und WILLIAMS [60]; die arabischen Ziffern die Ordnungszahlen der Metalle, vgl. Text)

Ergebnisse und Besprechung

1. Die vergleichende Prüfung von Durchtritt und Bindung von Eisen, Mangan, Kobalt, Nickel, Kupfer und Zink an isolierten, nichtdurchbluteten Jejunum-Segmenten in vitro von normalen und anämischen Ratten

Wir führten die ersten Versuche zum Studium des Durchtritts und der Aufnahme ins Gewebe mit verschiedenen Metallen an isolierten, nichtdurchbluteten „everted sac"-Präparaten nach WILSON und WISEMAN [70] durch [20, 28]. Diese Versuchsanordnung bietet übersichtliche Verhältnisse. Die damit gewonnenen Ergebnisse sind unabhängig von der *Darmpassagezeit* und von der *Durchblutung*; schließlich wird unter diesen Versuchsbedingungen die Resorption nicht durch die in vivo weitgehend unkontrollierbaren *intraluminalen Faktoren* beeinträchtigt, da die Darmstücke einer Lösung bekannter Zusammensetzung ausgesetzt sind.

a) *Der Durchtritt der Metalle von der Mucosa- zur Serosaseite.* Die Konzentration der Metalle war auf der Mucosa- und der Serosaseite des Präparates gleich, sie betrug $5 \cdot 10^{-6}$ M/l; die Tracer wurden jedoch nur auf der Mucosaseite angeboten. Am Versuchsende wurde die auf die Serosaseite übergetretene Radioaktivität gemessen, d.h. es wurde dabei der Teil der übergetretenen Metallmenge bestimmt, der aus der auf der Mucosaseite angebotenen Metallösung stammte. Das Volumen der Serosaflüssigkeit (0,5 ml) blieb während des Versuches konstant (vgl. Anhang: Methoden).

In einem Teil der Versuche mit ^{58}Co und ^{64}Cu wurden die beiden Metalle jeweils zugleich mit ^{59}Fe angeboten. Die Werte des Durchtritts durch und die Aufnahme der Metalle in die isolierten Darmabschnitte dieser Versuche unterschieden sich nicht von denen, die in anderen Versuchen ermittelt wurden, in denen die Metalle jeweils einzeln eingesetzt wurden. Deshalb wurden zur statistischen Beurteilung die Werte beider Versuchsgruppen zusammen genommen.

Der Durchtritt von ^{59}Fe zur Serosaseite ist bei Darmsegmenten *normaler* Ratten sehr gering (vgl. Tab. 1). Die ^{59}Fe-Konzentration in der Serosaflüssigkeit beträgt nur 3% der auf der Mucosaseite angebotenen. Die insgesamt durchgetretene Menge beträgt sogar nur 0,011% der angebotenen Menge. Der Durchtritt von ^{58}Co ist am größten; er ist bei Darmabschnitten normaler Ratten fast 13mal größer als der von ^{59}Fe; dann folgt ^{63}Ni, dessen Durchtritt 7mal größer als der ^{59}Fe-Durchtritt ist, und der von ^{54}Mn, dessen Durchtritt 2mal größer ist. Der ^{64}Cu-Durchtritt ist um ca. $^1/_3$ geringer als der von ^{59}Fe, und der von ^{65}Zn beträgt nur $^1/_{10}$ davon.

Tabelle 1. *Durchtritt von Mangan, Eisen, Kobalt, Nickel, Kupfer und Zink in vitro an isolierten, nichtdurchbluteten Jejunumsegmenten von normalen und anämischen Ratten*

Darmabschnitte: 2 je Ratte, von der Flexura duodenojejunalis an 3,5 cm lang. Konzentration der Metalle auf beiden Seiten des Präparates: $5 \cdot 10^{-6}$ M/l. Tracer wurden nur auf der Mucosaseite zugesetzt entsprechend 50 000 Imp/ml u. min. Inkubationslösung (Mucosaseite): 10 ml Tyrode-Lösung; Füllung der Darmsäckchen (Serosaflüssigkeit): 0,5 ml Tyrode-Lösung; am Versuchsende: 0.49 g ± 0,05. Versuchsdauer: 2 Std.; Temp. 37 °C. Die Tabelle enthält die Werte der zur Serosaseite übergetretenen Radioaktivität in Imp/min ± der Standardabweichung S_E. Index * bedeutet $p < 0{,}01$ gegen die Werte normaler Abschnitte.

Metall	Radioaktivität in der Serosaflüssigkeit			
	Normale Jejunumabschnitte		Eisenarme Jejunumabschnitte	
	Imp/min	n	Imp/min	n
^{54}Mn	**1500** ± 400	8	**4 650** ± 1 500	4
^{59}Fe	**750** ± 250	24	**8 150*** ± 2 200	16
^{58}Co	**9 500** ± 2 150	8	**28 500*** ± 5 500	8
^{63}Ni	**5 200** ± 1 850	14	**21 000*** ± 7 500	10
^{64}Cu	**450** ± 250	8	**400** ± 200	8
^{65}Zn	**70** ± 30	16	**130*** ± 45	4

Bei *eisenverarmten* Segmenten ist der Durchtritt aller Metalle mit Ausnahme des von ^{64}Cu angestiegen. Die Reihenfolge der nach der Höhe der durchgetretenen Menge geordneten Metalle lautet bei normalen Darmsegmenten: Co > Ni > Mn > Fe > Cu > Zn; sie ist bei eisenverarmten Segmenten nur geringfügig verändert: Co > Ni > Fe > Mn > Cu > Zn. Nur Fe und Mn haben ihren Platz getauscht, und zwar deshalb, weil die Zunahme des Durchtritts an eisenverarmten Darmsegmenten für Fe viel größer ist als für Mn.

Ein Bild über die *relative Steigerung* der Resorption der verschiedenen Metalle bei eisenverarmten Darmabschnitten gegenüber normalen vermittelt die Darstellung des Quotienten a/n, den man erhält, wenn man die zur Serosaseite durchgetretene Metallmenge bei eisenarmen (a) Darmsegmenten durch die normaler (n) dividiert (vgl. Abb. 2). Der Quotient a/n ist 1, wenn der Durchtritt bei eisenarmen und normalen Abschnitten gleich groß ist. Der Abbildung kann entnommen werden, daß für ^{59}Fe die relative Zunahme des Durchtritts am größten ist; der Quotient a/n beträgt für ^{59}Fe 11, für ^{63}Ni 4, für ^{58}Co und ^{54}Mn etwa 3 und für ^{65}Zn fast 2. Der Durchtritt von ^{64}Cu hat bei eisenarmen Segmenten gegenüber dem bei normalen nicht zugenommen, der Quotient a/n beträgt rund 1.

b) *Die Aufnahme der Metalle ins Darmgewebe.* Berücksichtigt man die Streuung der Ergebnisse, dann ist die Aufnahme ins Darmgewebe auf g Trockengewicht bezogen bei Segmenten normaler Ratten für alle Metalle etwa gleich groß; eine Ausnahme bildet ^{63}Ni, dessen Wert rund $^1/_{10}$ dessen der anderen Metalle beträgt (vgl. Tab. 2).

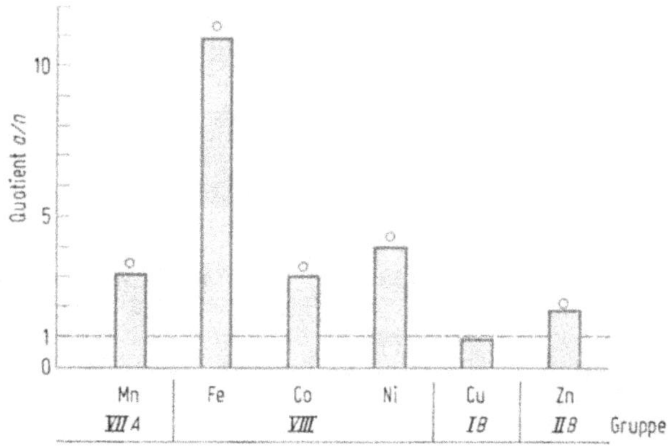

Abb. 2. Die relative Steigerung des Durchtritts von Mangan, Eisen, Kobalt, Nickel, Kupfer und Zink an isolierten eisenarmen Jejunumsegmenten in vitro im Vergleich mit normalen. Die Säulen repräsentieren die Quotienten a/n, die durch Division der zur Serosaseite penetrierten Metallmengen bei eisenarmen (a) und normalen (n) Jejunumabschnitten berechnet wurden. Die römischen Ziffern kennzeichnen die Gruppen der Metalle im periodischen System. Der Index ° kennzeichnet die statistisch gesicherte Steigerung des Durchtritts. Die unterbrochene Linie markiert die Höhe des Quotienten $a/n = 1$, d. h. einen gleich großen Wert für den Durchtritt der Metalle bei eisenarmen und normalen Segmenten (vgl. Tab. 1)

Bei den *eisenarmen* Segmenten ist die aufgenommene Menge von ^{59}Fe am größten; ihr folgen die von ^{54}Mn, ^{58}Co, ^{65}Zn, ^{64}Cu, und am Ende steht die von ^{63}Ni. Die *relative* Steigerung der Metallaufnahme eisenarmer Darmabschnitte gegenüber normalen ist wie die des Durchtritts für ^{59}Fe am größten. Legt man die pro g Trockengewicht berechneten Werte zugrunde, dann ergibt sich für ^{59}Fe ein Wert für die relative Steigerung von fast 4, für ^{58}Co von 2,9, für ^{54}Mn von 2,6, für ^{63}Ni von 2,5 und für ^{65}Zn von 1,7. ^{64}Cu, dessen Durchtritt bei eisenarmen Darmabschnitten im Vergleich mit normalen gleich geblieben ist, zeigt auch keine gesteigerte Aufnahme ins Gewebe. Angesichts der Streuung der Werte ist der geringfügigen Erniedrigung des ^{64}Cu-Gehaltes pro g Trockengewicht des Darmgewebes anämischer Ratten gegenüber dem von normalen wohl keine besondere Bedeutung beizumessen.

Aus der Berechnung des Metallgehaltes pro g Feuchtgewicht geht hervor (vgl. Tab. 2), daß die Metalle offenbar im Darmgewebe *angereichert* werden: im Vergleich zur Konzentration der markierten Metalle in der Inkubationslösung ist der Metallgehalt pro g Feuchtgewicht im Gewebe mit Ausnahme von ^{63}Ni für alle Metalle höher. Die Versuche geben keinen Aufschluß darüber, ob die Metalle nur von der Mucosa aufgenommen und dort gebunden wurden, oder ob sie gleichmäßig über das ganze Darmgewebe verteilt sind. Aus älteren Versuchen mit ^{59}Fe ist bekannt, daß 75% der Radioaktivität in der Mucosa vorliegen [8, 20, 54]. Geht man davon aus, daß auch die anderen Metalle wie Eisen überwiegend in der Mucosa gebunden werden und veranschlagt man ca. 50% des Feuchtgewichtes für das Gewicht der Mucosa (eigene unveröffentlichte Ergebnisse), dann ergibt sich ein noch höherer Anreicherungsgrad der Metalle im Gewebe. Für ^{59}Fe läßt sich auf diese Weise bei normalen Segmenten ein Anreicherungsgrad von 3,3 berechnen, bei eisenarmen Abschnitten sogar ein Anreicherungsgrad um das 9,2fache.

Tabelle 2. *Aufnahme ins Gewebe von Mangan, Eisen, Kobalt, Nickel, Kupfer und Zink in vitro an isolierten, nichtdurchbluteten Jejunumsegmenten von normalen und anämischen Ratten*

Darmabschnitte: 2 je Ratte, von der Flexura duodenojejunalis an 3,5 cm lang. Normale Abschnitte: Feuchtgewicht: 0,17 g ± 0,03; Trockengewicht: 0,03 g ± 0,02. Eisenarme Segmente: Feuchtgewicht: 0,24 g ± 0,04; Trockengewicht: 0,032 g ± 0,001. Inkubationslösung (Mucosaseite): 10 ml Tyrode-Lösung; Füllung der Darmsäckchen (Serosaflüssigkeit): 0,5 ml Tyrode-Lösung. Konzentration der Metalle auf beiden Seiten des Präparates: $5 \cdot 10^{-6}$ M/l; Tracer wurden nur auf der Mucosaseite zugesetzt entsprechend 50 000 Imp/ml u. min. Versuchsdauer: 2 Std.; Temp. 37 °C. Die Tabelle enthält die Werte der Radioaktivität pro 100 mg Trockengewicht des Darmgewebes ± der Standardabweichung S_E, sowie die pro 100 mg Feuchtgewicht des Darmgewebes und der Mucosa berechneten Werte. Index * bedeutet $p < 0{,}01$ gegen die Werte normaler Abschnitte.

Metall	Radioaktivität im Darmgewebe							
	Normale Jejunumabschnitte				Eisenarme Jejunumabschnitte			
	Imp/min u. 100 mg Gewebe (trocken) $\cdot 10^3$	n	Imp/min u. 100 mg Gewebe (feucht) $\cdot 10^3$	Imp/min u. 100 mg Mucosa (feucht) $\cdot 10^3$	Imp/min u. 100 mg Gewebe (trocken) $\cdot 10^3$	n	Imp/min u. 100 mg Gewebe (feucht) $\cdot 10^3$	Imp/min u. 100 mg Mucosa (feucht) $\cdot 10^3$
^{54}Mn	**55** ± 11	8	**97**	**146**	**143*** ± 47	4	**190**	**286**
^{59}Fe	**61** ± 15	24	**108**	**164**	**230*** ± 62	16	**310**	**470**
^{58}Co	**46** ± 3,6	8	**81**	**118**	**314*** ± 43	8	**178**	**279**
^{63}Ni	**6,7** ± 2,3	14	**12**	**18**	**16,5*** ± 3,7	10	**22**	**33**
^{64}Cu	**73** ± 33	8	**128**	**188**	**53** ± 25	8	**71**	**108**
^{65}Zn	**36** ± 6,6	16	**64**	**98**	**61*** ± 17	4	**81**	**122**

Das wichtigste Ergebnis dieser Versuche ist darin zu sehen, daß an nichtdurchbluteten Darmsegmenten, die von anämischen Tieren stammten, in vitro nicht nur Durchtritt und Aufnahme ins Gewebe von Eisen gesteigert ist, sondern auch der von den dem Eisen chemisch nahestehenden Metallen Kobalt und Nickel sowie der von Mangan und Zink; nicht dagegen der von Kupfer. Nimmt man die Zunahme von Durchtritt und Bindung im Gewebe bei eisenverarmten Segmenten gegenüber normalen als Maß, dann ist die Spezifität des eisenbindenden Systems in der Mucosa, das für die Durchschleusung des Eisens verantwortlich ist, für Eisen am größten und für die dem Eisen chemisch nahestehenden Nickel, Kobalt und Mangan größer als für das ferner stehende Zink.

2. Die Prüfung der Resorption von Eisen, Mangan, Kobalt, Kupfer und Zink aus durchbluteten Darmschlingen in situ bei normalen und anämischen Ratten

Zur Überprüfung der in vitro gewonnenen Ergebnisse wurden Versuche in vivo mit durchbluteten Darmschlingen in situ bei normalen und anämischen Ratten durchgeführt. Die resorbierten Mengen wurden mit Hilfe einer Kleintiermeßanlage nach Entfernung der Darmschlingen bestimmt; in diesen Versuchen wurden nur Metalle geprüft, deren Isotope γ-Strahlen emittieren. Die Ergebnisse dieser Versuche sind in Tab. 3 zusammengefaßt. Die relative Steigerung der Metallresorption bei anämischen Ratten gegenüber normalen ist als Quotient a/n in der Abb. 3 graphisch dargestellt.

Tabelle 3. *Resorption von Mangan, Eisen, Kobalt, Kupfer und Zink bei durchbluteten Jejunumschlingen in situ von normalen und anämischen Ratten*

Jejunumschlingen: durchschnittlich 12 cm lang von der Flexura duodenojejunalis an. Füllung: 2 ml physiol. NaCl-Lösung, pH 2,3. Metall-Konzentration: 0,09 μM/ml. Die Tiere hatten eine Urethraligatur. Versuchsdauer: 1 Std. Die resorbierte Menge der radioaktiv markierten Metalle wurde nach Entfernung der Darmschlinge durch Messung der Radioaktivität der Kadaver in einer Kleintiermeßanordnung bestimmt. Die Zahlen sind die Mittelwerte von n Versuchen, ausgedrückt in % der Dosis ± der Standardabweichung S_E. Index * bedeutet $p < 0{,}01$ gegen die Gruppe normaler Tiere.

	Normale Ratten		Anämische Ratten	
Metall	Resorbierte Menge	n	Resorbierte Menge	n
^{54}Mn	**5,7** ± 4,0	5	**9,5** ± 2,4	5
^{59}Fe	**7,1** ± 3,4	15	**37,1** ± 14,0	15
^{58}Co	**20,8** ± 6,2	10	**65,5** ± 14,2	10
^{64}Cu	**9,3** ± 3,4	10	**11,4** ± 4,0	10
^{65}Zn	**7,7** ± 2,3	10	**16,5*** ± 4,7	10

Die Metalle wurden in einer Konzentration von 0,09 μM/ml in 2 ml physiologischer NaCl-Lösung appliziert, deren pH-Wert auf 2,3 eingestellt worden war. (Die Erörterung der Frage, inwieweit eine sauer in die Darmschlingen injizierte Lösung die Resorptionsleistung beeinträchtigt, ist dem Anhang: Methoden zu entnehmen.)

Aus den Jejunumschlingen *normaler* Ratten wurde Kobalt im Vergleich zu den anderen Metallen am meisten resorbiert (20,8%). Die nach 1 Stunde inkorporierte ^{58}Co-Menge ist ca. 2—4mal größer. Die Streuungen der Werte von ^{54}Mn, ^{59}Fe, ^{64}Cu und ^{65}Zn sind so groß, daß keine ausgeprägten Unterschiede hinsichtlich der Resorbierbarkeit dieser Metalle zu erkennen sind. Ihre Resorptionsquoten bewegen sich zwischen 6 und 9% der angebotenen Dosis.

Die Unterschiede werden bei *anämischen* Ratten deutlicher. Auch hier wurde ^{58}Co am meisten resorbiert, nämlich zu 65% der Dosis. Für ^{59}Fe beträgt die resorbierte Menge 37% der Dosis. Die resorbierte ^{65}Zn-Menge beträgt weniger als die Hälfte derjenigen von ^{59}Fe, die von ^{64}Cu nur $^1/_3$ und die von ^{54}Mn nur $^1/_4$.

Die Steigerung der Resorptionsquoten für ^{54}Mn, ^{58}Co, ^{59}Fe und ^{65}Zn bei anämischen Ratten gegenüber normalen ist statistisch gesichert. An der Höhe der in Abb. 3 dargestellten Säulen, die den Quotienten a/n aus den Resorptionsquoten von anämischen und normalen Ratten wiedergeben, läßt sich ablesen, daß die

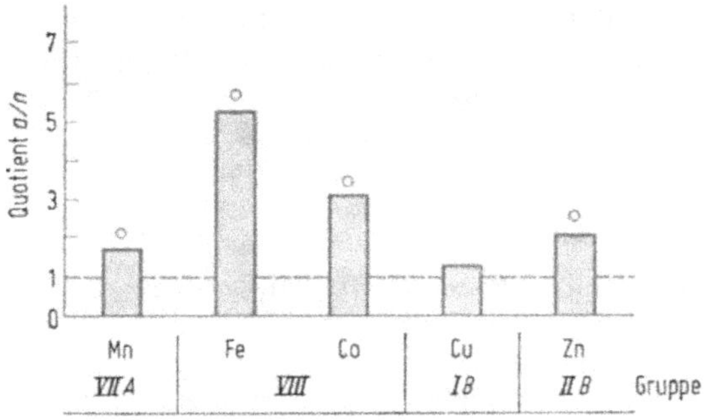

Abb. 3. Die relative Steigerung der Resorption von Mangan, Eisen, Kobalt, Kupfer und Zink an durchbluteten Jejunumschlingen in situ bei anämischen Ratten im Vergleich zu normalen. Die Säulen repräsentieren die Quotienten a/n, die durch Division der resorbierten Metallmengen bei anämischen (a) und normalen (n) Ratten berechnet wurden. Die römischen Ziffern kennzeichnen die Gruppierung der Metalle im periodischen System. Der Index ° bezeichnet eine statistisch gesicherte Steigerung der Resorption. Die unterbrochene Linie markiert die Höhe des Quotienten $a/n = 1$, d. h. einen gleich großen Wert der Resorption bei normalen und eisenarmen Ratten (vgl. Tab. 3)

Resorption von ^{59}Fe am stärksten zugenommen hat; der Quotient beträgt mehr als 5. Für ^{58}Co erreicht er nur einen Wert von 3 und für ^{54}Mn und ^{65}Zn von etwa 2. Für ^{64}Cu liegt auch in vivo der Quotient a/n bei 1; die geringgradige Steigerung der Resorptionsquote ist statistisch nicht gesichert (vgl. Tab. 3).

Das Ergebnis dieser Versuchsreihe besagt, daß auch in vivo — genau wie in vitro — bei anämischen Tieren nicht nur die Resorption von Eisen gesteigert ist — — was längst bekannt ist [2, 12, 23, 27, 31, 32, 34, 38, 39, 40, 48, 55, 57, 58, 62, 66, 67] —, sondern auch die anderer Metalle. Unter diesen zeichnet sich besonders Kobalt aus, das aufgrund seiner Stellung in der Gruppe VIII des periodischen Systems dem Eisen besonders nahe steht. Die Zunahme der Resorption bei anämischen Tieren betrifft auch die dem Eisen weniger nahe stehenden Metalle Mangan und Zink, nicht aber Kupfer. Gemessen an der relativen Steigerung der Resorptionsquoten bei anämischen Tieren gegenüber normalen hat das für die Resorption verantwortliche System in der Mucosa für Eisen den höchsten Grad an Spezifität. Dann folgen Kobalt und Zink bzw. Mangan. Der naheliegende Einwand gegen diese Interpretation der Ergebnisse, daß sich die anämischen Tiere genau wie für Eisen auch hinsichtlich der anderen Metalle in einem Mangelzustand befinden, ist deshalb gegenstandslos, weil dem Rattenfutter diese Metalle zugesetzt worden waren (vgl. Anhang: Methoden).

Das Resultat dieser Versuche steht in Übereinstimmung mit den Ergebnissen von VALBERG und OLATUNBOSUN [68] und denjenigen der Arbeitsgruppe von CROSBY [61]. Neben den von uns geprüften Metallen wurde von der letztgenannten

Gruppe an abgebundenen Darmschlingen in situ bei normalen und anämischen Ratten Cs, Mg, Hg und Ca geprüft; für diese Metalle konnte im Eisenmangel keine Steigerung der Resorptionsquote nachgewiesen werden.

In diesem Zusammenhang müssen die älteren Befunde von MANIS und SCHACHTER [54] erwähnt werden; diese Autoren konnten in vitro (everted sac) mit steigender Ca-Konzentration den Durchtritt von Fe durch den Dünndarm der Ratte hemmen und umgekehrt den von Ca mit steigender Fe-Konzentration. Es lag daher nahe, daran zu denken, daß für Fe und Ca ein gemeinsamer Durchtrittsmechanismus existiert. Wenn dies zutrifft, würde man allerdings auch für die Ca-Resorption in vivo im Eisenmangel eine Steigerung erwarten dürfen, was aber nicht der Fall ist [61].

Zur Erklärung des sich aus dieser Überlegung ergebenden Widerspruchs sind noch weitere Experimente erforderlich. Hier sei nur soviel gesagt, daß zur Deutung der wechselseitigen Beeinflussung von Ca und Fe beim Durchtritt durch die Mucosa des isolierten Darmes auch daran zu denken ist, daß Ca und Fe Mischkomplexe bilden, die möglicherweise für den Durchtritt schlecht geeignet sind. Als Liganden für derartige Mischkomplexe kämen z. B. Phosphate in Frage. Die Tatsache, daß MANIS und SCHACHTER [54] die Darmstücke in PO_4-freiem Milieu inkubierten, schließt diese Möglichkeit keineswegs aus, da das Gewebe bekanntlich beträchtliche Phosphatmengen bei der Inkubation verliert. Die Wechselwirkung von Fe und Ca wird übrigens auch für die Interpretation einer bei Bantunegern verbreiteten Osteoporose herangezogen [53]. Dabei ist jedoch ungeklärt, ob Fe und Ca sich bei der Resorption kompetitiv behindern oder ob der hohe Fe-Gehalt in der Nahrung der Bantus durch Erzeugung eines Phosphatmangels mittelbar einen Ca-Mangel verursacht. Für die letztgenannte Möglichkeit spricht die alte Beobachtung, daß experimentell eine Rachitis dadurch erzeugt werden kann, daß der Nahrung Fe oder Al zugesetzt wird [14]. Der hohe Metallgehalt in der Nahrung hat nämlich zur Folge, daß Phosphat in eine für die Resorption ungeeignete Form überführt wird.

3. Die wechselseitige Hemmung der Resorption von Eisen, Kobalt und Mangan an durchbluteten Jejunumschlingen in situ bei anämischen Ratten

Wenn die Arbeitshypothese zutrifft, daß die Akzeptorstellen für Eisen in der Mucosa auch mit Mangan oder Kobalt reagieren können und deren Durchschleusung durch die Darmwand besorgen, dann ist zu fordern, daß bei der Resorption dieser Metalle Verdrängungsreaktionen nachweisbar sind. Mit anderen Worten, es sollte gelingen, die Resorption von Eisen in Gegenwart von Kobalt oder Mangan zu hemmen, und umgekehrt sollte deren Resorption durch Eisen hemmbar sein.

Zu diesen Versuchen wurden nur anämische Ratten benutzt, weil eine inhibitorische Wirkung bei der gesteigerten Resorptionsleistung der Därme für diese Metalle besonders deutlich in Erscheinung treten sollte. Die Metalle wurden in 1 ml physiologischer NaCl-Lösung (pH 2,3) in die Jejunumschlingen injiziert; 1 Stunde später wurde die inkorporierte Radioaktivität mit Hilfe einer Kleintiermeßanlage bestimmt. Zunächst wurde die ^{59}Fe-Resorption in Gegenwart steigender Co-Konzentrationen bzw. die ^{60}Co-Resorption in steigenden Fe-Konzentrationen geprüft. Die Dosis der radioaktiv markierten Metalle war 0,05 μM. Das Ergebnis dieser Untersuchungen ist in Abb. 4 und Tab. 4 zusammengefaßt.

Aus der Abb. 4 geht hervor, daß die *Resorption* von ^{59}Fe durch eine 10fach höhere Co-Konzentration um fast $^2/_3$ vermindert wird, eine weitere Steigerung der Co-Konzentration auf den 100fachen Überschuß bringt die ^{59}Fe-Resorption völlig zum Erliegen. Prinzipiell das gleiche ergibt sich für die ^{60}-Resorption in Gegenwart von Fe. Auch hier senkt erst der 10fache Überschuß von Fe die Resorption von

^{60}Co derart, daß die Differenz statistisch zu sichern ist. Unter diesen Bedingungen beträgt die resorbierte ^{60}Co-Menge die Hälfte und in Gegenwart eines 100fachen Überschusses weniger als $^{1}/_{5}$ des Kontrollwertes. Durch Interpolation kann abge-

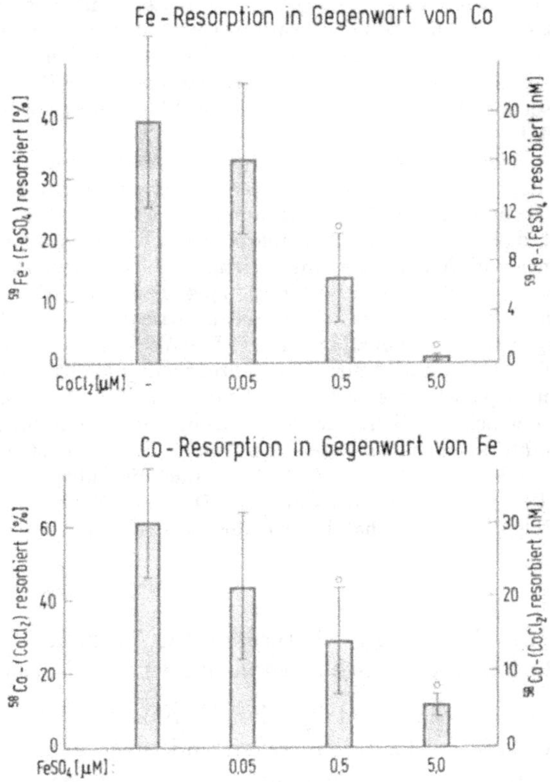

Abb. 4. Die Resorption von Eisen in Gegenwart von Kobalt und umgekehrt die von Kobalt in Gegenwart von Eisen an durchbluteten Jejunumschlingen in situ bei anämischen Ratten. Jejunumschlingen: 12 cm lang von der Flexura duodenojejunalis an. Darmfüllung: 1 ml physiol. NaCl-Lösung, pH 2,3. Metallangebot: 0,05 µM ^{59}Fe-($FeSO_4$) bzw. ^{60}Co-($CoCl_2$). Versuchsdauer: 1 Stunde. Die resorbierten Mengen der markierten Metalle wurden nach Entfernung der Jejunumschlingen durch Messung der Radioaktivität in einer Kleintiermeßanlage bestimmt. Linke Ordinate: resorbierte Metallmenge in % der Dosis, rechte Ordinate: in nM. Die Säulen repräsentieren die Mittelwerte aus je 5 Versuchen; die senkrechte Linie in den Säulen veranschaulicht die Standardabweichung S_E. Index ° bedeutet $p < 0{,}01$ gegen die Kontrollgruppe

schätzt werden, daß zur Hemmung der Resorption von ^{59}Fe und ^{60}Co um die Hälfte gerade etwa der 10fache Überschuß von Co bzw. Fe benötigt wird.

Wie die Werte der Tab. 4 besagen, ist auch die *Aufnahme der Metalle ins Gewebe* wechselseitig hemmbar. Jeweils die höchste Konzentration von Co bzw. Fe, d.h. der 100fache Überschuß hemmt die Aufnahme von ^{59}Fe bzw. ^{60}Co gerade etwa um die Hälfte.

Tabelle 4. *Aufnahme von Eisen ins Darmgewebe durchbluteter Jejunumschlingen anämischer Ratten in Gegenwart von Kobalt und umgekehrt von Kobalt in Gegenwart von Eisen*

Jejunumschlingen: 12 cm lang von der Flexura duodenojejunalis an. Darmfüllung: 1 ml physiol. NaCl-Lösung, pH 2,3. Metallangebot: 0,05 μM ^{59}Fe-($FeSO_4$) bzw. ^{60}Co-($CoCl_2$). Versuchsdauer: 1 Std. Die Zahlen sind die Mittelwerte von 5 Versuchen ± der Standardabweichung S_E. Sie geben den Gehalt an Radioaktivität im Gewebe der in EDTA-haltigen (10^{-2} M/l) physiol. NaCl-Lösung gewaschenen Darmschlingen in % der Dosis wieder. Index * bedeutet $p < 0{,}01$ gegen die Kontrolle.

Aufnahme von ^{59}Fe-($FeSO_4$) in Gegenwart von $CoCl_2$:				
$CoCl_2$ μM/ml:	—	0,05	0,5	5,0
Aufnahme ins Gewebe %:	**13,2** ± 3,5	**20,8** ± 8,8	**11,7** ± 5,5	**7,2*** ± 2,2
Aufnahme von ^{60}Co-($CoCl_2$) in Gegenwart von $FeSO_4$:				
$FeSO_4$ μM/ml:	—	0,05	0,5	5,0
Aufnahme ins Gewebe %:	**6,9** ± 1,6	**8,2** ± 2,2	**7,9** ± 3,6	**3,2*** ± 0,4

Eine wechselseitige Beeinflussung der Resorption kann auch für ^{59}Fe und Mn bzw. umgekehrt für ^{54}Mn und Fe nachgewiesen werden (vgl. Tab. 5). In diesen Versuchen wurden die radioaktiv markierten Metalle in einer Konzentration von 0,05 μM/ml in 1 ml physiologischer NaCl-Lösung (pH 2,3) zusammen mit einem 100fachen Überschuß der hemmenden Metalle angeboten. Sowohl die Resorption als auch die Aufnahme ins Gewebe konnte zwischen Fe und Mn wechselweise verringert werden.

Tabelle 5. *Wechselseitige Hemmung der Resorption von Eisen durch Mangan und umgekehrt der von Mangan durch Eisen an durchbluteten Jejunumschlingen in situ bei anämischen Ratten*

Jejunumschlingen: 12 cm lang von der Flexura duodenojejunalis an. Darmfüllung: 1 ml physiol. NaCl-Lösung, pH 2,3. Metallangebot: 0,05 μM ^{59}Fe-($FeSO_4$) bzw. ^{54}Mn-($MnCl_2$) in Gegenwart von 5 μM $MnCl_2$ bzw. $FeSO_4$. Versuchsdauer: 1 Std. Die Zahlen sind die Mittelwerte aus 5 Versuchen ± der Standardabweichung S_E. Sie geben die mit Hilfe einer Ganzkörpermessung ermittelte Radioaktivität im Kadaver der Ratten bzw. die Radioaktivität der in EDTA-haltigen (10^{-2} M/l) NaCl-Lösung gewaschenen Darmschlingen jeweils in % der Dosis an. Index * bedeutet $p < 0{,}01$ gegen die Kontrolle.

Resorption von:	^{59}Fe-($FeSO_4$)		^{54}Mn-($MnCl_2$)	
In Gegenwart von:	—	$MnCl_2$	—	$FeSO_4$
	43 ± 7	**25*** ± 2,4	**6,5** ± 3,3	**3,5** ± 1,8

Aufnahme ins Darmgewebe von:	^{59}Fe-($FeSO_4$)		^{54}Mn-($MnCl_2$)	
In Gegenwart von:	—	$MnCl_2$	—	$FeSO_4$
	17 ± 4	**8*** ± 2	**22,4** ± 11	**8,5*** ± 2,5

In der gleichen Versuchsanordnung konnte auch die Resorption von ^{59}Fe durch Zn und umgekehrt die von ^{65}Zn durch Fe gehemmt werden. Da jedoch Zn bereits in sehr niedrigen Konzentrationen die Resorptionsleistung des Darmes beeinträchtigt, sind diese Ergebnisse mit Zurückhaltung zu beurteilen. 5 µM/ml Zn bringen die Wasserresorption vollständig zum Erliegen, die Resorption von Glucose wird allerdings nur um $^1/_{10}$ verringert.

Die Hemmung der Resorptionsleistung des Darmes durch die Metalle, auf die im Anhang näher eingegangen wird, dürfte die Folge einer unspezifischen Schwermetallwirkung sein, die unumgänglich in Kauf genommen werden muß, wenn man die wechselseitige Beeinflussung der Metalle bei der Resorption studieren will. Die Konzentrationen der mit Tracern markierten Metalle sind in diesen Versuchen 100mal und die höchste Konzentration der nicht markierten Metalle ist 8mal niedriger als die von Pollack et al. [61] verwendeten (vgl. Anhang: Methoden).

Das Ergebnis dieser Versuchsreihe besagt, daß mit hoher Wahrscheinlichkeit die Metalle, deren Resorption bei anämischen Tieren gegenüber normalen gesteigert ist, von dem für die Eisen-Resorption verantwortlichen System der Mucosa mit Eisen verwechselt werden. Diese Metalle konkurrieren mit dem Eisen um die Akzeptorstellen in der Mucosa.

4. Die Retention von Mangan, Eisen, Kobalt und Zink durch normale und anämische Ratten in vivo nach oraler und intravenöser Verabfolgung

Die im Organismus verbleibende Metallmenge wird nicht nur von der *Resorption* bestimmt, sondern auch von der *Ausscheidung*. Deshalb prüften wir in einer letzten Versuchsreihe die *Retention* von Mangan, Kobalt und Zink im Vergleich mit derjenigen von Eisen. Zu diesem Zweck wurden 0,36 µM der mit Tracern vermischten Metalle in 2 ml physiologischer NaCl-Lösung (pH 2,3) mit der Schlundsonde an nüchternen normalen und anämischen Ratten verabreicht. Die Ergebnisse sind in Tab. 6 zusammengefaßt.

Die Retention der Metalle wurde durch Messung der Radioaktivität der Ratten in einer Kleintiermeßanordnung bestimmt. 2 Tage nach der oralen Gabe wurde die Radioaktivität erstmals gemessen. Zu diesem Zeitpunkt sind jedoch erst ca. 90% des nicht resorbierten ^{59}Fe mit dem Kot wieder ausgeschieden worden [33]. Deshalb kann für die Beurteilung der Retention nach der oralen Verabreichung frühestens der 3-Tage-Wert herangezogen werden.

3 Tage nach der oralen Gabe ist bei anämischen Tieren im Vergleich zu normalen die Retention aller geprüften Metalle höher. An der Spitze steht wiederum Eisen: die anämischen Ratten enthalten fast 4mal mehr ^{59}Fe als die normalen, 2,5mal mehr ^{54}Mn, 2mal mehr ^{58}Co und 1,6mal mehr ^{65}Zn. 21 Tage nach der oralen Gabe ist der Gehalt der normalen und anämischen Ratten an ^{54}Mn und ^{58}Co gleich. Der ^{65}Zn-Gehalt der anämischen Tiere ist zu diesem Zeitpunkt zwar statistisch gesichert, aber nur geringfügig größer als der normaler Ratten. Dagegen beträgt das Verhältnis des ^{59}Fe-Gehaltes anämischer und normaler Ratten unverändert 4. Das bedeutet, daß im Gegensatz zu Eisen die anderen Metalle, die im Eisenmangel ebenfalls vermehrt resorbiert wurden, den Organismus rasch wieder verlassen.

Die Ausscheidung scheint vor allem für ^{58}Co sehr rasch zu sein. Bereits 2 Tage nach der oralen Gabe fand sich in den normalen und anämischen Ratten nur noch 3–4% der ursprünglich verabreichten Dosis. Es sei daran erinnert, daß die Resorptionsquote für ^{58}Co an der abgebundenen Jejunumschlinge in situ bei anämischen Ratten während 1 Stunde 60–70% der Dosis betrug. Wie schnell gerade ^{58}Co ausgeschieden wird, geht aus den Daten der Tab. 7 hervor, in der die Ergebnisse

Tabelle 6. *Retention von Mangan, Eisen, Kobalt und Zink durch normale und anämische Ratten nach oraler Applikation mit der Schlundsonde*

n=normale und a=anämische Ratten: Metall-Dosis: 0,36 μM ^{54}Mn-($MnCl_2$), ^{59}Fe-($FeSO_4$), ^{58}Co-($CoCl_2$) und ^{65}Zn-($ZnCl_2$), gelöst in 2 ml physiol. NaCl-Lösung pH, 2,3. Die Zahlen sind die Mittelwerte von 6 Tieren ± der Standardabweichung S_E. Die Radioaktivität der Tiere wurde in einer Kleintiermeßanlage bestimmt und in % der Dosis umgerechnet. Index * bedeutet $p < 0{,}01$ jeweils gegen die Gruppe der normalen Ratten.

		Tage nach der oralen Applikation					
		2	3	4	7	13	21
^{54}Mn	n	**20,0** ±10,4	**11,2** ±7,5	**7,0** ±1,9	**4,4** ±1,4	**2,3** ±0,4	**1,4** ±0,3
	a	**45,7*** ±21,2	**28,2*** ±8,2	**16,0*** ±9,0	**7,5*** ±3,4	**1,7** ±0,9	**0,8** ±0,3
^{59}Fe	n	**21,4** ±11,0	**19,2** ±8,7	**16,5** ±6,7	**15,2** ±6,9	**15,5** ±6,9	**15,0** ±6,2
	a	**75,0*** ±9,1	**73,0*** ±10,7	**70,0*** ±9,6	**66,0*** ±9,3	**67,0*** ±10,2	**65,0*** ±10,0
^{58}Co	n	**2,8** ±0,2	**1,0** ±0,2	**0,9** ±0,3	**0,8** ±0,4	**0,4** ±0,2	**0,2** ±0,1
	a	**3,9*** ±0,6	**2,1*** ±0,5	**1,7*** ±0,4	**1,3*** ±0,5	**1,0*** ±0,4	**0,2** ±0,1
^{65}Zn	n	**35,0** ±4,3	**31,0** ±4,6	**29,0** ±5,0	**27,0** ±5,0	**19,0** ±1,7	**18,0** ±1,0
	a	**69,0*** ±12,6	**51,1*** ±8,9	**46,0*** ±8,8	**34,0*** ±3,3	**25,4*** ±1,5	**23,6*** ±1,1

Tabelle 7. *Retention von Mangan, Eisen, Kobalt und Zink durch normale und anämische Ratten nach intravenöser Applikation*

n = normale und a = anämische Ratten. Metall-Dosis: 0,18 μM ^{54}Mn-($MnCl_2$), ^{59}Fe-($FeSO_4$), ^{58}Co-($CoCl_2$) und ^{65}Zn-($ZnCl_2$), gelöst in 1 ml physiol. NaCl-Lösung, pH 2,3. Injektion in die Schwanzvene. Die Zahlen sind die Mittelwerte von 5 Tieren ± der Standardabweichung S_E. Die Radioaktivität der Tiere wurde in einer Kleintiermeßanlage bestimmt und in % der Dosis umgerechnet.

		Tage nach der iv.-Applikation	
		2	6
^{54}Mn	n	**42** ± 4	**24** ± 2
	a	**42** ± 5	**19** ± 2
^{59}Fe	n	—	**103** ± 8
	a	—	**98** ± 6
^{58}Co	n	**10** ± 3	**4** ± 0,8
	a	**8** ± 3	**3** ± 0,6
^{65}Zn	n	**58** ± 5	**38** ± 4
	a	**57** ± 4	**35** ± 3

der Retention der Metalle nach *intravenöser* Verabfolgung zusammengefaßt sind. 2 Tage nach der iv.-Applikation von ^{58}Co enthalten die normalen und anämischen Tiere gleichermaßen nur noch 10% der verabreichten Dosis.

Das wichtigste in der Tab. 7 enthaltene Ergebnis ist darin zu sehen, daß die Retention der iv. verabreichten Metalle bei normalen und anämischen Ratten gleich groß ist. Mit anderen Worten heißt das, daß für den im Eisenmangel gesteigerten Gehalt der Tiere an ^{54}Mn, ^{58}Co und ^{65}Zn nicht etwa eine erhöhte Retention, i. e. verminderte Ausscheidung, sondern nur die Steigerung der Resorption ursächlich in Frage kommt. Diese wird offenbar durch das System in der Mucosa vermittelt, das für die Resorption von Eisen verantwortlich ist.

Schlußfolgerung und Zusammenfassung

Im Eisenmangel resorbieren Ratten im Vergleich zu Tieren, deren Eisen-Bilanz ausgeglichen ist, aus dem Jejunum nicht nur vermehrt Eisen, sondern auch das ihm chemisch verwandte Kobalt, das ihm nahestehende Mangan und sogar Zink; nicht jedoch Kupfer. Die Retention dieser Metalle, die vermehrt resorbiert wurden, ist bei normalen und anämischen Ratten gleich. In vitro, an nicht durchbluteten Jejunumsegmenten anämischer Ratten ist der Durchtritt für Eisen und die ihm nahestehenden Metalle Nickel, Kobalt und Mangan, aber auch für Zink gesteigert. Auch hier macht Kupfer eine Ausnahme. Die Aufnahme ins Darmgewebe anämischer Ratten ist im Vergleich mit Segmenten normaler Tiere für die Metalle, deren Durchtritt im Eisenmangel gesteigert war, ebenfalls gesteigert. Schließlich ist zu erwähnen, daß sich Eisen und Kobalt bzw. Eisen und Mangan bei der Resorption und Aufnahme in die Mucosa gegenseitig behindern.

Die Befunde lassen sich plausibel durch die Annahme eines eisenbindenden Systems in der Mucosa erklären, das für die Resorption verantwortlich ist. Gemessen an der Zunahme der Resorption der Metalle im Eisenmangel hat dieses System in der Mucosa den höchsten Grad an Spezifität für Eisen; dann folgen Kobalt, Nickel, Mangan und am Ende steht Zink. Dieses System reagiert offenbar nicht mit Kupfer.

Wahrscheinlich ist das eisenbindende System nicht einheitlich, sondern aus mehreren Gliedern zusammengesetzt, wobei sicherlich Proteinen als Liganden eine besondere Bedeutung zukommt. Möglicherweise spielen als Akzeptoren für Eisen auch membranständige Proteine eine Rolle, vergleichbar etwa denjenigen der Reticulocytenmembran, die für die Aufnahme und Inkorporation von Eisen verantwortlich sind [1, 29, 45, 46, 47, 59]. Auch Reticulocyten nehmen nämlich nicht nur Eisen auf; sie binden und inkorporieren im Vergleich mit normalen Zellen ebenfalls mehr Kobalt, Mangan und Zink. Außerdem behindern sich auch an Reticulocyten Eisen und Kobalt gegenseitig bei der Aufnahme [30]. Die Übereinstimmungen dieser Vorgänge lassen daran denken, daß bei den Mucosaepithelien und bei den Reticulocyten ähnliche Mechanismen für den transzellulären Transport bzw. für die Aufnahme in die Zelle verantwortlich sind.

Meinem Lehrer, Herrn Prof. Dr. W. Rummel, möchte ich an dieser Stelle für die Unterstützung dieser Untersuchungen, seine Kritik und Anregung, sowie die ungezählten hilfreichen Diskussionen besonders herzlich danken.

Anhang: Methoden

1. Tiere und Futter

Weibliche Wistar-Ratten wurden sechs Wochen mit einer besonderen Diät gefüttert, die im wesentlichen aus Vollmilchpulver und Maisstärke bestand [21]. Die Mineralzusätze entsprachen den Vorschriften von WESSON [69], der Vitamingehalt denen von DEUEL [15]. Außerdem enthielt das Futter 20 mg/kg Vitamin K_3, 1,5 mg/kg Cholinchlorid und 0,66 mg/kg Inosit. Dem Futter waren folgende Metalle zugesetzt: $FeSO_4$ $3\cdot10^{-3}$ M, $MnCl_2$, $CoCl_2$, $NiSO_4$ und $ZnCl_2$ je $1{,}2\cdot10^{-3}$ M und $CuSO_4$ $2\cdot10^{-5}$ M (Angaben jeweils pro kg Futter).

Der Hämoglobin-Gehalt des Blutes der normalen Tiere betrug 14,6 g/100 ml Blut ($S_E = 1{,}5$; $n = 80$); das Gewicht dieser Tiere lag zwischen 170 und 190 g. Bei den Ratten, bei denen eine Eisenarmut erzeugt wurde, wurde der Fe-Zusatz zum Futter weggelassen. Der Fe-Gehalt dieses Futters betrug noch $6\cdot10^{-5}$ M/kg ($S_E = 5\cdot10^{-6}$ M/kg; $n = 11$); Fe-Bestimmung nach SCHADE [65]. Außerdem wurden 4—5 Aderlässe aus der angeschnittenen Zungenvene bei diesen Tieren vorgenommen; auf diese Weise wurden den Ratten im Mittel 1,6 mg Eisen entzogen ($S_E = 0{,}25$; $n = 80$). Der Hämoglobin-Gehalt des Blutes sank unter diesen Bedingungen auf 5,6 g/100 ml Blut ab. Einzelheiten zur Methode sind bereits veröffentlicht [21].

2. Versuche in vitro

Die im Mittel 3,5 cm langen Dünndarmsegmente (everted sac) wurden in Anlehnung an die von WILSON und WISEMAN beschriebene Methode [70] präpariert (2 Segmente pro Tier; beginnend an der Flexura duodenojejunalis). Die Darmsäckchen wurden mit 0,5 ml Tyrode-Lösung gefüllt und in Erlenmeyer-Kolben (50 ml) in 10 ml Tyrode-Lösung inkubiert. Die Tyrode-Lösung hatte folgende Zusammensetzung: NaCl 8,0; KCl 0,2; $CaCl_2$ (2 H_2O) 0,2; $MgCl_2$ (6 H_2O) 0,1; $NaHCO_3$ 1,25; NaH_2PO_4 0,05; Glucose 2,7 g/l. Die Inkubationsflüssigkeit wurde mit einem Gasgemisch aus 95% O_2 und 5% CO_2 mit 1,5 atü belüftet. Der pH-Wert dieser Lösung betrug 7,2, die Versuchsdauer 2 Std., die Temperatur 37 °C. Die Inkubationsflüssigkeit wurde durch Schaukelbewegungen (32/min; Ausschlag 20 Bogengrade) ständig durchmischt.

Die Metalle wurden auf der Mucosa- und der Serosaseite der Darmsäckchen in einer Konzentration von $5\cdot10^{-6}$ M/l angeboten. Diese Konzentration der Metalle hatte keinen Einfluß auf den Bergauftransport von Glucose. Der Zusatz der radioaktiv markierten Metalle erfolgte nur auf der Mucosaseite als ^{54}Mn-($MnCl_2$), ^{59}Fe-($FeSO_4$), ^{58}Co-($CoCl_2$) bzw. ^{60}Co-($CoCl_2$), ^{63}Ni-($NiSO_4$), ^{64}Cu-($CuSO_4$) und ^{65}Zn-($ZnCl_2$).

Zu Beginn und am Versuchsende wurde die Radioaktivität in der Inkubationsflüssigkeit gemessen. Dabei zeigte sich, daß die Konzentration der verschiedenen Metalle während der Versuchszeit unterschiedlich abfiel (vgl. Tab. 8). Am meisten fiel Eisen, Kupfer und Zink ab; der Abfall der Co-Konzentration betrug nur die Hälfte derjenigen des Eisens. Für Mangan und Nickel nahm die Konzentration am wenigsten ab. Wahrscheinlich sind die Unterschiede darauf zurückzuführen, daß, vom pH-Wert der Inkubationslösung und der O_2-Sättigung begünstigt, die Metalle in unterschiedlichem Maße oxydiert werden, wobei schwerlösliche Metall-Hydroxyde ausfallen bzw. adsorbiert werden.

Tabelle 8. *Der Abfall der Radioaktivität in der Inkubationslösung 10 Min. und 2 Std. nach Versuchsbeginn*

Die Werte geben den Abfall der Radioaktivität in % der angebotenen Aktivität an; sie sind Mittelwerte aus 24—32 Messungen. Die Standardabweichung S_E war nicht größer als 5% des Mittelwertes.

	Zusatz auf der Mucosaseite					
Zeit	^{54}Mn	^{59}Fe	^{58}Co	^{63}Ni	^{64}Cu	^{65}Zn
10 Min.	5	20	5	15	14	35
2 Std.	14	55	28	17	36	46

Am Versuchsende wurde die Radioaktivität in der Serosaflüssigkeit (0,49 g durch Wägung bestimmt. $S_E = 0{,}05$; $n = 360$) gemessen. Vor der Bestimmung der Radioaktivität im Darmgewebe (Feuchtgewicht der normalen Segmente 0,17 g; $S_E = 0{,}03$; $n = 118$. Trockengewicht 0,03 g; $S_E = 0{,}002$. Feuchtgewicht der Segmente anämischer Ratten 0,24 g; $S_E = 0{,}04$; $n = 50$. Trockengewicht 0,032 g; $S_E = 0{,}001$) wurden die Darmstücke 3mal in je 100 ml EDTA-haltiger (10^{-2} M/l) physiol. Kochsalzlösung gewaschen. In Übereinstimmung mit den Ergebnissen anderer Autoren [8, 54] wurden 75—80% der im Darmgewebe gemessenen ^{59}Fe-Aktivität in der abgeschabten Mucosa gefunden [20].

Bei gleichzeitigem Angebot von ^{59}Fe und ^{64}Cu bzw. ^{58}Co wurde die Messung der Radioaktivität in einem NaJ-Bohrlochkristall gemessen, das mit einem Vielkanal-Impulshöhenanalysator verbunden war (RCL 256 Channel-Analyzer). In den Versuchen, in denen die Metalle einzeln angeboten wurden, wurde die γ-Aktivität mit einem NaJ-Bohrlochkristall der Fa. Friesecke und Höpfner (Erlangen-Bruck) vom Typ FH421 gemessen. ^{63}Ni, das nur β-Strahlen aussendet, wurde mittels Flüssigkeitsscintillationsmessung („Tricarb“ der Fa. Packard, USA) in einer Scintillationsflüssigkeit auf Dioxanbasis gemessen.

3. Versuche an durchbluteten Dünndarmsegmenten in situ

In Äthernarkose wurden bei den Ratten, die 16 Stunden nüchtern waren, an der Flexura duodenojejunalis beginnend im Mittel 12,3 cm lange ($S_E = 1{,}2$; $n = 130$) Jejunumschlingen präpariert. Die Schlingen wurden mit 2 ml, in einigen Versuchen auch nur mit 1 ml physiol. Kochsalzlösung gefüllt, deren pH auf 2,3 gebracht worden war. Diese Lösung enthielt die mit Tracern vermischten Metalle in den angegebenen Konzentrationen. Das Abdomen wurde schichtweise verschlossen. Zur Vermeidung der Ausscheidung der resorbierten Metalle mit dem Urin wurde eine Urethra-Ligatur angelegt. Nach einer Stunde wurden die Tiere in Äthernarkose getötet, die Darmschlingen entfernt und die inkorporierte Radioaktivität der Kadaver in einer Kleintiermeßanordnung (NaJ-Bohrlochkristall der Fa. Friesecke und Höpfner, Typ FH 421) gemessen. Die Meßwerte wurden mit

Hilfe eines Standards in % der verabreichten Dosis umgerechnet. Zur Bestimmung der Radioaktivität im Darmgewebe wurde wie oben bei den in vitro-Versuchen beschrieben verfahren.

Hier muß die Frage erörtert werden, welchen Einfluß die Applikation der Metalle in saurer Lösung auf die Resorptionsleistung des Darmes hat. Die Ratte produziert in einer halben Stunde etwa 2 ml Magensekret mit insgesamt 80 mÄq/l Salzsäure [63]; das entspricht einem pH-Wert des Sekretes von 2,9. Der Magensaft gelangt in das Duodenum und wird dort nach Einmündung von Galle und Pankreassaft normalerweise neutralisiert. In der Versuchsanordnung der abgebundenen Jejunumschlingen, die hier verwendet wurde, ist die Pufferung einer sauren Lösung durch die Sekrete der Verdauungsdrüsen nicht möglich, die Pufferung hängt nur von der Pufferkapazität des Darmgewebes ab. Fünf Minuten nach Füllung der Jejunumschlingen mit einer Lösung von pH 2,4 wurde im Schlingeninhalt ein pH-Wert von 4,8 gemessen ($S_E = 0{,}7$; $n = 5$); weitere 10 Minuten später war der pH-Wert auf 6,7 ($S_E = 0{,}5$) angestiegen. Auf diesem Wert blieb der pH des Schlingeninhaltes stehen; mit anderen Worten ist die Pufferung des sauren Inhaltes zwischen 5 und 15 Minuten nach der Injektion abgeschlossen.

Normalerweise beträgt der Schlingeninhalt nach Injektion von 2 ml physiol. NaCl-Lösung (pH 7) nach 1 Std. noch 0,6 ml ($S_E = 0{,}2$; $n = 5$); nach Injektion saurer NaCl-Lösung (pH 2,3) betrug sie jedoch 1,72 ml ($S_E = 0{,}12$). Wird der Schlingenfüllung Glucose zugesetzt (15 µM/ml), dann nimmt die Konzentration so stark ab, daß nach 1 Stunde die Glucose-Konzentration im Schlingeninhalt niedriger ist als im Blut, d.h. die Glucose wird bergauf transportiert. Die Resorption von Glucose, deren Abhängigkeit vom pH-Wert der Inkubationsflüssigkeit in vitro nachgewiesen wurde [44], ist unter den hier gewählten Versuchsbedingungen in vivo offensichtlich nicht beeinträchtigt; sie betrug in beiden Gruppen bei einem Angebot von insgesamt 30 µM 98—99% der Dosis.

Das histologische Bild (wir danken Herrn Prof. Dr. H. KULENKAMPFF für die Anfertigung und Beurteilung der Schnitte) der so behandelten Jejunumsegmente zeigt in der ersten Zeit (5, 10 und 20 Min. nach Injektion der sauren Lösung) eine Auflockerung des Gewebes, ohne daß die Kontinuität der Epithelschicht unterbrochen gewesen wäre; im Darmlumen fanden sich im Vergleich mit Kontrollschnitten mehr abgeschilferte Epithelien. Nach 1 Stunde unterschieden sich jedoch die Bilder der einer sauren Lösung ausgesetzten Jejunumsegmente nicht mehr von den Kontrolldärmen. Diese Beobachtung deckt sich mit derjenigen anderer Autoren, die den Einfluß der Acidität auf das morphologische Bild und die Funktion des Dünndarmes untersuchten [42]. In Anbetracht der Tatsache, daß nur in saurem Medium die zu untersuchenden Metalle in Lösung gehalten und vor der Bildung schwer löslicher Hydroxyde geschützt werden können, entschlossen wir uns dazu, die unter diesen Bedingungen unvermeidbare Beeinträchtigung der Funktion der Dünndarmschlingen in Kauf zu nehmen.

Auch in die Versuchsreihe zur Prüfung der wechselseitigen Hemmung der Resorption wurden die Metalle in saurer Lösung (pH 2,3) in die Darmschlingen injiziert. Wie aus Tab. 9 hervorgeht, ist in diesen Versuchen, in denen nur die Hälfte der Säuremengen der oben beschriebenen Experimente appliziert wurde, keine Wirkung der Säure auf die Wasserresorption zu verzeichnen. Dagegen hemmen die Metalle die Wasser- und Glucoseresorption in unterschiedlichem Maße. Selbst in der höchsten Konzentration von 5 µM/ml blieb Fe ohne Wirkung. Mn hemmte in einer Konzentration von 0,5 µM/ml die Wasserresorption um ca. $^1/_3$; durch eine weitere Steigerung der Mn-Konzentration um das 10fache wurde die inhibitorische Wirkung auf die Wasserresorption nicht weiter verstärkt. In dieser Konzentration wirkt sich Mn aber — wenn auch nur geringfügig — auf die Resorption von Glucose aus.

Co hat in einer Konzentration von 0,5 µM/ml noch keinen statistisch zu sichernden Effekt auf die Wasser- und Glucoseresorption. 5 µM/ml Co hemmen die Wasserresorption um $^1/_3$, die Glucoseresorption ist geringfügig eingeschränkt.

Tabelle 9. *Der Einfluß von Säuregrad und Metallkonzentration auf die Wasser-Resorption und Bergauftransport von Glucose an der durchbluteten Jejunumschlinge in situ bei anämischen Ratten*

Jejunumschlingen: 12 cm lang von der Flexura duodenojejunalis an. Darmfüllung: 1 ml physiol. NaCl-Lösung. Versuchsdauer: 1 Std. Die Zahlen sind die Mittelwerte aus *n* Versuchen $\pm$ der Standardabweichung S_E. Index * bedeutet $p < 0{,}01$ gegen die zugehörige Kontrolle.

Zusatz	Konz. µM/ml	pH	Restflüssigk. g	Netto-Flüssigk. Resorption % des Angebots	Glucose-Konz. i. d. Restflgk. µM/ml	Netto-Glucose Resorption % des Angebots	*n*
		6	**0,23** ± 0,12	**77**	**1,0** ± 0,11	**98,5**	5
		2,2	**0,25** ± 0,11	**75**	**1,4*** ± 0,17	**97,6**	25
$FeSO_4$	5,0	2,2	**0,28** ± 0,15	**72**	**1,6*** ± 0,12	**95,5**	5
$MnCl_2$	0,05	2,2	**0,24** ± 0,16	**76**	**0,95** ± 0,20	**98,5**	5
	0,5	2,2	**0,49*** ± 0,3	**51**	**1,3*** ± 0,14	**95,5**	5
	5,0	2,2	**0,44*** ± 0,19	**56**	**1,4*** ± 0,17	**96**	5
$CoCl_2$	0,05	2,2	**0,36** ± 0,15	**64**	**1,6*** ± 0,05	**96,1**	5
	0,5	2,2	**0,32** ± 0,18	**68**	**1,6*** ± 0,11	**96,4**	5
	5,0	2,2	**0,56*** ± 0,08	**44**	**2,3*** ± 0,7	**92**	20

4. Die Messung der Retention der Metalle bei Ratten in vivo

In dieser Versuchsreihe wurden die Metalle teils mit der Schlundsonde verabreicht (in 2 ml physiol. Kochsalzlösung, pH-Wert 2,3), teils iv. appliziert (in 1 ml physiol. Kochsalzlösung, pH-Wert 2,3). Nach bestimmten Zeitpunkten wurde die jeweils retinierte Menge der Metalle in einer Kleintiermeßanordnung bestimmt und in % der verabreichten Dosis umgerechnet. Vor der oralen Verabreichung der Metalle waren die Tiere 16 Stunden nüchtern, nicht jedoch vor der iv.-Applikation. In beiden Fällen erhielten die Tiere 3 Stunden nach der Eisen-Gabe ihr gewohntes Futter wieder. Die erste Messung der Radioaktivität wurde 48 Stunden nach der oralen Verabreichung durchgeführt; aus früheren Untersuchungen war bekannt, daß zu diesem Zeitpunkt über 90% der nicht resorbierten ^{59}Fe-Aktivität mit dem Kot wieder ausgeschieden worden ist [33].

Literatur

1. Allen, D. W., Jandl, I. H.: Kinetics of intracellular iron in rabbit reticulocytes. Blood **15**, 71 (1960).
2. Bannerman, R. M., O'Brien, I. R. P., Wilts, L. J.: Studies on iron metabolism IV. Iron absorption in experimental iron deficiency. Blood **20**, 532 (1962).
3. Blanc, B.: Etude isotopique des protéines impliquées dans la résorption et dans le métabolisme du fer. Vortrag gehalten auf der 3. Jahrestagung der Ges. f. Nuclearmedizin in Lausanne, 21.—23. X. 1965.
4. — Issliker, H.: Répartition du fer dans diverses fractions tissulaires de la muqueuse intestinale au cours de l'absorption. Mise en évidence et caractérisation partielle d'une mucoprotéine fixatrice du fer. Helv. physiol. acta **23**, 145 (1965).

5. BLANC, B.: Rôle des protéines impliquées dans la résorption et le métabolisme du fer. In: Radioisotope in der Gastroenterologie (VANNOTTI u. HOFFMANN, Hgb.). Stuttgart: F. K. Schattauer 1967, S. 113.
6. BOTHWELL, TH. H.: Iron overload in the Bantu. In: Iron metabolism (F. GROSS, Ed.). Berlin-Göttingen-Heidelberg: Springer 1964, p. 362.
7. — PIRZIO-BIROLI, G., FINCH, C.: Iron absorption I. Factors influencing absorption. J. Lab. Clin. Med. **51**, 24 (1958).
8. BROWN, E. B., ROTHER, M. L.: Studies of the mechanism of iron absorption I. Iron uptake by the normal rat. J. Lab. Clin. Med. **62**, 357 (1963).
9. — —: Studies of the mechanism of iron absorption II. Influence of iron deficiency and other conditions of iron uptake by rats. J. Lab. Clin. Med. **62**, 804 (1963).
10. CHARLEY, P. J., STITT, C., SHORE, E., SALTMAN, P.: Studies in the regulation of intestinal iron absorption. J. Lab. Clin. Med. **61**, 397 (1963).
11. CHARLTON, R. W., LACOBS, P., TORRANCE, J. D., BOTHWELL, T. H.: The role of intestinal mucosa in iron absorption. J. Clin. Invest. **44**, 543 (1965).
12. CHIRASIRI, L., IZAK, G.: The effect of acute haemorrhage and acute haemolysis on the intestinal iron absorption in the rat. Brit. J. Haemat. **12**, 611 (1966).
13. CONRAD, M. E., CROSBY, W. H.: Intestinal mucosal mechanisms controlling iron absorption. Blood **22**, 406 (1963).
14. DEOBALD, H. J., ELVEHJEM, C. A.: The effect of feeding high amounts of soluble iron and aluminium salts. Amer. J. Physiol. **111**, 118 (1935).
15. DEUEL: Zit. nach: Biochemisches Taschenbuch (H. M. RAUEN, Hgb.). Berlin-Göttingen-Heidelberg: Springer 1958, S. 791.
16. DOWDLE, E. B., SCHACHTER, D., SCHENKER, H.: Active transport of Fe^{59} by everted segments of rat duodenum. Amer. J. Physiol. **198**, 609 (1960).
17. DUTHIE, H. L.: The relative importance of the duodenum in the intestinal absorption of iron. Brit. J. Haemat. **10**, 59 (1964).
18. FINCH, C. A., LODEN, B.: Body iron exchange in man. J. Clin. Invest. **38**, 392 (1959).
19. FORTH, W.: In: Radioisotope in der Hämatologie. I. Intern. Symp. Freiburg 1962 (W. KEIDERLING, G. HOFFMANN, Hgb.). Stuttgart: F. K. Schattauer 1963, S. 61.
20. —: Untersuchungen über die Resorption von Eisen und chemisch verwandten Schwermetallen an Därmen normaler und anämischer Ratten in vivo und in vitro; ein Beitrag zur Frage der Spezifität des Eisenbindenden Systems in der Mucosa. Habil.-Schrift Homburg/Saar 1966.
21. — ANDRES, H.: Methode zur Erzeugung eines Eisenmangels bei Ratten. Arzneimittelforschung **19**, 363 (1969).
22. — LEOPOLD, G., RUMMEL, W.: Eisendurchtritt von der Mucosa- zur Serosaseite und umgekehrt an isolierten eisenarmen und normalen Segmenten von Jejunum und Ileum. Naunyn-Schmiedebergs Arch. Pharmak. exp. Path. **261**, 434 (1968).
23. — PFLEGER, K., RUMMEL, W.: Zur Wirkung von Komplexbildnern auf die Fe-Resorption der Ratte. Naunyn-Schmiedebergs Arch. Pharmak. exp. Path. **245**, 89 (1963).
24. — RUMMEL, W.: Eisenresorption an isolierten Dünndarmpräparaten von normalen und anämischen Ratten. Naunyn-Schmiedebergs Arch. Pharmak. exp. Path. **252**, 205 (1965).
25. — —: Beziehungen zwischen Eisen-Konzentration und -Resorption am isolierten Dünndarm normaler und anämischer Ratten. Med. Pharmacol. exp. **14**, 289 (1966).
26. — —: Abhängigkeit der Eisenresorption von der Eisenbindung durch den Darm. Med. Pharmacol. exp. **14**, 384 (1966).
27. — —: Zur Frage der Regulation der Eisenresorption durch Gastroferrin, ein eisenbindendes Protein des Magensaftes. Klin. Wschr. **46**, 1003 (1968).
28. — — BECKER, P. J.: Die vergleichende Prüfung von Bindung und Durchtritt von Eisen, Kobalt und Kupfer durch isolierte Jejunumsegmente normaler und anämischer Ratten. Med. Pharmacol. exp. **15**, 179 (1966).
29. — — CRÜSEMANN, D., PFLEGER, K.: Die Bedeutung der Bindung in der Zellmembran für die Fe-Aufnahme der Reticulozyten. Naunyn-Schmiedebergs Arch. Pharmak. exp. Path. **257**, 275 (1967).

30. FORTH, W., RUMMEL, W., CRÜSEMANN, D., SIMON, J.: Die Bindung von Eisen und verwandten Schwermetallen in der Membran und ihre Aufnahme in Reticulozyten. In: I. Intern. Sympos. über Stoffwechsel und Permeabilität von Erythrozyten und Thrombozyten (DEUTSCH, GERLACH, MOSER, Eds.). Stuttgart: G. Thieme 1968.
31. — — PFLEGER, K.: Der Einfluß von Liganden auf die Retention von Eisen nach oraler Verabfolgung an normalen und anämischen Ratten. Naunyn-Schmiedebergs Arch. Pharmak. exp. Path. **261**, 225 (1968).
32. — — SEIFEN, E.: Der Einfluß verschiedener Liganden auf Eisendurchtritt und -bindung an isolierten Darmpräparaten normaler und anämischer Ratten. Naunyn-Schmiedebergs Arch. Pharmak. exp. Path. **252**, 224 (1965).
33. — PFLEGER, K., RUMMEL, W., SEIFEN, E., RICHMOND, SP. J.: Der Einfluß verschiedener Liganden auf Resorption, Verteilung und Ausscheidung von Eisen nach oraler Verabfolgung. Naunyn-Schmiedebergs Arch. Pharmak. exp. Path. **252**, 242 (1965).
34. — SEIFEN, E.: Der Einfluß von Komplexbildnern auf die ^{59}Fe-Resorption am isolierten Darm der Ratte. Naunyn-Schmiedebergs Arch. Pharmak. exp. Path. **241**, 556 (1961).
35. GRANICK, S.: Iron metabolism and hemochromatosis. Bull. N. Y. Acad. Med. **25**, 403 (1949).
36. —: Structure and physiological functions of ferritin. Physiol. Rev. **31**, 489 (1951).
37. HAHN, P. F., BALE, W. F., LAWRENCE, E. O., WHIPPLE, G. H.: Radioactive iron and its metabolism in anemia. J. exp. Med. **69**, 739 (1939).
38. — — ROSS, J. F., BALFOUR, W. M., WHIPPLE, G. H.: Radioactive iron absorption by gastrointestinal tract. J. exp. Med. **78**, 169 (1943).
39. — CAROTHERS, E. L., DARBY, W. I., MARTIN, M., SHEPPARD, C. W., CANNON, R. O., BEAM, A. S., DENSEN, P. M., PETERSEN, J. C., MCCLELLAN, G. S.: Iron metabolism in human pregnancy as studied with the radioactive isotope ^{59}Fe. Amer. J. Obst. Gyn. **61**, 477 (1951).
40. HALLBERG, L., SÖLVELL, L.: Iron absorption studies. Acta med. Scand. **168** (Suppl. 358), 71 (1960).
41. HELBOCK, H. J., SALTMAN, P.: The transport of iron by rat intestine. Biochem. Biophys. Acta **135**, 979 (1967).
42. HELLER, M. D., KERN, F.: Absorption of lactic acid from an isolated intestinal segment in the intact rat. Proc. Soc. Exp. Biol. Med. **127**, 1103 (1968).
43. HEVESY, G. v., LOCKNER, D.: Iron metabolism in health and in the neoplastic state. Arkiv för Kemi **19**, Nr. 26, 303 (1961).
44. IIDA, H., MOORE, E. W., BROITMAN, S. A., ZAMCHECK, N.: Effect of pH on active transport of D-glucose in the small intestine of hamsters. Proc. Soc. exp. biol. med. **127**, 730 (1968).
45. JANDL, J. H., INMAN, I. K., SIMMONS, R. L., ALLEN, D. W.: Transfer of iron from serum iron-binding protein to human reticulocytes. J. clin. Invest. **28**, 161 (1959).
46. — KATZ, J. H.: The plasma-to-cell cycle of transferrin. J. clin. Invest. **42**, 314 (1963).
47. KATZ, J.H., JANDL, J. H.: The role of transferrin in the transport of iron into the developing red cell. In: Iron metabolism (F. GROSS, Ed.). Berlin-Göttingen-Heidelberg: Springer 1964, p. 103.
48. KAUFMAN, R. M., POLLACK, S., CROSBY, W. H.: Iron-deficient diet: effect in rats and humans. Blood **28**, 726 (1966).
49. KRANTZ, S., GOLDWASSER, E., JACOBSON, L. O.: Studies on erythropoiesis XIV. The relationship of humoral stimulation to iron absorption. Blood **14**, 654 (1959).
50. LAFFERTY, F. W., SPENCER, G. E., PEARSON, O. E.: Effects of androgens, estrogens and high calcium intakes on bone formation and resorption in osteoporosis. Amer. J. Med. **36**, 514 (1964).
51. LAURELL, C. B.: Studies on the transportation and metabolism of iron in the body with special reference to the iron binding component in human plasma. Acta physiol. Scand. Suppl. **46**, 14 (1947).
52. —: Iron transportation. In: Iron in clinical medicine (R. O. WALLERSTEIN, S. R. METTIER, Eds.). Berkeley-Los Angeles: Univ. of California Press 1958, p. 8.

53. LYNCH, S. R., BERELOWITZ, I., SEFTEL, H. C., MILLER, G. B., KRAWITZ, P., CHARLTON, R. W., BOTHWELL, T. H.: Osteoporosis in Johannesburg bantu males, its relationship to siderosis and ascorbic acid deficiency. Amer. J. clin. Nutr. **20**, 799 (1967).
54. MANIS, J. G., SCHACHTER, D.: Active transport of iron by intestine: features of the two-step mechanism. Amer. J. Physiol. **203**, 73 (1962).
55. — —: Active transport of iron by intestine: effects of oral iron and pregnancy. Amer. J. Physiol. **203**, 81 (1962).
56. MCCANCE, R. A., WIDDOWSON, E. M.: Absorption and excretion of iron. Lancet **2**, 680 (1937).
57. MCGOWAN, J. P., CRICHTON, A.: Iron deficiency in pigs. Biochem. J. **18**, 265 (1928).
58. MENDEL, G. S., WELLER, R. I., MANGALIK, A.: Studies on iron absorption II. The absorption of iron in experimental anemias of diverse etiology. Blood **22**, 450 (1963).
59. MORGAN, E. H., LAURELL, C. B.: Studies on the exchange of iron between transferrin and reticulocytes. Brit. J. Haemat. **9**, 471 (1963).
60. PHILLIPS, C. S. G., WILLIAMS, R. J. P.: Inorganic chemistry I and II. Oxford: Clarendon 1965.
61. POLLACK, S., GEORGE, J. N., REBA, R. C., KAUFMANN, R. M., CROSBY, W. H.: The absorption of nonferrous metals in iron deficiency. J. clin. Invest. **44**, 1470 (1965).
62. REIMANN, F., FRITSCH: Zit. nach E. STARKENSTEIN: Eisen. In: Handbuch der exp. Pharmakologie. 3. Band, 2. Teil (A. HEFFTER, W. HEUBNER, Hgb.). Berlin: Springer 1934, S. 859.
63. ROE, J. H., DWYER, H. M.: Proc. Soc. Exp. Biol. Med. **41**, 603 (1939). Zit. nach: The rat (E. J. FARRIS, J. Q. GRIFFITH, Eds.). Philadelphia-London-Montreal: Lippincott 1949, p. 168.
64. RULIFFSON, W. S., HOPPING, J. M.: Maturation, iron deficiency and ligands in enteric radioiron transport in vitro. Amer. J. Physiol. **204**, 171 (1963).
65. SCHADE, A. L., OYAMA, S., REINHART, R. W., MILLER, J. R.: Bound iron and unsaturated iron-binding capacity of serum: rapid and reliable determination. Proc. Soc. Exp. Biol. Med. **87**, **443** (1954).
66. SMITH, N. J., SCHULZ, J.: In: Iron in clinical medicine (R. O. WALLERSTEIN, S. R. METTIER, Eds.). Berkeley-Los Angeles: Univ. of California Press 1958, p. 65.
67. STURGEON, P.: Studies on iron requirements in infants and children I. Normal values for serum iron, copper and free erythrocyte protoporphyrin. Pediatrics **13**, 107 (1954).
68. VALBERG, L. S., OLATUNBOSUN, D.: Alteration in cobalt absorption in patients with disorders of iron metabolism. Gastroenterology **54**, 1279 (1968).
69. WESSON: Zit. nach: Biochemisches Taschenbuch (H. M. RAUEN, Hgb.). Berlin-Göttingen-Heidelberg: Springer 1958, S. 793.
70. WILSON, T. H., WISEMAN, G.: The use of sacs of everted small intestine for the study of the transferrence of substances from the mucosal to the serosal surface. J. Physiol. (London) **123**, 116 (1954).

Theoretische Analyse der Kinetik der Verteilung und Ausscheidung von Radium-Isotopen mit Hilfe eines Compartmentmodells

Theoretical Analysis of the Kinetics of Distribution and Excretion of Radium Isotopes by Means of a Compartment Model

A. KAUL

Nuklearmedizinische Abteilung des Klinikums Steglitz der Freien Universität Berlin

Summary

The contrast medium Thorotrast (^{232}Th) was often used in the period 1930–1945; data obtained from the fairly large number of surviving patients enable radiation doserate and somatic radiation effects to be correlated but give no information about radiation dosage over a period of time. To enable the cumulative dose to be calculated, experiments over 530 days were carried out on adult male rabbits which had 2 cm^3 of contrast medium injected into the ear vein. To establish the theoretical basis of the daily excretion of the disintegration products, ^{228}Ra and ^{224}Ra, a three-compartment model was constructed: skeleton excluding bone marrow (K); reticulo-endothelial system RES (R); the remaining intra and extra-cellular spaces (G). By comparing the curves calculated by means of the model with the measured daily ^{228}Ra excretion, it is possible to evaluate the total skeletal retention and excretion from the body of radium in ionic form.

Einleitung

Die weitverbreitete Anwendung von kolloidalem ThO_2 als Röntgenkontrastmittel in den Jahren zwischen 1930 und 1945 und die daraus resultierende verhältnismäßig große Zahl dieser heute noch lebenden sog. Thorotrastpatienten ermöglicht es u.U., eine Korrelation zwischen Strahlendosis und somatischer Strahlenwirkung unmittelbar am Menschen und mit statistisch gesicherter Aussagekraft aufzustellen.

Die Analyse der Patienten ist aber keineswegs so problemlos, wie man zunächst annehmen möchte: Neben der Auswahl des strahlenbedingten Effekts sind Zeitpunkt des Auftretens und quantitative Angabe der Höhe einer Schädigung entscheidend für eine *klinische* Beurteilung der Fälle.

Nicht weniger problematisch ist die *strahlenbiophysikalische* Analyse der Patienten, nämlich die Ermittlung der Strahlendosis, die mit einem bestimmten Effekt korreliert werden soll.

Obwohl bereits eine größere Zahl von relevanten Ergebnissen über Verteilung und Ausscheidung von Th-232 und seinen Folgeprodukten vorliegen, war bisher eine Aussage über die Höhe der gesamten, während der Dauer der Thorotrastspeicherung an ein Organ abgegebenen Strahlendosis nur bedingt möglich. Der Grund hierfür ist, daß die Mehrzahl der Untersuchungen 20–30 Jahre nach

Inkorporierung durchgeführt wurden, d.h. zu einem Zeitpunkt, als bereits physiologisches Gleichgewicht zwischen Thoriumfolgeprodukten herrschte.

Da jedoch die Kenntnis des zeitlichen Verlaufs der Strahlendosisleistung Voraussetzung ist zur Berechnung der akkumulierten Strahlendosis, derartige Untersuchungen aber aus naheliegenden Gründen nicht am Menschen selbst durchgeführt werden können, mußte der Frage nach der Kinetik der Verteilung und Ausscheidung von Th-232 und seinen Folgeprodukten im Tierexperiment nachgegangen werden. Darüberhinaus mußte ein Modell entwickelt werden, das die Translokation von Thoriumfolgeprodukten im Organismus und deren Ausscheidung aus dem Organismus in Übereinstimmung mit den experimentellen Befunden zu beschreiben gestattet und damit die Voraussetzung bietet, das biologische und physikalische Geschehen im Organismus als Grundlage zur Berechnung der zeitlichen Änderung der Strahlendosisleistung in einzelnen Organen mit Hilfe von geschlossenen mathematischen Ausdrücken zu formulieren.

Die experimentellen Untersuchungen wurden an männlichen Kaninchen der Rasse „Gelbsilber" (Alter der Tiere zu Versuchsbeginn: 9 Monate; mittleres Gewicht: 3080 $\pm$ 300 g) durchgeführt. Der Versuchszeitraum erstreckte sich über 530 Tage, das Kontrastmittel (2 cm^3 entsprechend 0,7 cm^3/kg) wurde den Tieren im Alter zwischen 9 und 12 Monaten in die Ohrvene injiziert.

Ergebnisse der theoretischen und experimentellen Untersuchungen*

Zur Berechnung der täglichen Gesamtausscheidung der beiden Radiumisotope Ra-228 und Ra-224 wurde der Organismus als ein System der folgenden 3 Compartments aufgefaßt:

1. Markfreies Skelett (K),
2. Organe des RES (R),
3. „übriger" Intra- und Extrazellularraum („Gewebe", G).

Diese verhältnismäßig wenig differenzierte Gliederung des Organismus in funktionelle Einheiten ist wegen der spezifischen Affinität des Kolloids bzw. der Th-232-Folgeprodukte Ra-228 und Ra-224 in Ionenform zu bestimmten Organen möglich:

Während Th-232 und Th-228 als Komponenten der Teilchenfraktion praktisch ausschließlich auf die Organe des RES Leber, Milz und Knochenmark beschränkt sind, ist das markfreie Skelett das „kritische" Organ für die „osteotropen" Radionuklide Ra-228 und Ra-224 in Ionenform. Die Translokation der beiden Radiumisotope in das Skelett sowie deren Ausscheidung aus dem Organismus aufgrund der Dauerzufuhr von Radium aus dem RES und dem markfreien Skelett laufen über den „übrigen" Intra- und Extrazellularraum ab.

Auf Ra-228 angewandt, beschreibt das Modell die Kinetik der Verteilung und Ausscheidung von Ra-228 mit der Teilchenfraktion und in Ionenform wie folgt (s. Abb. 1):

* Auf die Ableitung der formalen Ausdrücke zur theoretischen Beschreibung der Kinetik der Verteilung und Ausscheidung von Radium muß im Rahmen der vorliegenden, zusammenfassenden Darstellung verzichtet werden. Einzelheiten werden an anderer Stelle ausführlicher behandelt (KAUL, 1969).

Der Anteil an der Ra-228-Gesamtaktivität der Suspension (S), der an das ThO_2-Kolloid (T) gebunden ist ($A_2^{S,T}$) (t_i), wird aus dem Blut retiniert und im RES abgelagert. Mit der Ausscheidung der Teilchenfraktion wird Ra-228 entsprechend

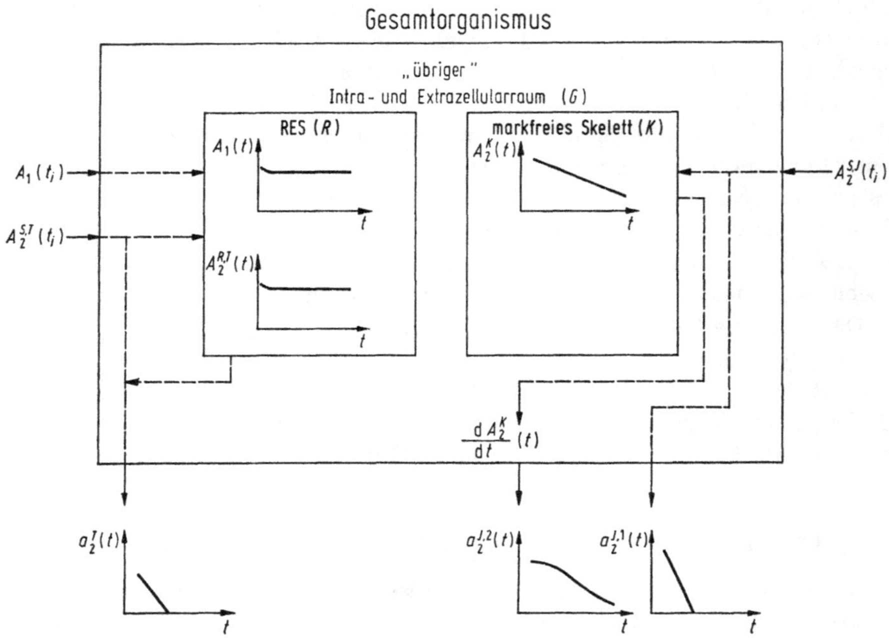

Abb. 1. Schematische Darstellung des Compartmentmodells zur Berechnung der täglichen Ausscheidung von Ra-228

dem Aktivitätsverhältnis $a_{2,1}^{S,T}(t_i)$ zwischen Ra-228 und Th-232 in den ThO_2-Teilchen der Suspension zum Zeitpunkt t_i der Injektion aus dem Organismus eliminiert:

$$a_2^T(t) = a_{2,1}^{S,T}(t_i)\, a_1(t_i)\, e^{-\lambda_T t}. \tag{1}$$

Der dem Organismus zum Zeitpunkt der Inkorporierung in Ionenform (J) zugeführte Anteil der Ra-228-Aktivität der Suspension $A_2^{S,J}(t_i)$ wird zu einem bestimmten Bruchteil (F) in das Skelett translokalisiert und von dort nach einer Potenzfunktion in das Gewebe ausgeschieden. Die daraus resultierende Ra-228-Aktivität $A_{2,2}^{G,J}$ des Gewebes wird durch Gl. (2) beschrieben:

$$A_{2,2}^{G,J}(t) = F A_2^{S,J}(t_i) e^{-\lambda_{2,J} t} b \int_1^t e^{\lambda_{2,J} t} t^{-(1+b)}\, dt. \tag{2}$$

$\lambda_{2,J}$ sei die Ausscheidungskonstante für Ra-228 in Ionenform.

Da die pro Zeiteinheit aus dem Gewebe durch Ausscheidung eliminierte Ra-228-Aktivität in Ionenform proportional der vorhandenen ist, folgt für den zeitlichen Verlauf der Ra-228-Ausscheidung in Ionenform aus dem Organismus aufgrund der Translokation von *einmalig zum Zeitpunkt der Inkorporierung im Skelett abgelagertem Radium* in das Gewebe:

$$a_2^{J,2}(t) = \lambda_{2,J} F A_2^{S,J}(t_i)\, e^{-\lambda_{2,J} t}\, b \int_1^t e^{\lambda_{2,J} t}\, t^{-(1+b)}\, dt. \tag{2a}$$

Der Bruchteil $(1 - F)$ der in Ionenform zum Zeitpunkt der Inkorporierung zugeführten Ra-228-Aktivität, der nicht in das Skelett translokalisiert wurde, verbleibt im Gewebe und wird von dort aus dem Organismus ausgeschieden:

$$a_2^{J,1}(t) = \lambda_{2,J}(1 - F)\, A_2^{S,J}(t_i)\, e^{-\lambda_{2,J} t}. \tag{3}$$

Da im Beobachtungszeitraum Zerfall und Nachbildung von Ra-228 wegen dessen Halbwertzeit von 5,8 Jahren zu vernachlässigen sind, findet keine nennenswerte Dauerausscheidung von Ra-228 aus dem RES und demzufolge keine dauernde Translokation von Radium in das Skelett statt. Die tägliche Gesamtausscheidung an Ra-228 wird damit durch den folgenden Ausdruckformal beschrieben:

$$a_2(t) = \lambda_{2,J} A_2^{S,J}(t_i) e^{-\lambda_{2,J} t} \left[(1 - F) + F b \int_1^t e^{\lambda_{2,J} t} t^{-(1+b)}\, dt\right] + a_{2,1}^{S,T}(t_i)\, a_1(t_i)\, e^{-\lambda_T t}. \tag{4}$$

Die Gl. (4) enthält 2 Unbekannte, die der unmittelbaren experimentellen Bestimmung nicht zugänglich sind. Es sind dies die Ausscheidungskonstante $\lambda_{2,J}$ für Ra-228 in Ionenform und die Größe F, die der Retention von Radium im gesamten markfreien Skelett zum Zeitpunkt 1 Tag nach Inkorporierung entspricht.

Dagegen sind aus dem Experiment bekannt:
Die Ra-228-Aktivität $A_2^{S,J}(t_i)$ der Suspension in Ionenform, der Exponent b in der nach einer Potenzfunktion verlaufenden Retention von Radium im Skelett, das Aktivitätsverhältnis $a_{2,1}^{S,T}(t_i)$ zwischen Ra-228 und Th-232 in der Teilchenfraktion der Suspension und die „fiktive" Th-232-Ausscheidung zum Zeitpunkt der Injektion sowie die Ausscheidungskonstante λ_T für die Teilchenfraktion. Darüber hinaus ist die Größe F/M^K, d.h. die auf die Gesamtmasse M^K des markfreien Skeletts bezogene Retention bekannt; ihr Wert ergab sich zu $1{,}5 \cdot 10^{-3}\,g^{-1}$ (KAUL und HEYDER, 1969).

Zur Bestimmung von $\lambda_{2,J}$ und F und damit der Masse des markfreien Skeletts wurden 14 F-Werte (zwischen 0,105 entsprechend 70 g Skelettmasse und 0,525 entsprechend 350 g Skelettmasse) mit 13 $\lambda_{2,J}$-Werten (zwischen $0{,}64\,d^{-1}$ entsprechend 26 h Halbwertzeit und $2{,}07\,d^{-1}$ entsprechend 8 h Halbwertzeit) kombiniert, so daß für insgesamt 182 Wertepaare der jeweilige theoretische Verlauf der täglichen Gesamtausscheidung an Ra-228 im Zeitintervall zwischen Injektion und 30 Tagen danach bestimmt werden konnte. (Bedingt durch das in Gl. (4) aufgetretene Exponentialintegral und den mit 182 Wertepaaren von F und $\lambda_{2,J}$ verbundenen Rechenaufwand wurde die Berechnung des theoretischen Verlaufs von a_2 als Funktion von F und $\lambda_{2,J}$ mit Hilfe einer elektronischen Rechenanlage durchgeführt.)

Bei Variation von F und festgehaltenem $\lambda_{2,J}$ ändern sich die Steigung von $a_2(t)$ im Zeitintervall zwischen einem und etwa 10 Tagen nach Inkorporierung sowie der Betrag von $a_2(t)$ dergestalt, daß mit wachsendem F die zeitliche Änderung der Ra-228-Gesamtausscheidung kleiner wird (s. Abb. 2). Hält man dagegen F fest und variiert $\lambda_{2,J}$, dann ändert sich im wesentlichen die Steigung von $a_2(t)$ im Zeitraum bis zu etwa 10 Tagen nach Injektion (s. Abb. 3).

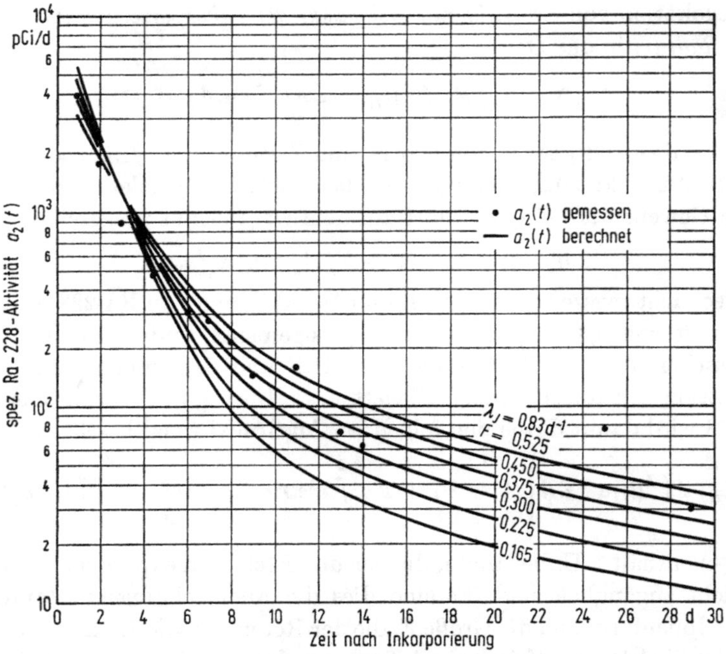

Abb. 2. Änderung des theoretischen Verlaufs der täglichen Ausscheidung von Ra-228 bei Variation von F

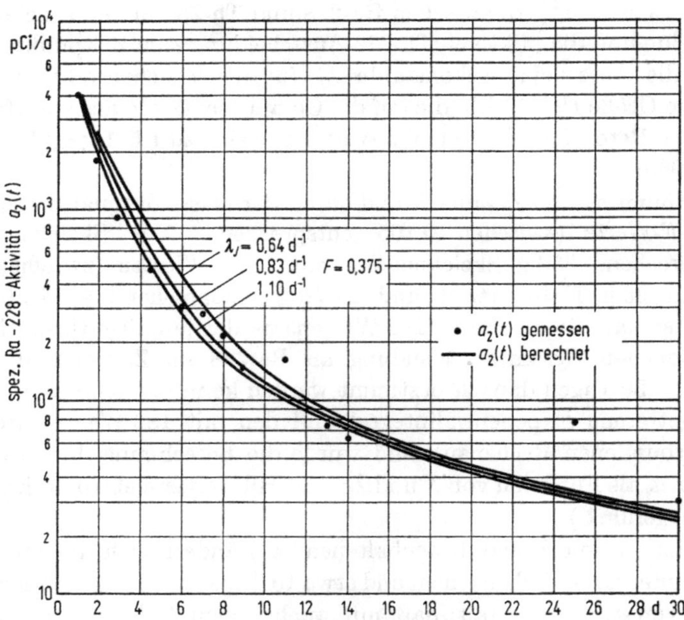

Abb. 3. Änderung des theoretischen Verlaufs der täglichen Ausscheidung von Ra-228 bei Variation von λ_J

Aus der Gegenüberstellung der durch Permutation von F und $\lambda_{2,J}$ gewonnenen theoretischen Ausscheidungskurven und des gemessenen Verlaufs von $a_2(t)$ ergab sich das folgende Wertepaar als beste Annäherung an das Experiment:

$$F = 0{,}375 \pm 0{,}075 \text{ (entsprechend } M = 250 \pm 50 \text{ g)},$$
$$\lambda_{2,J} = 0{,}83 \pm 0{,}23 \text{ (entsprechend } T^J_{1/2} = 20 \pm 5 \text{ h)}.$$

Mit diesen Werten für F und $\lambda_{J,2}$ wurde der theoretische Verlauf der einzelnen Komponenten (a_2^T, $a_2^{J,1}$ und $a_2^{J,2}$) der täglichen Gesamtausscheidung an Ra-228 sowie die Gesamtausscheidung selbst berechnet. Die Ergebnisse sind in Abb. 4 dargestellt und zeigen, daß nach etwa 10 Tagen nach der Inkorporierung der Suspension von kolloidalem ThO_2 die Ra-228-Ausscheidung nur noch durch die Translokation der einmalig zum Zeitpunkt der Injektion im markfreien Skelett abgelagerten Ra-228-Aktivität aus dem Skelett und dem Organismus bestimmt wird.

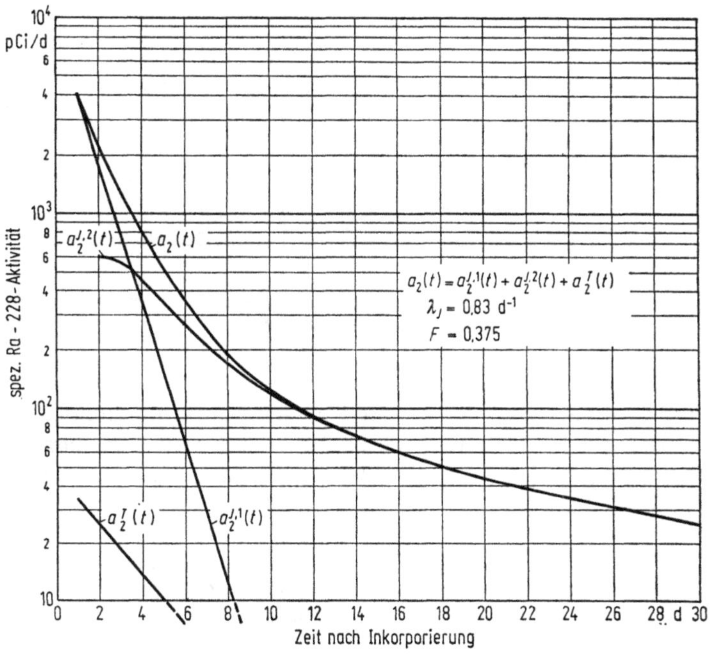

Abb. 4. Zeitlicher Verlauf der theoretisch berechneten Gesamtausscheidung von Ra-228 und deren Komponenten

Auf die Kinetik der Verteilung und Ausscheidung von Ra-224 angewandt, mußte das Modell dahingehend geändert werden, daß es auch die *Dauerausscheidung* von Ra-224 aus dem RES durch Zerfall von Th-228 zu beschreiben gestattet. Diese Dauerausscheidung von Ra-224 hat eine *kontinuierliche Translokation* von Ra-224 in das Skelett zur Folge. Durch den Abfluß von Ra-224 aus dem Skelett und die Dauerausscheidung von Radium aus dem RES in den übrigen Intra- und

Extrazellularraum entstehen 2 weitere Komponenten der Gesamtausscheidung, so daß im Gegensatz zu Ra-228 die tägliche Ausscheidung an Ra-224 aus 5 Komponenten resultiert (s. Abb. 5).

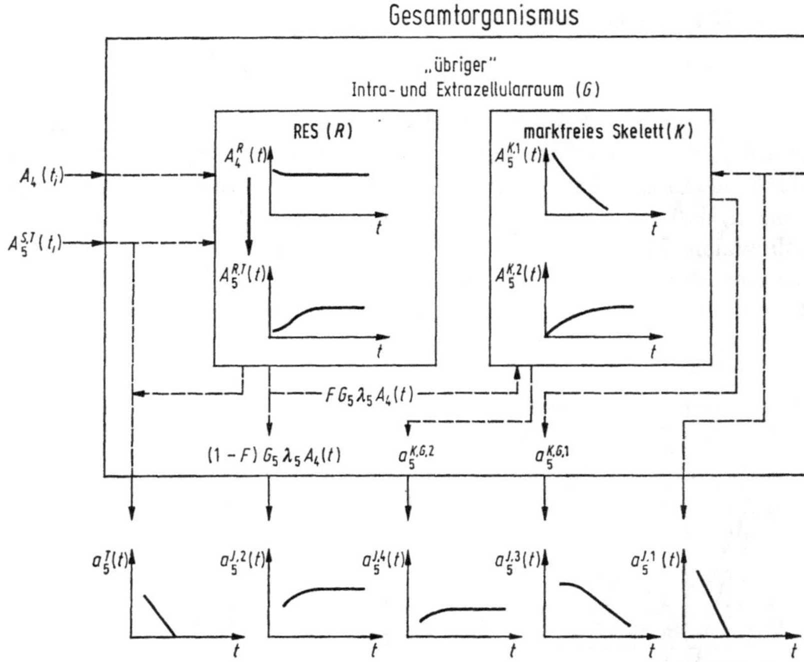

Abb. 5. Schematische Darstellung des Compartmentmodells zur Berechnung der täglichen Ausscheidung von Ra-224

Da sich die Ausscheidung der an die Teilchenfraktion gebundenen Ra-224-Aktivität wie die der Th-232- und Th-228-Aktivität vollzieht, kann sie unter Berücksichtigung des Zerfalls von Ra-224 während der Ausscheidung wie folgt ausgedrückt werden:

$$a_5^T(t) = a_{5,1}^{S,T}(t_i)\, a_1(t_i)\, e^{-(\lambda_5+\lambda_T)t}. \tag{5}$$

Durch *einmalige Zufuhr von Ra-224* in Ionenform zum Zeitpunkt der Inkorporierung findet unter Berücksichtigung des Zerfalls von Ra-224 im Skelett (K) eine Ausscheidung von Radium in das Gewebe (G) statt $\left(a_5^{K,G}(t)\right)$.

Die daraus resultierende Ra-224-Aktivität $A_{5,2}^{G,J}(t)$ wird durch Gl. (6) beschrieben:

$$A_{5,2}^{G,J}(t) = F A_5^{S,J}(t_i)\, e^{-\lambda_{5\text{eff}}t}\left[\lambda_5 \int_1^t e^{\lambda_{5,J}t}\, t^{-b}\, dt + b \int_1^t e^{\lambda_{5,J}t}\, t^{-(1+b)}\, dt\right] \tag{6}$$

mit der Definition von $\lambda_{5\text{eff}}$: $\lambda_{5\text{eff}} = \lambda_5 + \lambda_{5,J}$.

Damit folgt für die Ausscheidung $a_5^{J,3}(t)$ aufgrund der Elimination des zum Zeitpunkt der Inkorporierung *einmalig in das Skelett translokalisierten* Ra-224:

$$a_5^{J,3}(t) = \lambda_{5,J} F A_5^{S,J}(t_i)\, e^{-\lambda_{5\text{eff}}t}\left[\lambda_5 \int_1^t e^{\lambda_{5,J}t}\, t^{-b}\, dt + b \int_1^t e^{\lambda_{5,J}t}\, t^{-(1+b)}\, dt\right]. \tag{6a}$$

Der Bruchteil $(1 - F)$ der in Ionenform zum Zeitpunkt der Inkorporierung zugeführten Ra-224-Aktivität, der nicht in das Skelett translokalisiert wurde, verbleibt im Gewebe und wird von dort aus dem Organismus ausgeschieden:

$$a_5^{J,1}(t) = \lambda_{5,J}(1 - F)\, A_5^{S,J}(t_i)\, e^{-\lambda_{5\text{eff}} t}. \tag{7}$$

Aus der im Gewebe in Ionenform vorhandenen Ra-224-Aktivität aufgrund der *Dauerzufuhr von Radium aus dem RES* unter Berücksichtigung der Translokation von Ra-224 in das Skelett und der Agglomerierung der ThO_2-Teilchen in den Organen des RES (G_5, KAUL und HEYDER, 1969) folgt als weitere Komponente an der täglichen Gesamtausscheidung an Ra-224:

$$a_5^{J,2}(t) = \lambda_{5,J}(1 - F)\,\frac{\lambda_5}{\lambda_{5\text{eff}}}\, G_5 \left[\left(A_5(t_i) - a_{4,1}^S(t_i)\,\frac{a_1(t_i)}{\lambda_T}\right)(1 - e^{-\lambda_{5\text{eff}} t}) + \frac{\lambda_{5\text{eff}}}{\lambda_T - \lambda_{5\text{eff}}}\, a_{4,1}^S(t_i)\,\frac{a_1(t_i)}{\lambda_T}\,(e^{-\lambda_{5\text{eff}} t} - e^{-\lambda_T t})\right]. \tag{8}$$

Aus der *Dauerzufuhr* von Ra-224 aus dem RES *in das markfreie Skelett* und von dort zurück in das Gewebe $(a_5^{K,G,2}(t))$ resultiert als 5. Komponente an der täglichen Ra-224-Gesamtausscheidung der folgende Ausdruck:

$$a_5^{J,4}(t) = \lambda_{5,J} F G_5 \lambda_5 e^{-\lambda_{5\text{eff}} t} \int_1^t \Bigg[\left(A_4(t_i) - a_{4,1}^S(t_i)\,\frac{a_1(t_i)}{\lambda_T}\right) \cdot \left(1 - \lambda_5 \int_0^t e^{-\lambda_5 t}\, t^{-b}\, dt - e^{-\lambda_5 t}\, t^{-b}\right) + a_{4,1}^S(t_i)\,\frac{a_1(t_i)}{\lambda_T} \cdot \left(e^{-\lambda_T t} - \lambda_5 \int_0^t e^{-(\lambda_5 + \lambda_T) t}\, t^{-b}\, dt - e^{-(\lambda_5 + \lambda_T) t}\, t^{-b}\right)\Bigg] e^{\lambda_{5\text{eff}} t}\, dt. \tag{9}$$

Die Zusammenfassung der einzelnen, die Gesamtausscheidung an Ra-224 bestimmenden Komponenten (Gln. (5), (6a), (7), (8) und (9)) hat die folgende Form:

$$\begin{aligned} a_5(t) = \lambda_{5,J} \Bigg(&(1 - F)\left(A_5^{S,J}(t_i)\, e^{-\lambda_{5\text{eff}} t} + \frac{\lambda_5}{\lambda_{5\text{eff}}}\, G_5 \left[\left(A_4(t_i) - a_{4,1}^S(t_i)\,\frac{a_1(t_i)}{\lambda_T}\right) \cdot (1 - e^{-\lambda_{5\text{eff}} t}) + \frac{\lambda_{5\text{eff}}}{\lambda_T - \lambda_{5\text{eff}}}\, a_{4,1}^S(t_i)\,\frac{a_1(t_i)}{\lambda_T}\,(e^{-\lambda_{5\text{eff}} t} - e^{-\lambda_T t})\right]\right) \\ &+ F \Bigg(A_5^{S,J}(t_i)\, e^{-\lambda_{5\text{eff}} t} \left[\lambda_5 \int_1^t e^{-\lambda_{5,J} t}\, t^{-b}\, dt + b \int_1^t e^{\lambda_{5,J} t}\, t^{-(1+b)}\, dt\right] \\ &+ G_5 \lambda_5 e^{-\lambda_{5\text{eff}} t} \int_1^t \Bigg[\left(A_4(t_i) - a_{4,1}^S(t_i)\,\frac{a_1(t_i)}{\lambda_T}\right)\left(1 - \lambda_5 \int_0^t e^{-\lambda_5 t}\, t^{-b}\, dt - e^{-\lambda_5 t}\, t^{-b}\right) \\ &+ a_{4,1}^S(t_i)\,\frac{a_1(t_i)}{\lambda_T}\left(e^{-\lambda_T t} - \lambda_5 \int_0^t e^{-(\lambda_5 + \lambda_T) t}\, t^{-b}\, dt - e^{-(\lambda_5 + \lambda_T) t}\, t^{-b}\right)\Bigg] e^{\lambda_{5\text{eff}} t}\, dt\Bigg)\Bigg) \\ &+ a_{5,1}^{S,T}(t_i)\, a_1(t_i)\, e^{-(\lambda_5 + \lambda_T) t}. \end{aligned} \tag{10}$$

Da der Organismus nicht zwischen den beiden Radiumisotopen Ra-228 und Ra-224 zu diskriminieren vermag, muß die Ausscheidungskonstante $\lambda_{2,J}$ für Ra-228 in Ionenform auch der für Ra-224 entsprechen:

$$\lambda_{2,J} = \lambda_{5,J} = \lambda_J.$$

Damit sind alle in Gl. (10) auftretenden und zur Berechnung des theoretischen Verlaufs von $a_5(t)$ notwendigen Größen bekannt. Ihre Zahlenwerte sowie die Angabe der experimentellen Untersuchungen, aus denen sie gewonnen wurden, sind in einer Übersicht zusammengestellt (s. Tab. 1).

In Abb. 6 sind die Ergebnisse der Berechnungen dargestellt. Neben dem theoretischen Verlauf der täglichen Gesamtausscheidung an Ra-224 ist in der Abbildung der zeitliche Verlauf der einzelnen Komponenten der Gesamtausscheidung wiedergegeben.

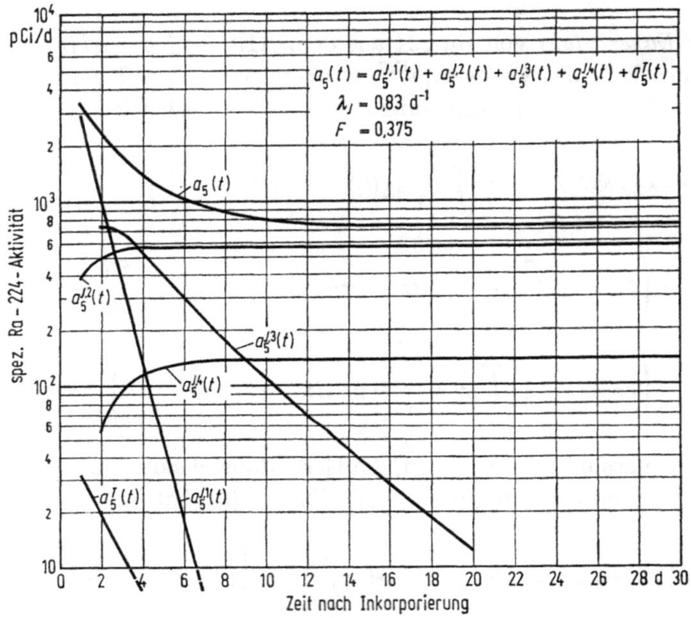

Abb. 6. Zeitlicher Verlauf der theoretisch berechneten täglichen Gesamtausscheidung von Ra-224 und deren Komponenten (Testagar)

Vergleicht man die experimentell bestimmte Ausscheidung von Ra-224 mit dem auf der Grundlage des Compartmentmodells berechneten zeitlichen Verlauf der Gesamtausscheidung nach Inkorporierung einer *Testagar-* oder *Heydensuspension* (s. Abb. 7), dann kann die Übereinstimmung von Experiment und Theorie als Bestätigung für die Gültigkeit des Modells gewertet werden.

Die Beobachtung, daß die Ra-224-Ausscheidung nach Inkorporierung der *Heydensuspension* im gesamten Beobachtungszeitraum konstant bleibt, spiegelt sich auch im theoretischen Verlauf der Gesamtausscheidung wieder. Der Grund hierfür liegt in der gegenüber der *Testagarsuspension* geringeren Ra-224-Aktivität

Tabelle 1. *Zusammenstellung der zur Berechnung des theoretischen Verlaufs der täglichen Ra-224-Gesamtausscheidung notwendigen Größen, ihrer Zahlenwerte sowie der experimentellen Untersuchungen, aus denen sie gewonnen wurden*

Gemessene/ Berechnete Größe	Zahlenwert	Dimension	Phys./biol. Bedeutung der Größe; Experimentelle Bestimmung der Größe
$\bar{\lambda}_J$ $\hat{=} \bar{T}^J_{1/2}$ $\bar{F}$	0,83 20 0,375	d^{-1} h —	$\bar{\lambda}_J$: Konstante für die Ausscheidung von Radium in Ionenform (Mittelwert); $\bar{F}$: Retention von Radium im ges. markfreien Skelett zum Zeitpunkt 1 Tag nach Inkorporierung (Mittelwert); Anpassung des theoretischen Verlaufs der Gesamtausscheidung von Ra-228 an den experimentellen Befund durch Permutation von frei gewählten Werten für λ_J und F
$\bar{\lambda}_T$ $\hat{=} \bar{T}^T_{1/2}$	0,31 56	d^{-1} h	$\bar{\lambda}_T$: Konstante für die Ausscheidung der Teilchenfraktion (Mittelwert); Messung des zeitlichen Verlaufs der Ausscheidung von Th-232
G_5	0,35	—	G_5: Verhältnis der aus den ThO_2-Agglomeraten im RES ausgeschiedenen zu den in den Agglomeraten entstehenden Ra-224-Kernen; Bestimmung aus dem Aktivitätsverhältnis zwischen Ra-224 und Th-228 im RES zum Zeitpunkt $t \geq 30$ d nach Inkorporierung
$\bar{b}$	0,26	—	$\bar{b}$: Exponent in der nach einer Potenzfunktion verlaufenden Retention von Radium im markfreien Skelett (Mittelwert); Bestimmung aus dem zeitlichen Verlauf der spezifischen Ra-228-Aktivität des markfreien Skeletts
$A_4(t_i)$ $A_5^{S,J}(t_i)$ $a_{4,1}^S(t_i)$ $a_{5,1}^{S,T}(t_i)$	$17{,}86 \cdot 10^3$ $15{,}55 \cdot 10^3$ 0,41 0,06	pCi/2 cm³ pCi/2 cm³ — —	$A_4(t_i)$: Th-228-Aktivität der inkorporierten Suspension (Testagar) $A_5^{S,J}(t_i)$: Ra-224-Aktivität der *Ionenfraktion* der inkorporierten Suspension (Testagar) $a_{4,1}^S(t_i)$: Aktivitätsverhältnis zwischen Th-228 und Th-232 in der *Teilchenfraktion* der Suspension zum Zeitpunkt der Inkorporierung (Testagar) $a_{5,1}^{S,T}(t_i)$: Aktivitätsverhältnis zwischen Ra-224 und Th-232 in der *Teilchenfraktion* der Suspension zum Zeitpunkt der Inkorporierung (Testagar) Bestimmung nach Ultrazentrifugation und/oder γ-Strahlen- bzw. α-Strahlenspektrometrie
$a_1(t_i)$	850	pCi/d	$a_1(t_i)$: „fiktive" Th-232-Ausscheidung zum Zeitpunkt der Inkorporierung Durch Extrapolation des gemessenen zeitlichen Verlaufs der Th-232-Ausscheidung auf den Zeitpunkt $t = t_i = 0$

in Ionenform zum Zeitpunkt der Inkorporierung (Testagar $A_5^{S,J}(t_i) = 15{,}55 \cdot 10^3$ pCi/2 cm³; Heyden $A_5^{S,J}(t_i) = 2{,}28 \cdot 10^3$ pCi/2 cm³), so daß der Anteil an der Gesamtausscheidung aufgrund der einmaligen Zufuhr von Ra-224 aus dem

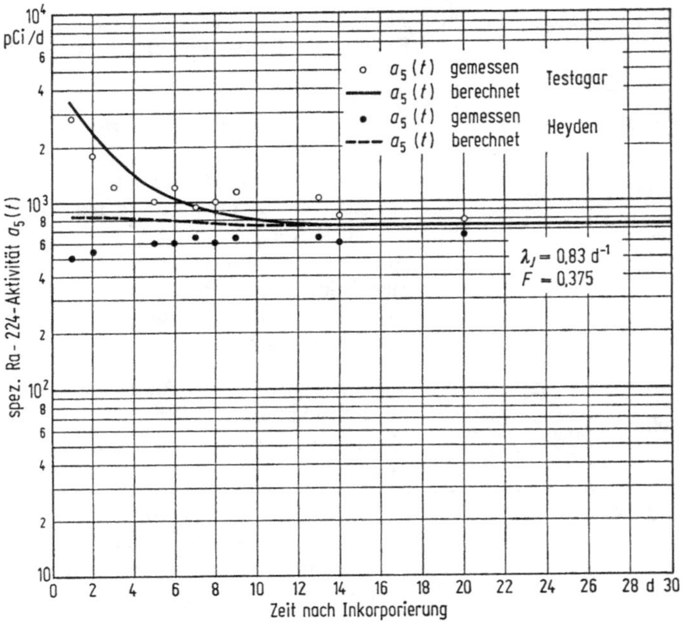

Abb. 7. Zeitlicher Verlauf der gemessenen und theoretisch berechneten täglichen Gesamtausscheidung von Ra-224 (Testagar und Heyden)

Gewebe bzw. dem markfreien Skelett gegenüber dem aufgrund der Dauerzufuhr durch Zerfall von Th-228 im RES bereits im Zeitraum unmittelbar nach Inkorporierung zu vernachlässigen ist.

Bezieht man die tägliche Ra-224-Ausscheidung auf die pro Tag dem Organismus durch Zerfall von Th-228 im RES und Elimination von Ra-224 aus den Organen des RES zugeführte Ra-224-Aktivität in Ionenform, dann sind dies im Zustand des stationären Gleichgewichts zwischen Ra-224 und Th-228 im RES ($t \geq 30$ d) 60% (KAUL, 1969).

Zusammenfassung

Zur Berechnung des theoretischen Verlaufs der täglichen Gesamtausscheidung an Ra-228 und Ra-224 wurde ein Compartmentsystem entwickelt, bestehend aus dem markfreien Skelett, den Organen des RES und dem übrigen Intra- und Extrazellularraum.

Mit Hilfe des Modells wurde die Kinetik der Verteilung und Ausscheidung von Ra-224 und Ra-228 beschrieben. Auf Ra-228 angewandt, mußten die folgenden, im Organismus ablaufenden Prozesse berücksichtigt werden:

Translokation von Ra-228 in das Skelett aufgrund der einmaligen Zufuhr von Ra-228 in Ionenform mit der Suspension zum Zeitpunkt der Inkorporierung;

Elimination von Ra-228 aus dem Skelett in den übrigen Intra- und Extrazellularraum und Ausscheidung aus dem Organismus;

Direkte Ausscheidung des nicht in das Skelett translokalisierten Anteils an Ra-228 in Ionenform aus der Suspension;

Ausscheidung von Ra-228 mit der Teilchenfraktion.

Zur Beschreibung der Kinetik von Ra-224 mußten darüber hinaus die folgenden Prozesse berücksichtigt werden:

Dauerzufuhr von Ra-224 in den übrigen Intra- und Extrazellularraum durch Zerfall von Th-228 in den Organen des RES unter Berücksichtigung der Agglomerierung der ThO_2-Teilchen;

Kontinuierliche Translokation von Ra-224 in das markfreie Skelett;

Zerfall von Ra-224 im Skelett und dauernder Abfluß von Ra-224 aus dem Skelett verbunden mit der Ausscheidung aus dem Organismus;

Direkte Dauerausscheidung des nicht in das Skelett translokalisierten Anteils an Ra-224 in Ionenform.

Aus dem Vergleich der theoretisch mit Hilfe des Compartmentmodells bestimmten Ausscheidungskurven mit dem gemessenen Verlauf der täglichen Gesamtausscheidung an Ra-228 wurden der Wert für die Retention von Radium zum Zeitpunkt 1 Tag nach Inkorporierung (F) und die Konstante für die Ausscheidung von Radium in Ionenform (λ_J) ermittelt:

$$F = 0{,}375 \pm 0{,}075,$$
$$\lambda_J = 0{,}83 \pm 0{,}23\ \mathrm{d}^{-1} \mathrel{\hat{=}} T^J_{1/2} = 20 \pm 5\ \mathrm{h}.$$

Literatur

KAUL, A.: Habilitationsschrift. Freie Universität Berlin 1969.
— HEYDER, J.: Biophysik **5**, 331—340 (1969).

Effects of Ionizing Radiation on the In Vivo Metabolism of Monocarbon Fragment Precursors to CO_2

T. M. Ngo and H. S. Winchell

Donner Laboratory, University of California, Berkeley, California, U.S.A.

Introduction

In the previous study, we have demonstrated that *in vivo* oxidation of the No. 2 carbon atom of the imidazole ring of histidine to CO_2 was inhibited following ionizing radiation [1]. This result postulated that ionizing radiation might interfere with monocarbon fragment transport by an *in vivo* inactivation of tetrahydrofolic acid or the enzymatic processes responsible for its production. The present study is a further evaluation of this finding. The rate of $^{14}CO_2$ excretion in the breath of rats given ^{14}C-labeled formate, glycine, serine, and methionine was studied in rats prior to, and subsequent to, exposure of various doses of X-irradiation.

Material and Methods

Inbred male Buffalo rats (Simonsen Laboratory, Gilroy, California) weighing 230 to 250 g were used in all experiments. These animals had free access to food and water during the period of study.

In the first series of studies, 7 male rats were divided into 2 groups of 3 sham irradiated rats and 4 irradiated rats. The irradiated rats were subdivided further into 2 pairs of rats exposured to 400 to 600 R of X-rays. The source of X-rays for these experiments was an X-rays machine operated at 150 kVp and 15 mA and with an inherent filtration of 1 mm of aluminum and 1 mm of copper. Tube-target-distance was 10 inches. The dose rates, measured in air, were 15 to 27 r/min. Animals were studied at 20 minutes and then at 2, 5, 9, 16, and 21 days after sham irradiation or irradiation procedures. Each rat received 14 μCi of L-methionine-CH_3-^{14}C intravenously (specific activity: 14.77 mCi/mM, New England Nuclear Corp., 575 Albany Street, Boston, Massachusetts 02118, U.S.A.), after a light anesthesia with diethyl ether.

In the second series of studies, 16 rats were divided into 2 groups of 7 sham irradiated rats and 9 irradiated rats studied in groups receiving 400 and 600 R of X-rays. Irradiation procedure was similar to that has been described previously. Sham irradiated rats and irradiated rats were studied at 20 minutes and then at 3, 5, 7, 12, and 20 days after sham irradiation or irradiation procedure. In each study, the rat received 2.5 μCi of L-glycine-1-^{14}C intravenously (specific activity: 5.0 mCi/mM, Nuclear Chicago, 333 H. Howard Ave., Des Plaines, Illinois 60018, U.S.A.).

In the third series of studies, 8 rats were divided into 2 groups of 4 sham irradiated rats and 4 rats given 400 and 600 R of X-rays. Sham irradiated and irradiated rats were studied in pairs after sham irradiation or irradiation and then studies were repeated at 3, 5, 9, 14, and 21 days later. In each study, the rat, after anesthetisation with diethyl ether, received 2.5 μCi of L-serine-3-^{14}C intravenously (specific activity: 8.5 mCi/mM, Nuclear Chicago).

In the fourth series of studies, 12 rats were divided into sham irradiated and irradiated groups. The first experiments were studied at 20 minutes and repeat studies were then performed at 2, 5, 9, 14, and 20 days after irradiation. In each study, the rat received 2.5 μCi of ^{14}C-formate intravenously (specific activity: 4.62 mCi/mM, New England Nuclear Corp.).

In each study, immediately after the intravenous administration of the ^{14}C-labeled material, $^{14}CO_2$ excretion was measured continuously by a breath analyzer (Fig. 1). The procedure for such measurement has been described previously [2—6].

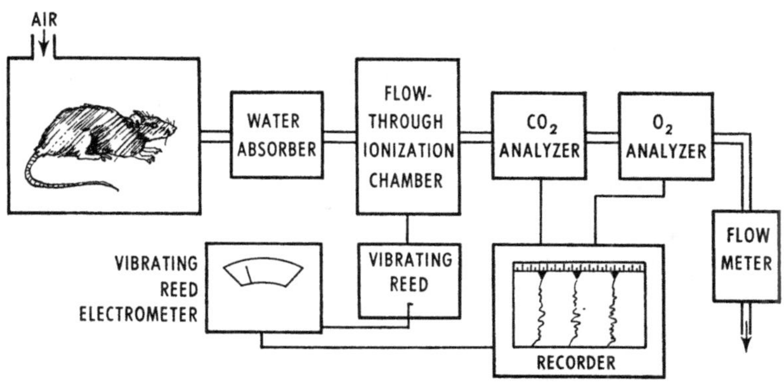

Fig. 1. Diagrammatic representation of the $^{14}CO_2$ breath analyzer. The expired air is serially passed through a water absorber, an ionization chamber, an infrared CO_2 analyzer, a paramagnetic O_2 analyzer, and a flowmeter

Results

Fig. 2 presents representative curves showing the rate and appearance of $^{14}CO_2$ in the breath of 3 control rats (upper curve) and 2 rats 2 days after exposure to 600 R (lower curve) subsequent to the intravenous administration of L-methionine-CH_3-^{14}C. The ordinate represents $^{14}CO_2$ excretion rate expressed as mμCi per minute and the abscissa as time in minutes following intravenous injection of the ^{14}C-labeled methionine. In the normal curve, each point represents the mean of the rate of $^{14}CO_2$ excretion of 3 control rats of the given time. The vertical bars through each point represent 1 standard error for the mean of this group. It is clear that there is a qualitative difference between the two $^{14}CO_2$ breath curves.

Fig. 3 presents composite data of the serial change in T_{max}, the time at which maximum rate of $^{14}CO_2$ excretion occurred in control and irradiated rats subsequent to the intravenous administration of the ^{14}C-labeled methionine. Irradiated rats showed a significant prolongation in T_{max} on the second and fifth post-irradiation days and the T_{max} returned to the normal by the ninth day.

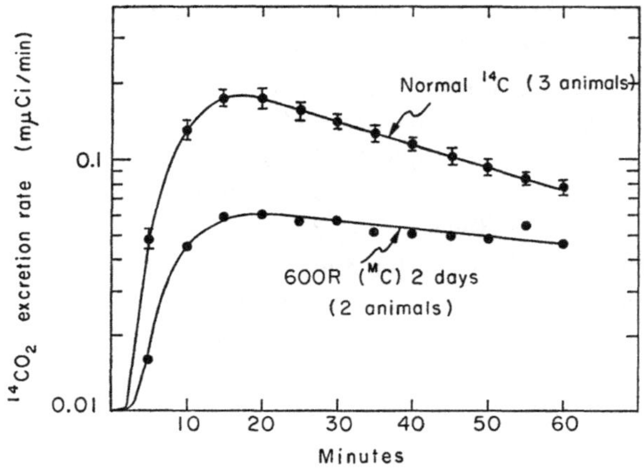

Fig. 2. Pattern of $^{14}CO_2$ excretion rates in the breath of 3 normal rats and in 2 irradiated rats 2 days after exposure to 600 R following intravenous administration of L-methionine-CH_3-^{14}C

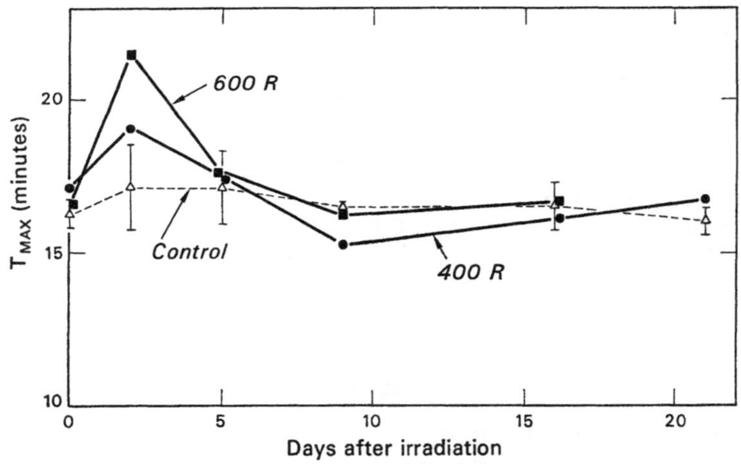

Fig. 3. Serial T_{max} of the $^{14}CO_2$ breath curves after intravenous administration of L-methionine-CH_3-^{14}C in irradiated rats and sham-irradiated rats

Fig. 4 presents the cumulative percent of $^{14}CO_2$ excretion of control and irradiated rats during the 60 minutes after intravenous injection of L-methionine-CH_3-^{14}C. There is a significant decrease of the cumulative excretion of $^{14}CO_2$ in rats given 400 R at 5 and 9 days, returning to the normal range by the sixteenth day. A markedly diminished $^{14}CO_2$ production in rats given 600 R was noted as early as 20 minutes after irradiation, and this decrease continued at least until the sixteenth postirradiation day. Table 1 summarizes data presented in Figs. 3 and 4.

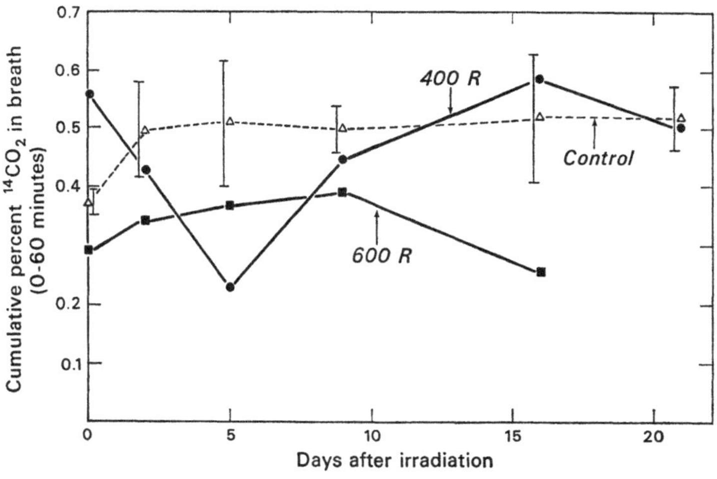

Fig. 4. Cumulative percentage of $^{14}CO_2$ excreted in the breath during the initial 60 minutes following intravenous administration of L-methionine-CH_3-^{14}C in irradiated rats and sham-irradiated rats

Table 1. *Serial T_{max} and integral ^{14}C excretion determined from $^{14}CO_2$ appearance in the breath following intravenous administration of L-methionine-CH_3-^{14}C in control and irradiated rats (the number of animals in each group of rats is noted in parentheses)*

Category	Time of study after irradiation or sham irradiation	Serial T_{max} (minutes)	Integral ^{14}C excretion in 60 minutes (%)
Control rats (3)	20 minutes, 2, 5, 9, 16, and 21 days	16.34 ($\pm$ S. E. = 0.49)	0.49 ($\pm$ S. E. = 0.04)
Irradiated rats			
400 R (2)	20 minutes	17.00	0.57
	2 days	18.25	0.43
	5 days	17.25	0.23
	9 days	15.75	0.45
	16 days	16.25	0.59
	21 days	16.75	0.50
600 R (2)	20 minutes	16.50	0.29
	2 days	21.75	0.34
	5 days	17.25	0.37
	9 days	16.25	0.39
	16 days	15.27	0.26

Fig. 5 presents serial T_{max} of $^{14}CO_2$ breath curves of control and irradiated rats following the intravenous injection of L-glycine-1-^{14}C at 20 minutes, and then at 3, 7, 12, and 20 days after irradiation.

Fig. 6 presents the cumulative excretion of $^{14}CO_2$ in the breath during the initial 60 minutes following administration of the ^{14}C-labeled glycine in irradiated

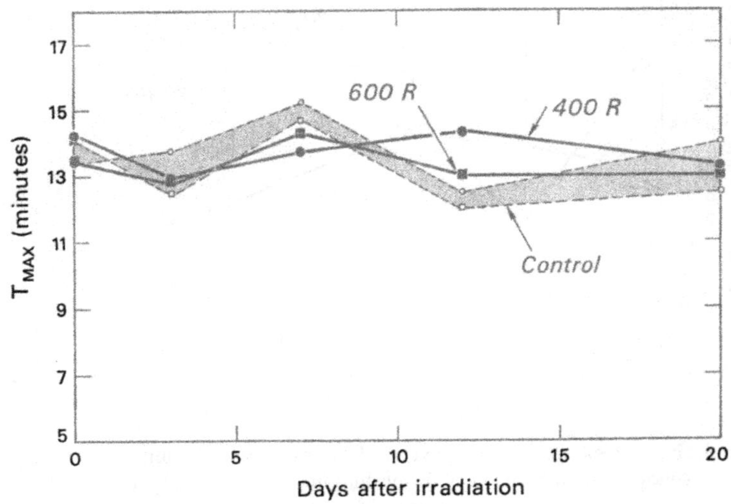

Fig. 5. Serial T_{max} of the $^{14}CO_2$ breath curves after intravenous administration of L-glycine-1-^{14}C in irradiated rats and sham-irradiated rats

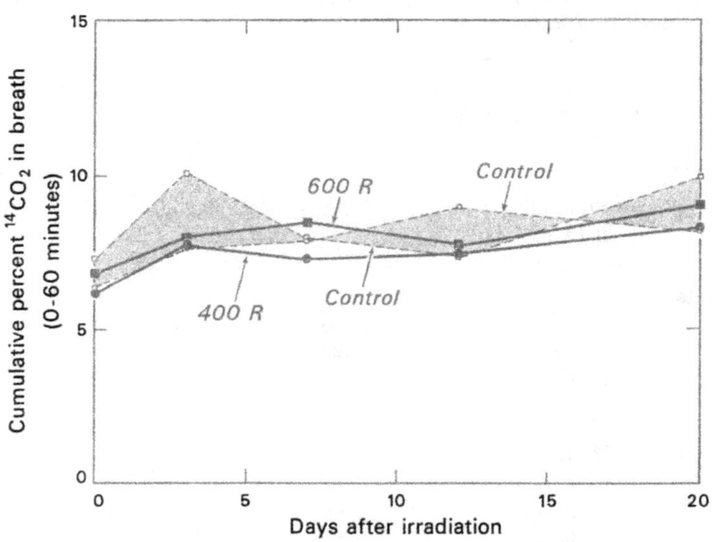

Fig. 6. Cumulative percentage of $^{14}CO_2$ excreted in the breath during the initial 60 minutes following intravenous administration of L-glycine-1-^{14}C in irradiated rats and sham-irradiated rats

rats. It can be seen that there is no significant change in T_{max} and cumulative excretion of $^{14}CO_2$ in irradiated rats. Table 2 summarizes data presented in Figs. 5 and 6.

Table 2. *Serial T_{max} and integral ^{14}C excretion determined from $^{14}CO_2$ appearance in the breath following intravenous administration of L-glycine-1-^{14}C in control and irradiated rats (the number of animals in each group of rats is noted in parentheses)*

Category	Time of study after irradiation or sham irradiation	Serial T_{max} (minutes) $\pm$ S. E.	Integral ^{14}C excretion in 60 minutes (%) $\pm$ S. E.
Control rats (4)	20 minutes	13.87 $\pm$ 0.77	6.85 $\pm$ 0.63
	3 days	13.12 $\pm$ 1.12	8.91 $\pm$ 0.73
	7 days	15.00 $\pm$ 0.46	7.98 $\pm$ 0.35
	12 days	12.25 $\pm$ 0.25	7.68 $\pm$ 1.14
	20 days	13.25 $\pm$ 0.94	9.25 $\pm$ 0.92
Irradiated rats			
400 R (2)	20 minutes	14.25	6.30
	3 days	13.00	7.87
	7 days	13.75	7.41
	12 days	14.25	7.59
	20 days	13.25	8.45
600 R (2)	20 minutes	13.50	6.98
	3 days	13.00	8.07
	7 days	14.25	8.55
	12 days	13.00	7.89
	20 days	13.00	9.17

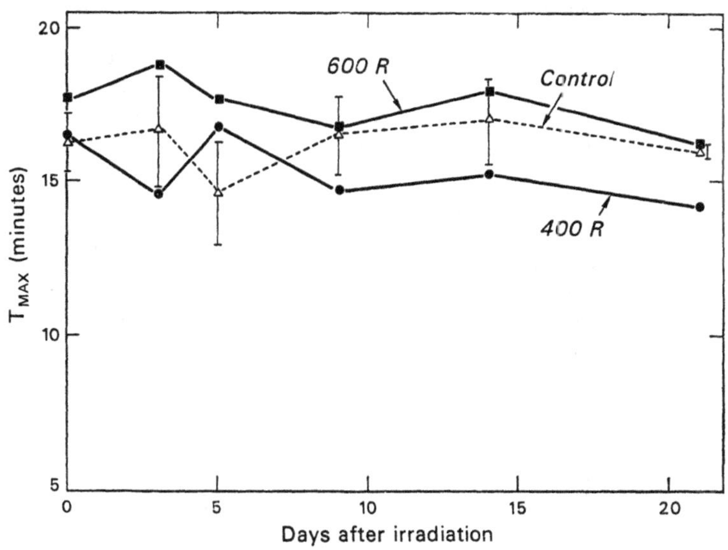

Fig. 7. Serial T_{max} of the $^{14}CO_2$ breath curves after intravenous administration of L-serine-3-^{14}C in irradiated rats and sham-irradiated rats

Fig. 7 presents serial T_{max} of $^{14}CO_2$ breath curves of control and irradiated rats subsequent to the administration of L-serine-3-^{14}C at 20 minutes, and 2, 5, 9, 13, and 19 days after radiation exposure. Fig. 8 presents results of the cumulative excretion of the $^{14}CO_2$ in the breath during the initial 60 minutes following the injection change in T_{max} and cumulative excretion of $^{14}CO_2$ in irradiated rats. Table 3 summarizes data presented in Figs. 7 and 8.

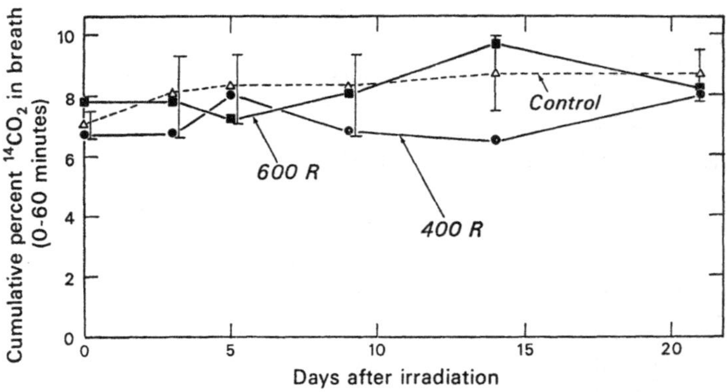

Fig. 8. Cumulative percentage of $^{14}CO_2$ excreted in the breath during the initial 60 minutes following intravenous administration of L-serine-3-^{14}C in irradiated rats and sham-irradiated rats

Table 3. *Serial T_{max} and integral ^{14}C excretion determined from $^{14}CO_2$ appearance in the breath following intravenous administration of L-serine-3-^{14}C in control and irradiated rats (the number of animals in each group of rats is noted in parentheses)*

Category	Time of study after irradiation or sham irradiation	Serial T_{max} (minutes) ± S. E.	Integral ^{14}C excretion in 60 minutes (%) ± S. E.
Control rats (4)	20 minutes	16.25 ± 0.92	7.14 ± 0.42
	3 days	16.75 ± 1.90	8.18 ± 0.90
	5 days	14.62 ± 1.25	8.36 ± 0.33
	9 days	16.50 ± 1.36	8.19 ± 1.06
	14 days	17.00 ± 1.43	8.79 ± 0.95
	21 days	16.00 ± 0.20	8.75 ± 0.64
Irradiated rats			
400 R (2)	20 minutes	16.50	7.44
	3 days	14.50	6.78
	5 days	16.75	8.10
	9 days	14.75	6.85
	14 days	15.25	6.60
	21 days	14.25	8.09
600 R (2)	20 minutes	17.75	7.90
	3 days	18.75	7.84
	5 days	17.75	7.33
	9 days	16.75	8.10
	14 days	18.00	9.70
	21 days	16.25	8.38

Fig. 9 presents the single exponential regression function with 95% confidence limits for the downslope of the $^{14}CO_2$ breath curves in 4 control rats (dotted line) and 2 irradiated rats after exposure to 600 R (solid line) following the intravenous administration of ^{14}C-formate. The ordinate presents the rate of $^{14}CO_2$ excretion expressed as mμCi per minute and the abscissa as time in minutes following the intravenous injection of ^{14}C-formate. Each point at 5, 15, 25, and 40 minutes represents the mean of excretion rates of $^{14}CO_2$ for the control group. Vertical

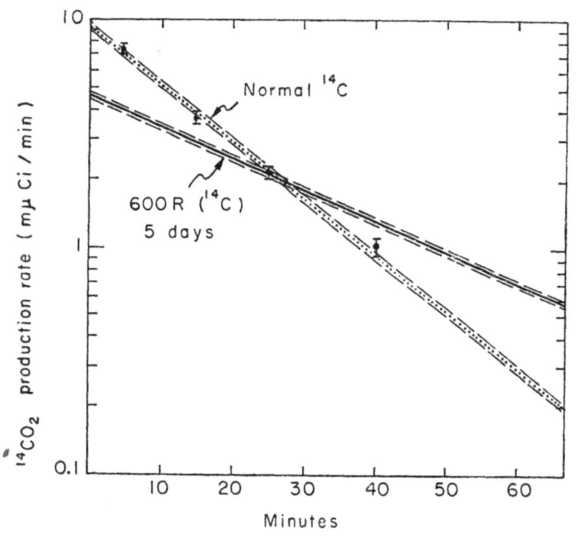

Fig. 9. Single exponential regression function with 95% confidence limits for the downslope of the $^{14}CO_2$ breath curves in normal rats and in 2 irradiated rats 5 days after exposure to 600 R following intravenous administration of ^{14}C-formate

bars at each point define precision of position of the mean of $^{14}CO_2$ excretion rates with 95% limits based on $t_{0.95} S_{y \cdot x}$ ($S_{y \cdot x} = 1$ standard error of the estimate). The zero-intercept (A), the slope of the regression function (B), and the standard error of the slope ($S_{b(x) \cdot (y)}$) were determined by least-squares best fit of the data to the function

$$Y = A + (B \pm S_{b(x)} \cdot {}_{(y)}) X .$$

The broken lines above and below each curve are regression lines of ± 1 standard error of the estimate. It is clear that there is significant difference between two single exponential curves.

Fig.10 presents the serial changes in the half-time of such curves subsequent to the administration of ^{14}C-formate to pairs of rats receiving various doses of X-irradiation at 20 minutes, and 2, 5, 9, 13, and 19 days after irradiation. Irradiated rats given 600 R showed early significant prolongation of $T_{1/2}$ within 20 minutes following irradiation. The maximum prolongation of $T_{1/2}$ was reached at 2 to 5 days with a return to near normal by the ninth day after irradiation. Fig. 11 presents the results of integral ^{14}C excretion (from 0 to ∞) after the intravenous injection of ^{14}C-formate in irradiated rats given 600 R decreased during the initial 2 to 5 days returning to near normal by ninth day after irradiation. Table 4 summarizes data presented in Figs. 10 and 11.

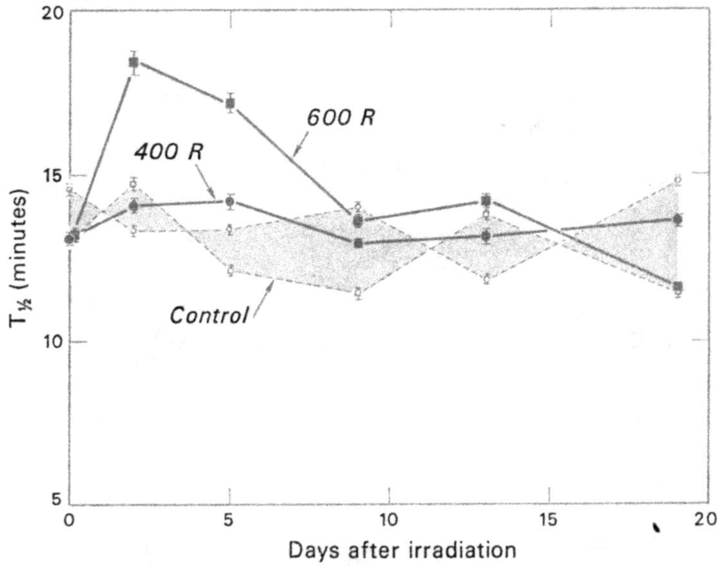

Fig. 10. Half-time $T_{1/2}$ of the $^{14}CO_2$ breath curves after intravenous administration of ^{14}C-formate in irradiated rats and sham-irradiated rats

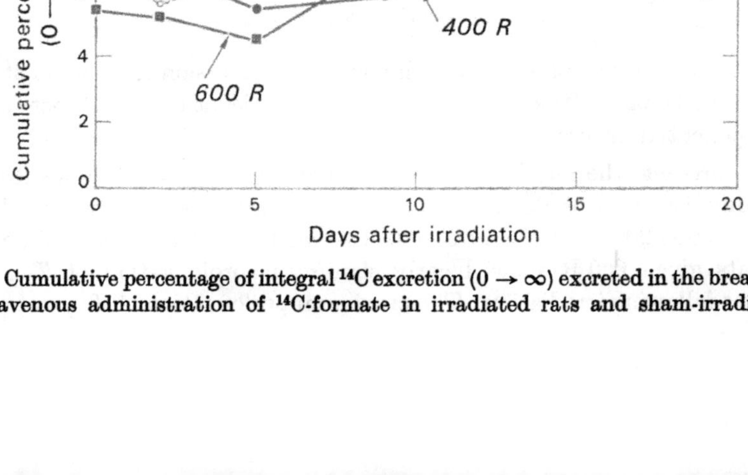

Fig. 11. Cumulative percentage of integral ^{14}C excretion ($0 \rightarrow \infty$) excreted in the breath following intravenous administration of ^{14}C-formate in irradiated rats and sham-irradiated rats

Table 4. *Half-time $T_{1/2}$ and integral ^{14}C excretion determined from $^{14}CO_2$ appearance in the breath following intravenous administration of ^{14}C-formate in control and irradiated rats (the number of animals in each group of rats is noted in parentheses)*

Category	Time of study after irradiation or sham irradiation	Half-time $T_{1/2}$ (minutes) ± S. E.	Integral ^{14}C excretion $(0 \rightarrow \infty)$ (%) ± S. E.
Control rats (4)	20 minutes, 2, 5, 9, 13, and 19 days	12.81 ± 0.13	6.92 ± 0.10
Irradiated rats			
400 R (2)	20 minutes	13.12 ± 0.16	6.00 ± 0.01
	2 days	14.14 ± 0.23	6.65 ± 0.01
	5 days	14.27 ± 0.20	5.42 ± 0.08
	9 days	12.98 ± 0.17	5.87 ± 0.07
	13 days	13.12 ± 0.24	6.92 ± 0.07
	19 days	12.38 ± 0.18	6.86 ± 0.07
600 R (2)	20 minutes	13.09 ± 0.17	6.86 ± 0.09
	2 days	18.37 ± 0.34	5.30 ± 0.08
	5 days	17.20 ± 0.29	4.58 ± 0.08
	9 days	13.69 ± 0.18	6.74 ± 0.09
	13 days	14.27 ± 0.20	6.73 ± 0.07
	19 days	11.60 ± 0.11	5.84 ± 0.06

Discussion

It is known that ^{14}C atoms from ^{14}C-formate, L-histidine (imidazole-2-^{14}C) and L-methionine-CH_3-^{14}C is readily incorporated into the monocarbon fragment pool attached to tetrahydrofolic acid. From this point, they may be either utilized in synthetic processes or rapidly oxidized to $^{14}CO_2$ (Fig. 12).

In the present studies, we demonstrate a difference in the pattern of $^{14}CO_2$ in the breath subsequent to the intravenous administration of L-methionine-CH_3-^{14}C and ^{14}C-formate between normal and irradiated rats. However, no significant effects were seen in the oxidation to $^{14}CO_2$ of L-glycine-1-^{14}C and L-serine-3-^{14}C in irradiated rats. At various time intervals following irradiation, the time at which the maximum rate of excretion of $^{14}CO_2$ in the breath (T_{max}) was prolonged, and there was a diminution of the total amount of the ^{14}C appearing in the breath during the initial 60 minutes of the study following the injection of the ^{14}C-labeled methionine. The half-time ($T_{1/2}$) of the single exponential regression function with 95% confidence limits for the downslope of $^{14}CO_2$ breath curves in irradiated rats was prolonged and there was a diminution of the integral ^{14}C excretion in the breath subsequent to the injection of ^{14}C-formate. The effect described is seen within 20 minutes after irradiation and therefore can not be ascribed to delayed non-specific effects of radiation on the animal's overall physiology. These results are consistent with the previous findings that *in vivo* oxidation of the No. 2 carbon atom of the imidazole ring of histidine to CO_2 was inhibited following ionizing radiation. These data led us to postulate that ionizing radiation interfered with monocarbon fragment transport possibly by inactivation of tetrahydrofolic acid or the enzymatic processes required for its production [1]. Recent study from this laboratory demonstrated that the *de novo* synthesis of thymine and adenine utilizing ^{14}C-formate was more radiosensitive than the incorporation of ^{3}H-thymidine into DNA thymine and both the synthesis of thymidine, thymine, and the

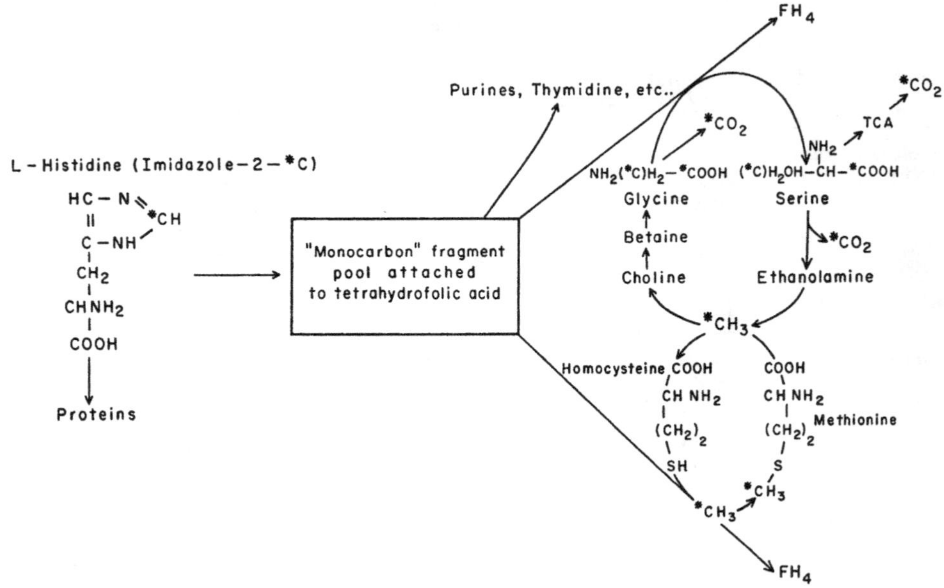

Fig. 12. Metabolic fate of the imidazole No. 2 Carbon atom of histidine

purine, adenine were significantly inhibited by radiation exposure [7]. In this work, they also postulated that the marked radiosensitivity of the synthesis of these bases, which require a common pathway via monocarbon transport, involved radiation inactivation of tetrahydrofolic acid or the enzymatic processes required for its production.

The No. 1 carbon atom of glycine or No. 3 carbon atom of serine may arise from monocarbon pool but their subsequent oxidation to CO_2 is essentially independent of any further passage of this carbon atom through the monocarbon pool.

A decreased pattern of $^{14}CO_2$ excretion in the breath of irradiated rats subsequent to the intravenous administration of ^{14}C-formate would not be expected to be due to any of the physical transport of this compound across the cell membrane because the oxidation of formate to CO_2 proceeds very rapidly and thus its transport across the cell membrane must be equally rapid. It is known that ^{14}C atoms from ^{14}C-formate enter the monocarbon pool attached to tetrahydrofolic acid. That radiation affects the pattern of $^{14}CO_2$ excretion in the breath of irradiated rats subsequent to the intravenous injection of ^{14}C-formate suggests that ^{14}C-formate might be useful in the study of folic acid deficiency in human subjects similar to L-histidine (imidazole-2-^{14}C) [8] and L-methionine-CH_3-^{14}C [9] which have been used previously for this purpose.

Summary

These results show a decrease of the initial rate and amount of $^{14}CO_2$ excretion in the breath of rats given various doses of X-rays following the intravenous administration of L-methionine-CH_3-^{14}C and ^{14}C-formate. No significant alteration in $^{14}CO_2$ production was noted in irradiated rats following the intravenous injection of either L-glycine-1-^{14}C or L-serine-3-^{14}C. The present study is a further evaluation of our previous finding that ionizing radiation suppresses the turnover rate of the monocarbon fragments regardless of the source of the monocarbon unit by inactivation of tetrahydrofolic acid or the enzymatic processes responsible for its production. The radiation effect on the monocarbon pool may present a clinically significant metabolic defect following exposure to ionizing radiation since the availability of monocarbon fragments is required for a large number of critical life processes, such as the production of purines and thymidine in nucleic acid synthesis.

References

1. Ngo, T., Winchell, H. S., Landaw, S.: Altered *in vivo* metabolism of histidine (imidazole-2-^{14}C) in irradiated rats. Rad. Research **34**, 390–395 (1968).
2. Tolbert, B. M., Kirk, M., Baker, E. M.: Continuous $^{14}CO_2$ excretion studies in experimental animals. Am. J. Physiol. **185**, 263–273 (1956).
3. Domingues, F. J., Gildner, K. J., Baldwin, R. R., Lowry, J. R.: An instrument and technique for the continuous measurement of respiratory pattern in metabolic tracer studies. Intern. J. Appl. Radiation Isotopes **7**, 77–86 (1959).
4. Ngo, T. M., Winchell, H. S., Williams, M. A., Finley, N. N., Lawrence, J. H.: Decreased $^{14}CO_2$ production in thiamine-deficient rats given pyruvate and acetate (-1-^{14}C): A possible means for early diagnosis of beriberi? J. Nucl. Med. **10**, 676–682 (1969).
5. — —: Alterations in histidine catabolism in normal rats given pharmacological doses of folic acid and cyanocobalamin. Proc. Soc. Exp. Biol. Med. **132**, 168–170 (1969).
6. —: Quantitation of certain alterations in intermediary metabolism *in vivo* associated with changes in body content of vitamins B_1, B_6, B_{12}, and folic acid using $^{14}CO_2$ breath analysis. Ph.D. thesis, Univ. of California, Berkeley 1969.
7. Winchell, H. S., Vimokesant, S. L., Raley, R.: Relative effect of radiation on *de novo* DNA base synthesis and thymidine incorporation into DNA. Semiannual Report, Biology and Medicine, Univ. of California, Berkeley, Spring 1969, p. 149–154.
8. Fish, M. B., Pollycove, M., Feichmeir, T. V.: Differentiation between vitamin B_{12}-deficient and folic acid-deficient anemias with ^{14}C-histidine. Blood **21**, 447–461 (1963).
9. — Pollycove, M.: San Francisco Medical Center and Donner Lab., Univ. of California, personal communication.

Pulmonary Microembolism

A. EVERETTE JAMES, Jr. and S. BOYD EATON, Jr.

Massachusetts General Hospital, Boston, Massachusetts, U.S.A.

The importance of pulmonary embolic disease needs no emphasis. Its frequency and potential seriousness are well known [2, 9, 12]. Most of the medical literature on this topic concerned with the problem of large pulmonary emboli [8, 11]. This paper is devoted to certain manifestations of pulmonary microembolism. The research of SMITH, DAMMIN and others [12, 13] in pulmonary pathological survey studies first emphasized that pulmonary microemboli were considerably more common than large emboli. Pre-mortem diagnosis of this entity has been difficult. Our interest in nuclear medicine led us to investigate pulmonary scintillography for diagnosis of pulmonary microembolism.

We have observed a scintigraphic pattern of circumferential hypoperfusion involving one or more lobes which we believe is due to multiple pulmonary microemboli.

Observation

All lung scintigrams of patients clinically suspected of having pulmonary embolic disease performed at the Massachusetts General Hospital between August 1966 and April 1969 were reviewed. Three hundred microcuries of high specific activity macroaggregated albumin labeled with ^{131}I was administered intravenously for each scintigram. A stationary scintillation "camera" with an 11.5 inch thallium-activated sodium iodide crystal and a parallel-hole collimator was employed for all examinations. Multiple views (e.g. both laterals, anterior and/or posterior projections) were invariably obtained. Both supine and upright studies were made.

The scintigraphic appearance of both the normal lung and the patterns associated with macroemboli have been carefully characterized [15]. Review of our scintigrams revealed an additional pattern commonly observed in patients with clinical and, occasionally, roentgenographic evidence of pulmonary embolism. This pattern consists of a circumferential zone of diminished radioactivity surrounding one or more lobes. Decreased radioactivity may be present around the circumference of a single lobe causing it to appear smaller in volume than one would expect (Figs. 1 and 2) or it may be apparent only as a linear area of diminished activity along an interlobar fissure. This has been described as the "fissure sign" (Figs. 3 and 4). Various combinations of these findings have been seen as well as concomitant occurrence of both microembolic and characteristic macroembolic scintigraphic patterns in the same patient (Fig. 5). Sequential scintigraphic appearance of both macro- and microemboli has also been observed.

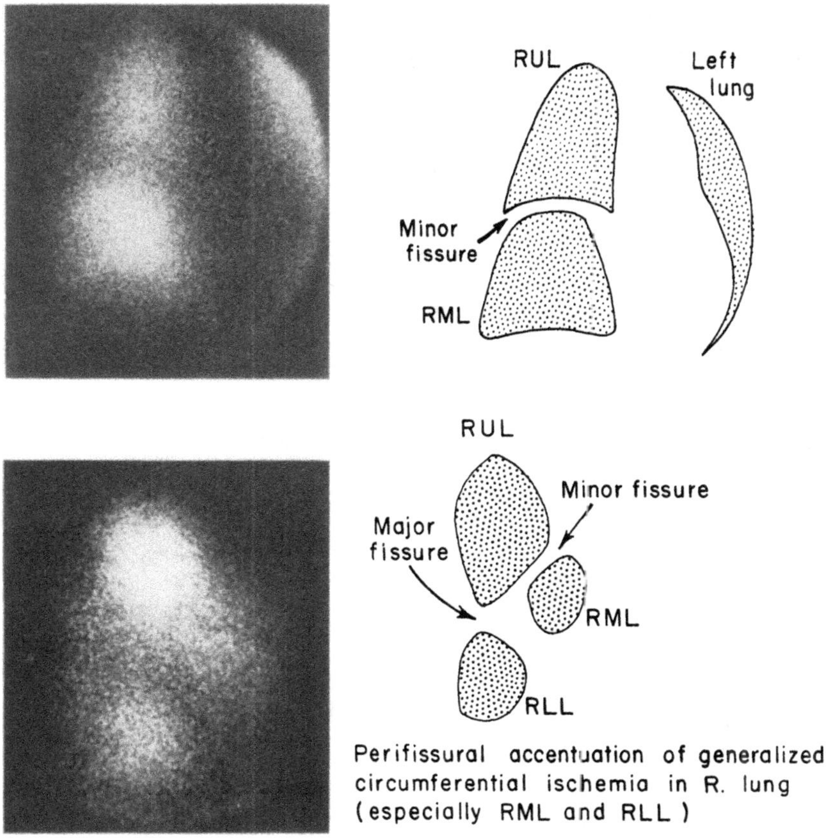

a

b

Fig. 1. Scintigram of patient with clinical evidence of pulmonary embolism. a and b Right anterior and lateral views: circumferential hypoperfusion is present in both the right upper, middle and lower lobes. The divisions of the lung appear diminished in volume. Linear areas of hypoperfusion correspond to the position and distribution of the fissures

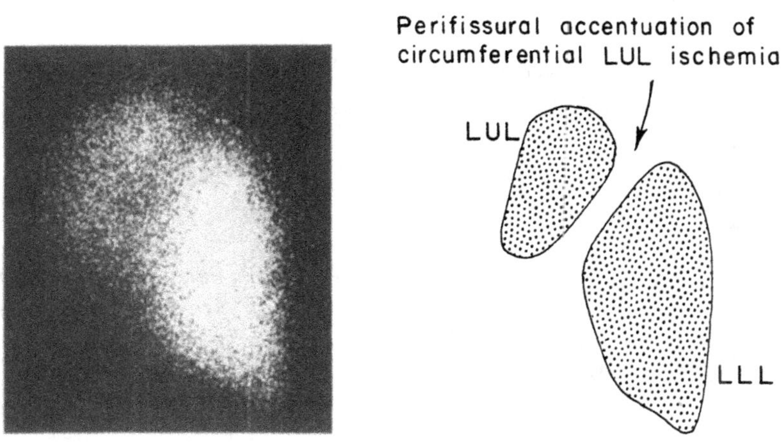

Fig. 2. Left lateral view: circumferential hypoperfusion surrounding the left upper lobe

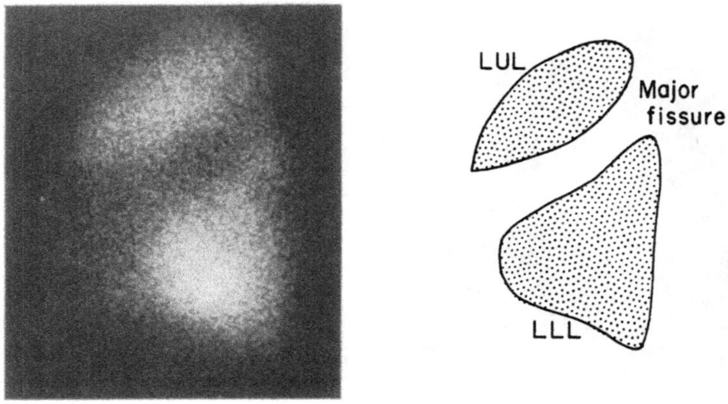

Fig. 3. Left lateral view: linear band of hypoperfusion in location of left major fissure

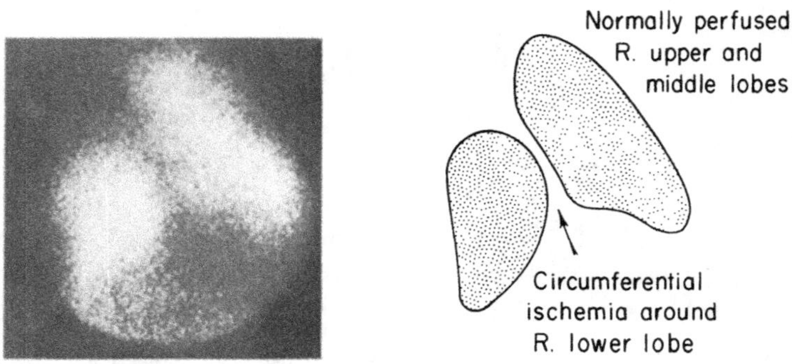

Fig. 4. Right lateral view: linear area of diminished radioactivity in position of right major fissure

a

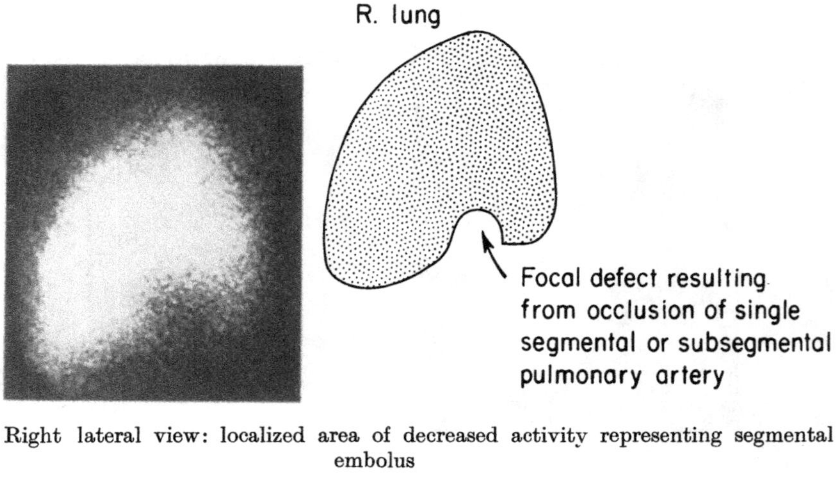

Fig. 5a. Right lateral view: localized area of decreased activity representing segmental embolus

b

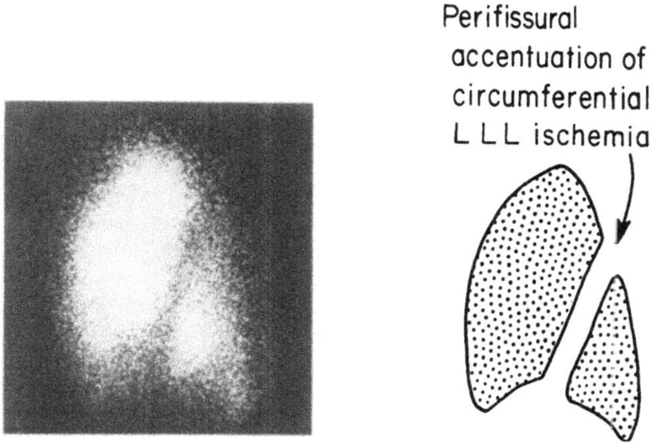

Fig. 5b. Left lateral view: (same patient) "fissure sign" associated with circumferential hypoperfusion involving left lower lobe

Theoretical Explanations

Since it has been documented that changes in pulmonary vascularity and differential perfusion occur with congestive heart failure, it is possible that the scintigraphic pattern of circumferential hypoperfusion might reflect these phenomena. To investigate this theory we reviewed the chest roentgenograms in patients with scintigraphic findings considered typical for microembolism and also performed pulmonary scintigrams on patients with congestive failure. By these two techniques we were able to prove to our satisfaction that congestive heart failure was not the basis for the pattern of circumferential hypoperfusion in our patients. Scintigraphic patterns typically associated with congestive heart failure are not the same as those associated with microembolism. In mild degrees of congestive failure there is increased perfusion in the upper lobes [16] and, therefore, "shunting" of radioactivity into these areas. Irregularity of perfusion becomes much more prominent in severe congestive failure assuming a mottled almost "Swiss-cheese" pattern (Fig. 6). The fissure sign is seen only if pleural fluid is present.

Another possible mechanism for circumferential hypoperfusion is pleural effusion. Dissection of pleural fluid into the fissure may result in the linear hypoperfusion defect noted on the lateral scintigrams. We specifically reviewed the roentgenographs in our patients with scintigrams demonstrating circumferential hypoperfusion to explore this possibility. In addition, we studied the scintigrams of patients with roentgenologically obvious pleural effusion. Characteristic scintigraphic findings of pleural effusion were observed (Fig. 7), but were not those of circumferential hypoperfusion. The scintigraphic findings associated with pleural effusion are flattening of the diaphragmatic surface and "blunting" or upward curvature of the posterior costophrenic angle area depicted on the lateral views (Fig. 8). An associated "fissure sign" may be present. Fluid in the fissures does not, on roentgenographs, appear to dissect superiorly nearly so far as the

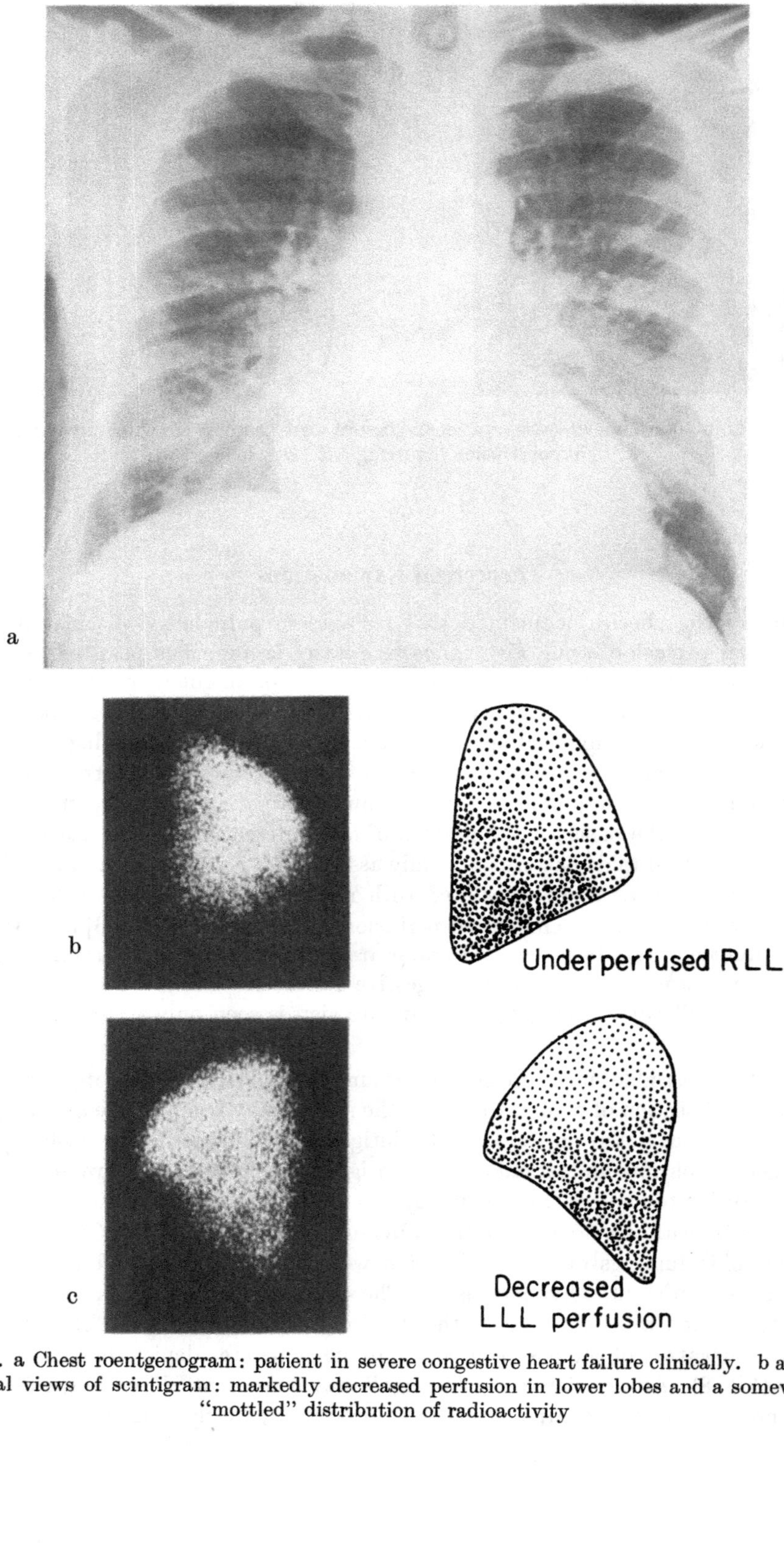

Fig. 6. a Chest roentgenogram: patient in severe congestive heart failure clinically. b and c Lateral views of scintigram: markedly decreased perfusion in lower lobes and a somewhat "mottled" distribution of radioactivity

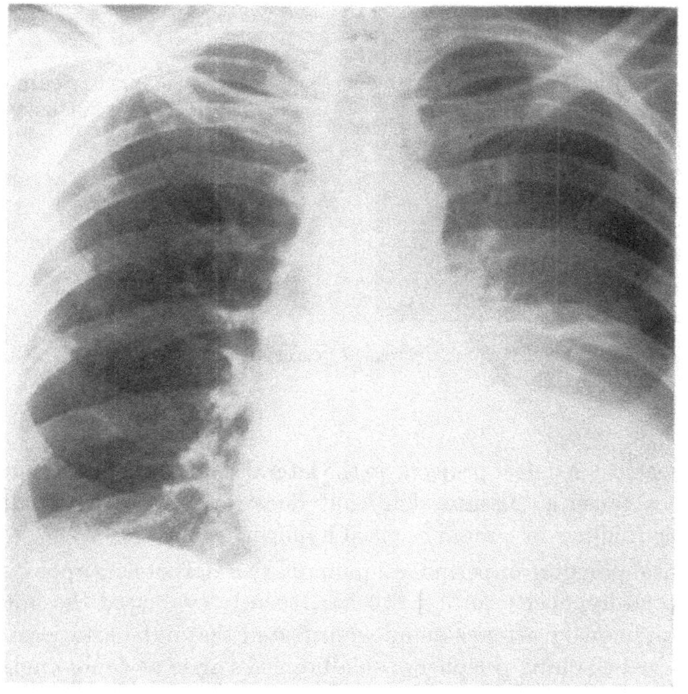

a

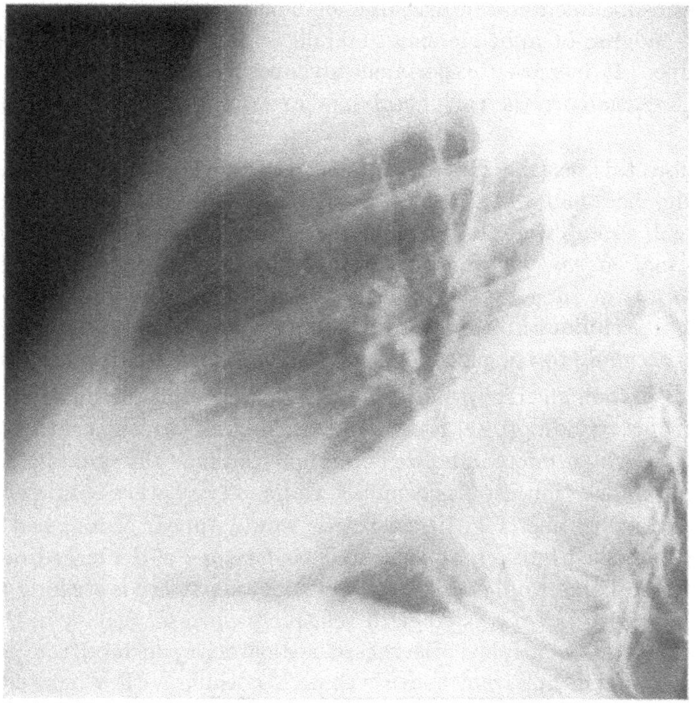

b

Fig. 7a and b. Chest roentgenograms: large pleural effusion on left; mainly posterior in location

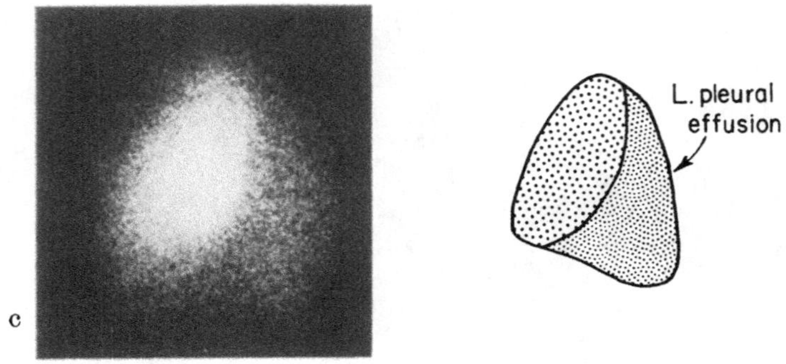

Fig. 7c. Left lateral scintigram: decreased perfusion in the area of the pleural effusion

linear hypoperfusion defect projects on the lateral scintigram. In summary, pleural effusion does cause a "fissure sign" but does not appear to produce all the scintigraphic findings of circumferential hypoperfusion.

Mechanical plugging of peripheral pulmonary axial arteries appears to produce circumferential hypoperfusion. REID has recently evaluated the branching pattern of the pulmonary arteries and demonstrated that axial arteries course rather directly toward the lung periphery; side branches arise at acute angles to supply the more central portions of each lobe [4, 10]. This anatomical situation coupled with the well documented laminar flow of blood in the pulmonary arteries predisposes to lodging of multiple small emboli in the periphery of the pulmonary vascular tree. It has been experimentally observed that small emboli tend to impact preferentially in the peripheral axial arteries after intravenous administration.

To explore this possibility we examined post-mortem lung specimens in those patients who had had scintigraphic patterns of circumferential hypoperfusion. Multiple small emboli were found in the region and distribution of the lung shown to be abnormal on the lung scintigrams. It must be emphasized that in a cursory examination of the lung by the pathologist, microemboli will usually be missed. The injection techniques outlined by DAMMIN and others [12] (Fig. 9) are often necessary as a guide to the pathologist in his search for small pulmonary emboli.

Since our pathological information suggested that multiple small emboli caused the scintigraphic findings we had noted, it became necessary to formulate a mechanism by which microembolization might produce this kind of pattern. If, on a Gough lung section used as a model, radioactivity were confined to an area within the arbitrary line (Fig. 10) each lobe would appear diminished in volume. With two diminished lobes in apposition, a linear zone of decreased radioactivity in the perifissural area would be seen (arrows). This pattern is striking because the diminished activity is contrasted with relatively normal activity in the adjacent central portions of each lobe. Decreased radioactivity around the periphery of each lobe may not be apparent because there is nothing with which to contrast it. (It is important to realize that scans do not necessarily show the outer border of the lung but demonstrate the outer margin of radioactivity.)

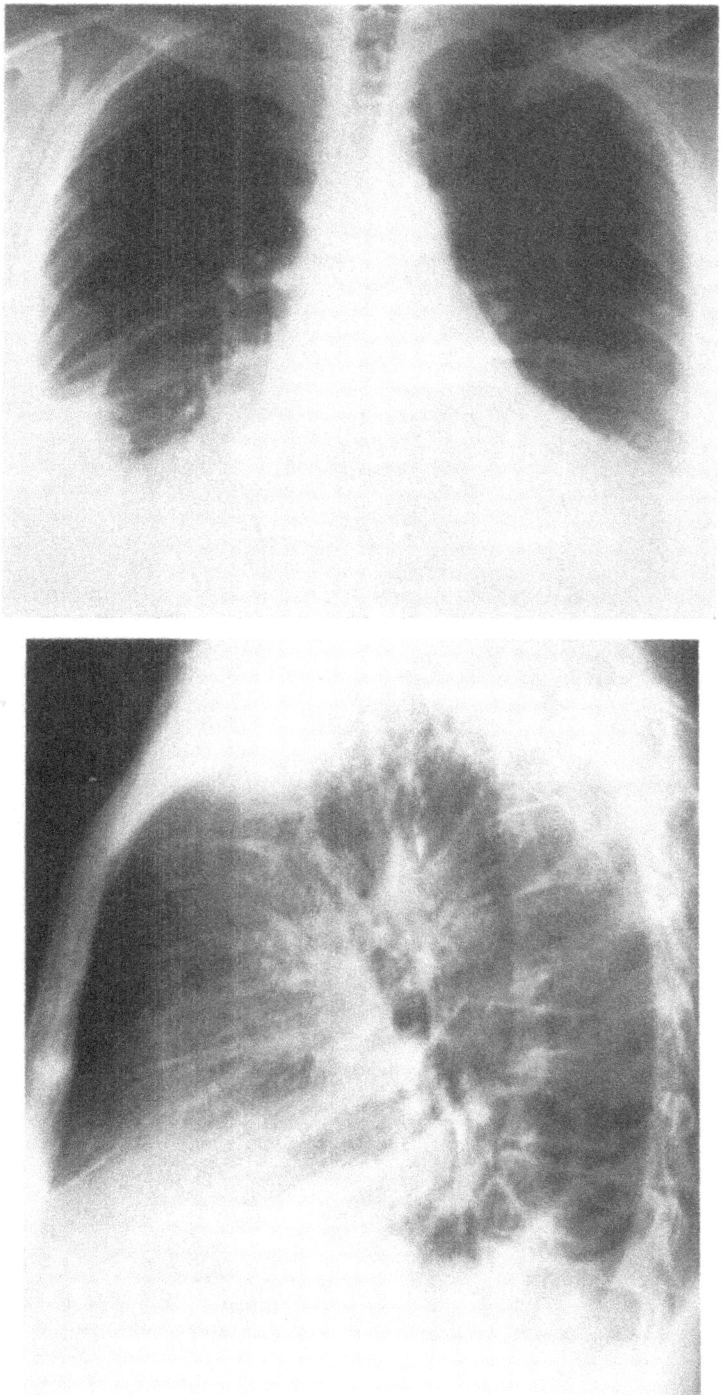

a

b

Fig. 8 a and b. Chest roentgenograms: bilateral pleural effusion with fluid dissecting into the fissure

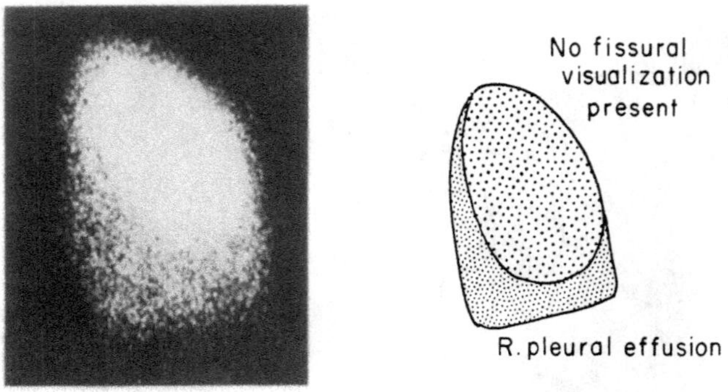

c

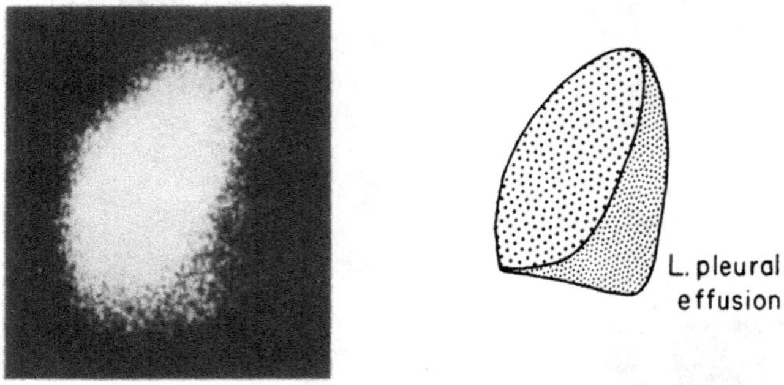

d

Fig. 8c and d. Lateral scintigrams: decreased activity in the area of the pleural effusion

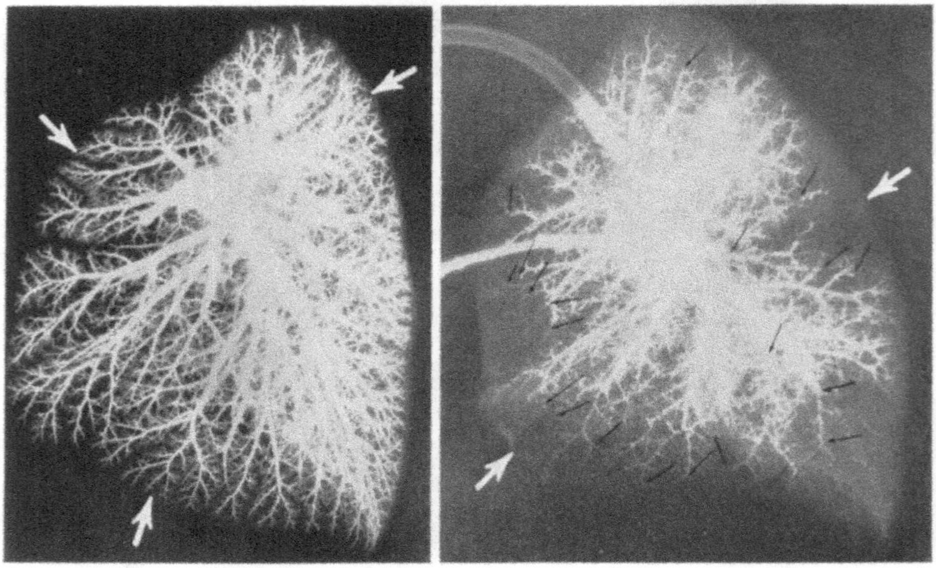

a b

Fig. 9. Specimen roentgenograms of human lung injected with Schlesinger mass.
a Normal lung: vessels fill to the periphery (large arrows) with the contrast material.
b Lung with multiple small emboli: vessels do not fill to periphery (large arrows) with Schlesinger mass

It should be emphasized that, in theory at least, peripheral hypoperfusion patterns are more easily depicted with the scintillation camera than a rectilinear scanner. The parallel-hole collimator isoresponse curves are such that objects assume a three-dimensional appearance, whereas the rectilinear scanner employs a focusing collimator which gives maximum detail at a given level. Therefore, the depth of the area of hypoperfusion will be optimally presented on the lateral view with the scintillation camera.

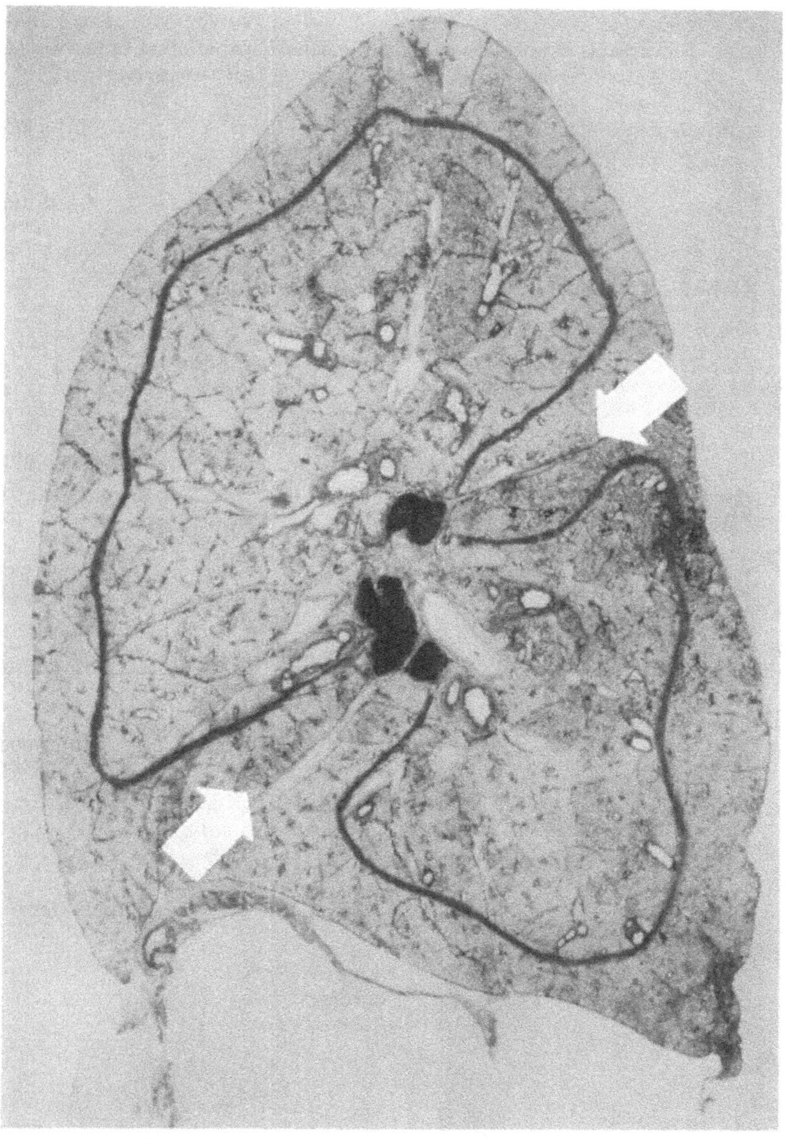

Fig. 10. Gough section of human lung: line drawing separates central from peripheral axial vessels. Large arrows enclose area of hypoperfusion that corresponds to position and configuration of the major fissure

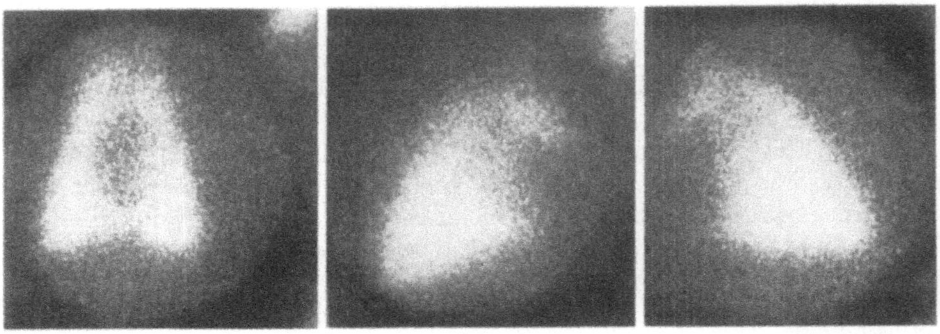

a b c

Fig. 11. Scintigrams of normal dog: homogeneous distribution of radioactivity.
a Anterior view. b Right lateral view. c Left lateral view

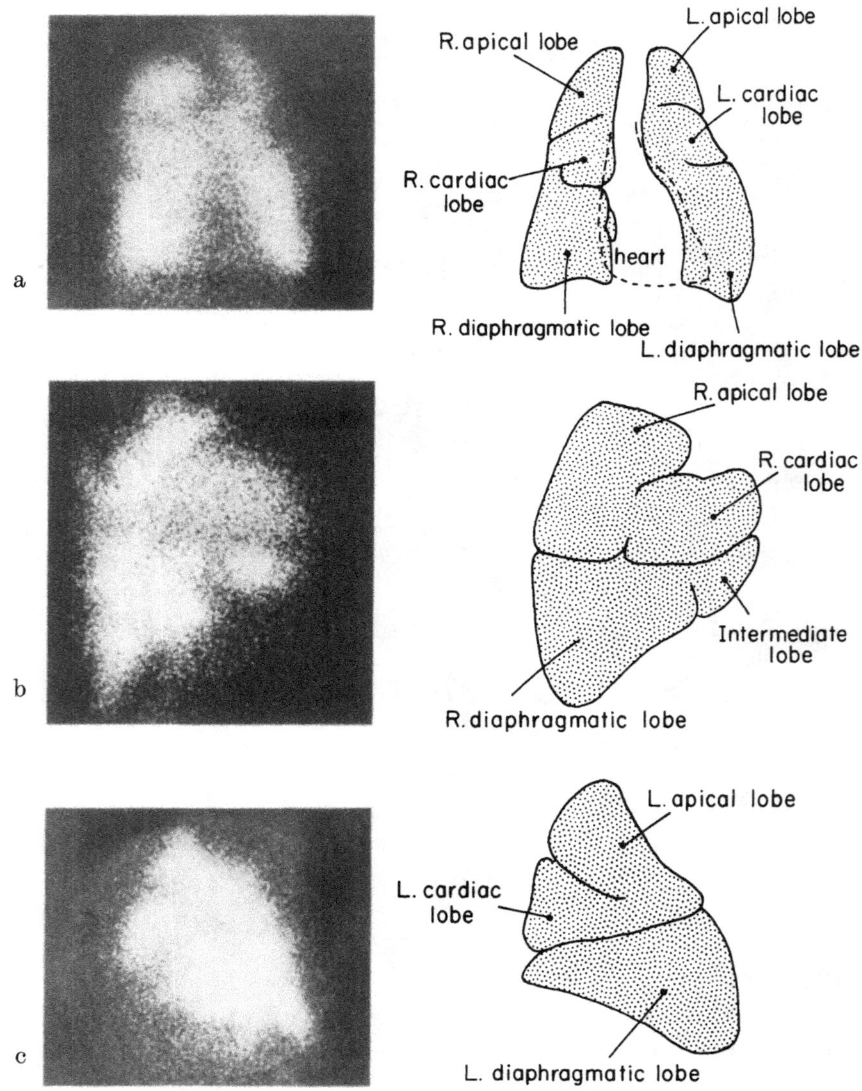

Fig. 12. Scintigrams of dog injected with microspheres: linear areas of hypoperfusion correspond to the anatomical divisions of the dog's lung (artist's drawing).
a Anterior view. b Right lateral view. c Left lateral view

Experimental Investigation

To reproduce the pathophysiology of multiple microembolism we administered polysterene microspheres of varying sizes [(0.147—3 mm)-preselected by passing the microspheres through sieves with various hole diameters] to large mongrel dogs. The microspheres were suspended in 2—3 cc. of an isopropyl alcohol-saline mixture and injected intravenously into a peripheral vein. After at least one circulation time, 125 microcuries of macroaggregated albumin labeled with ^{131}I was also injected intravenously. Scintigrams (Figs. 11 and 12) and roentgenograms were then performed (Fig. 13) [14]. Following this the animals were sacrificed and the heart and lungs removed intact. The pulmonary artery was cannulated and

a

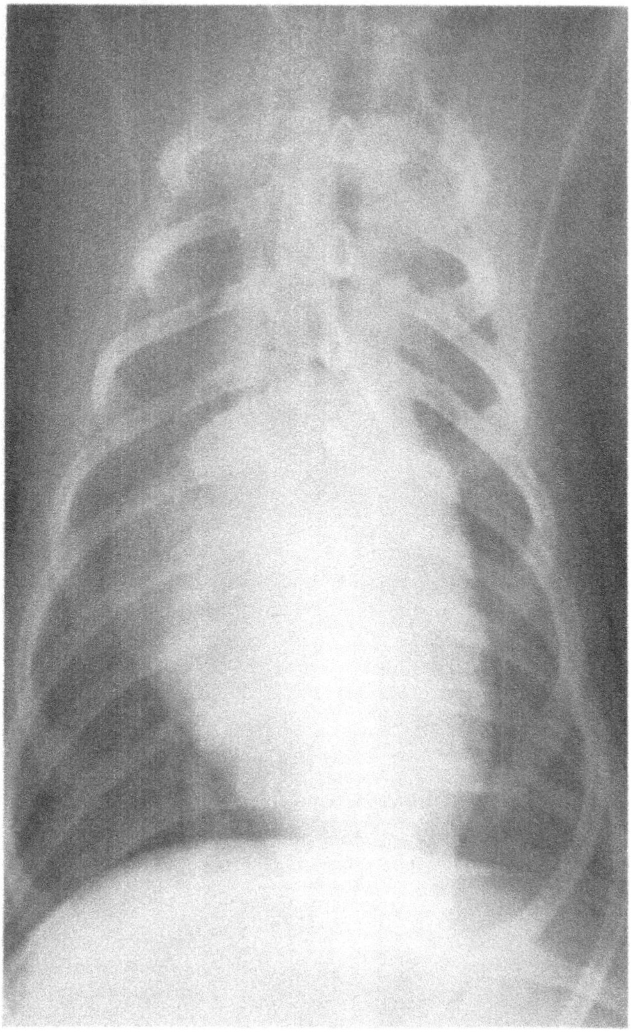

Fig. 13a and b (Fig. b on next page). Chest roentgenograms of dog injected with microspheres: appearance cannot be distinguished from normal

b

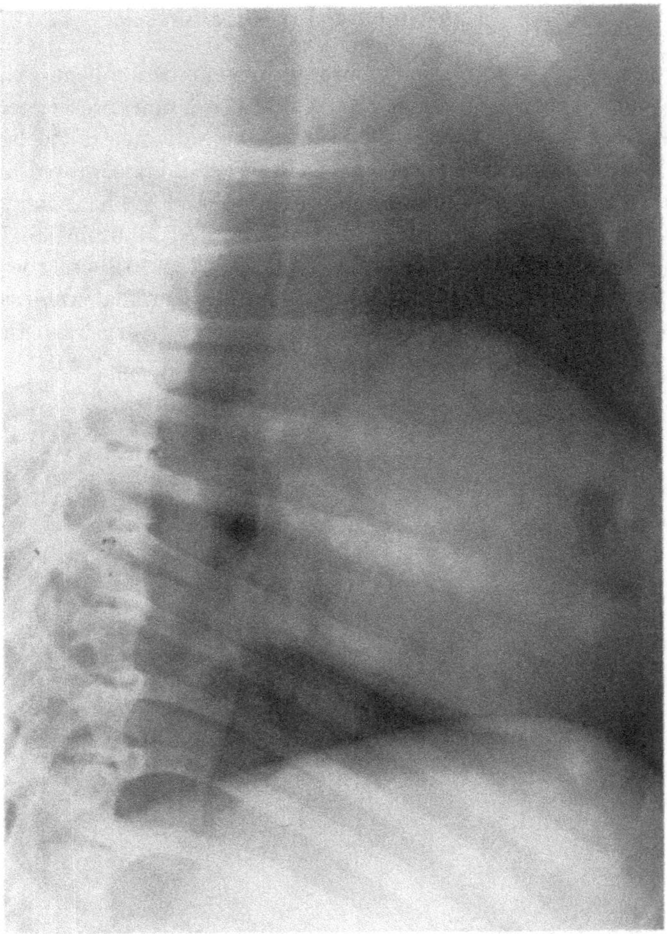

Fig. 13a and b (Fig. a on previous page). Chest roentgenograms of dog injected with microspheres. appearance cannot be distinguished from normal

---

A. *Part I*

Gelatin 130 gm

Potassium iodide 130 gm

Buffer

Monobasic sodium phosphate	30 ml 1 M
Dibasic sodium phosphate	15 ml 1 M

Pigment (Monastral Blue) 20 gm
Mix A 3.0 cc
H_2O 250 cc

B. *Part II*

Barosperse	600 gm
Water	200 cc

C. Mix Part I and Part II with Jet-Dry (a commercial preparation to reduce surface tension)

Fig. 14. Modified Schlesinger mass for pulmonary injection

the pulmonary vasculature injected with a modified Schlesinger mass (Fig. 14) [6, 7]. Roentgenograms of the injected specimens were obtained (Fig. 15). The lungs were then studied histologically; both hematoxylin and eosin and Verhoef-van Gieson (for elastica and collagen) preparations were made.

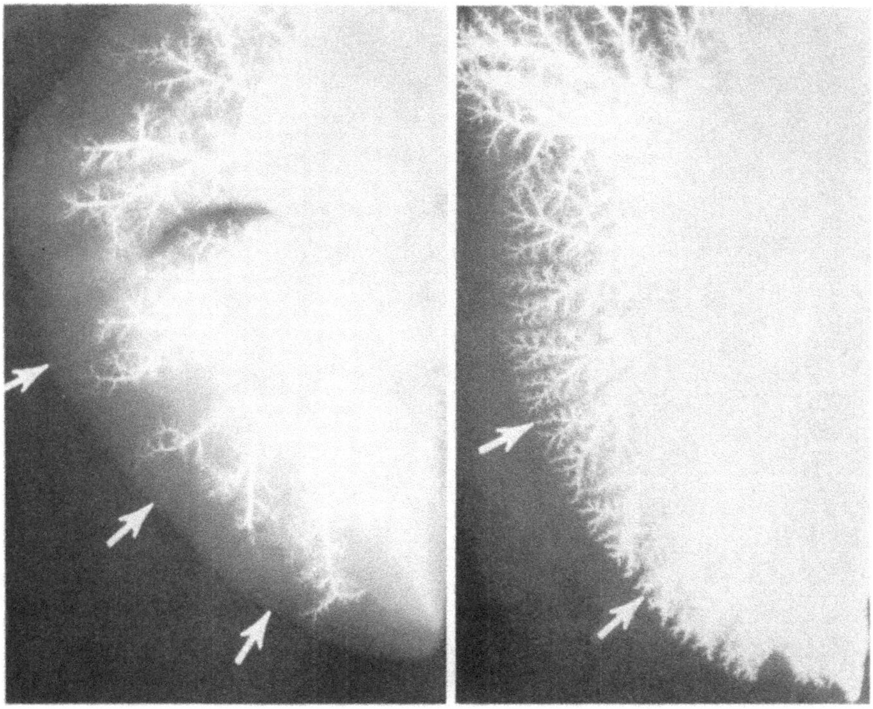

a b

Fig. 15. Roentgenogram of dog lung specimens injected with modified Schlesinger mass. a Normal dog: Schlesinger mass fills vessels to periphery (arrows). b Dog injected with microspheres: contrast does not fill vessels to periphery (arrows)

To study the appearance of both acute and chronic changes, scintigrams were performed at different time intervals (up to several months) following the injection of the microspheres. After the delayed scintigrams were obtained, the dogs were sacrificed and the subsequent studies performed in the usual manner.

Results

Injection of the microspheres resulted in a change in the dogs' respiratory rates but no episodes of coughing or hemoptysis were encountered in the immediate post-injection period. The chest roentgenograms appeared normal and no dilatation of the central pulmonary artery, pleural effusion, or density consistent with pneumonia was seen [5]. The scintigrams showed multiple bands of diminished perfusion which correspond quite well to the fissures that form the anatomical divisions of the dogs' lungs (Fig. 12). The specimens injected with the Schlesinger

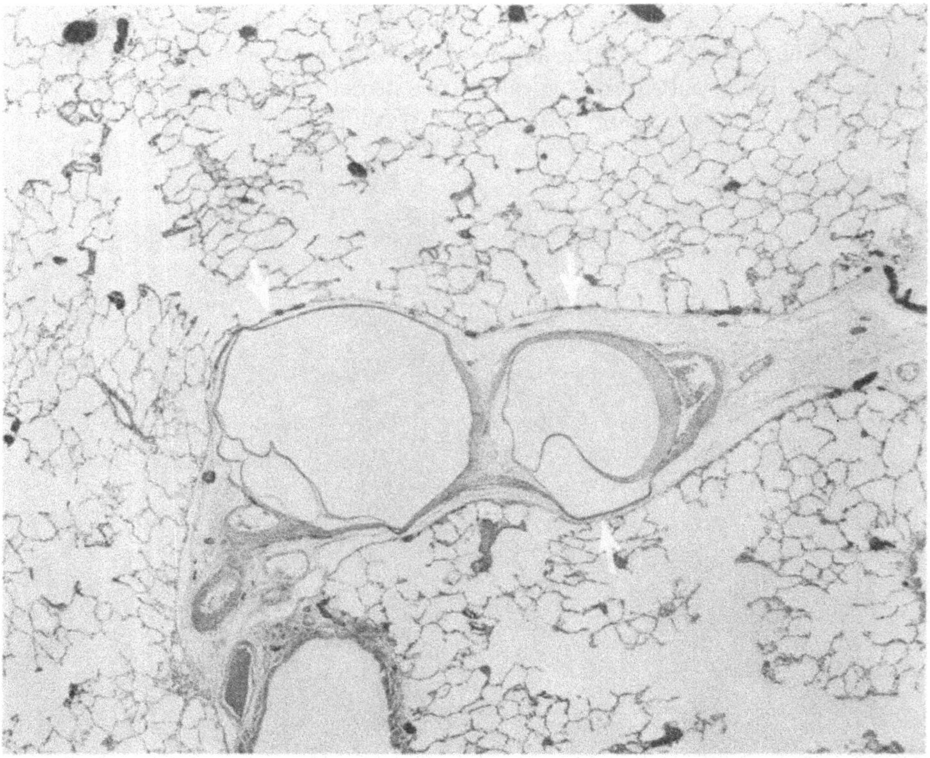

a

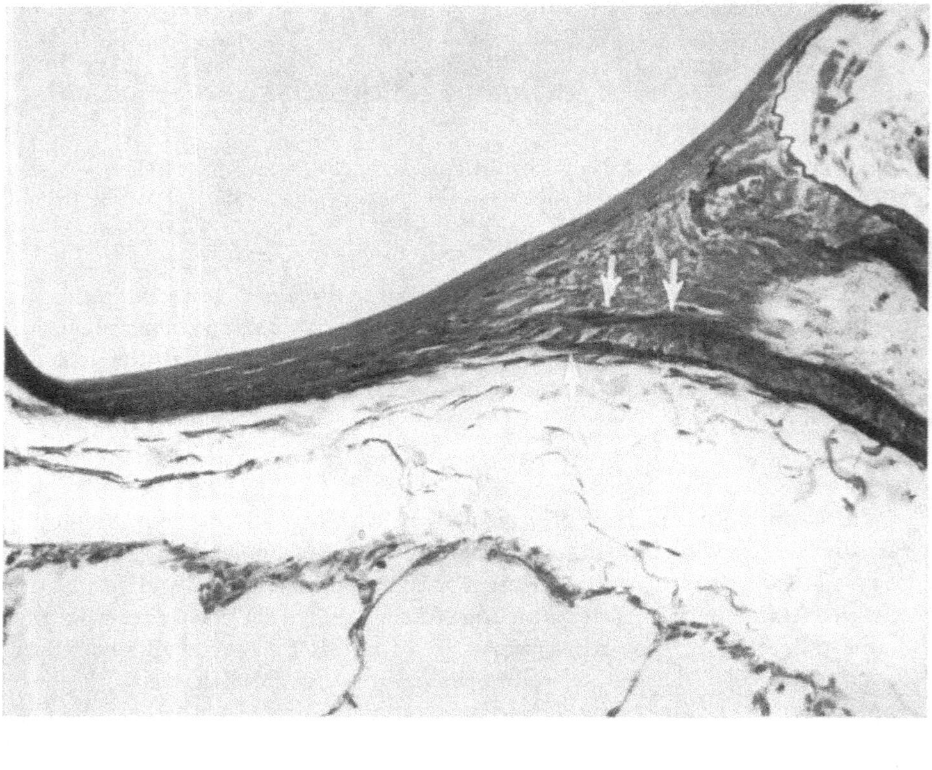

b

mass showed markedly decreased filling of the most peripheral pulmonary arteries giving a "leafless-tree" appearance [17]. No consistent bronchial collaterals or broncho-pulmonary shunts were demonstrated [3]. Corresponding histological studies showed small peripheral pulmonary arteries distended with the microspheres (Fig. 16a) and in the chronically injected dogs, replacement of the pulmonary arterial walls with fibrous tissue (Fig. 16b).

Conclusions

Several mechanisms can produce multiple microembolism. Thrombi in the extremities or pelvic veins may release showers of small emboli; larger emboli formed peripherally may fragment during passage through the heart. Large emboli lodged centrally in the pulmonary vasculature may fragment or partially lyse sending multiple microemboli to obstruct the smaller peripheral vessels. Any or all of the mechanisms may be active in the same patient thus explaining the changing patterns noted on serial pulmonary scintigrams.

The experimental model correlates with the pathological and roentgenographic manifestations of pulmonary microembolism [7]. Polystrene microspheres pass selectively to the peripheral pulmonary arteries much as the microemboli seem to do. Mechanical obstruction of the peripheral pulmonary vessels produces a scintigraphic pattern of peripheral hypoperfusion. Anatomically, the many fissures of the dogs' lungs result in multiple bands of diminished perfusion. Therefore, a linear pattern of many "fissural visualizations" is to be expected. The histological studies failed to reveal the presence of significant bronchopulmonary shunts [3], and instead emphasized the importance of the mechanical obstructive phenomena.

Summary

Pulmonary microembolism is a common clinical problem which has been difficult to diagnose by conventional laboratory and roentgenographic means. Scintillography performed in patients with multiple pulmonary microemboli results in a pattern of circumferential hypoperfusion involving one or several lobes. This pattern is optimally seen on the lateral pulmonary scintigram using the scintillation camera as the detection device.

Acknowledgements

The authors wish to thank Drs. M. S. Potsaid, L. L. Robbins, and the late Felix G. Fleischner for their initial encouragement and guidance as well as Dr. David L. Page for his assistance with the pathological specimens. Critical analysis of the causes of the fissure sign by Dr. Henry N. Wagner, Jr. is greatly appreciated.

←

Fig. 16. a Photomicrograph of histologic section of dog's lung injected with microspheres distended pulmonary arteries where microspheres were lodged (arrows). b Chronic embolized dog: the internal and external elastica (arrows) are lost and the muscular wall is replaced with collagenous tissue (Verhoef-van Gieson stain)

References

1. ANGER, H. O.: Scintillation camera with multichannel collimators. J. Nucl. Med. **5,** 515 (1964).
2. BLAISDELL, F. W., et al.: Pulmonary microembolism. Arch. Surgery **30,** 776 (1966).
3. DALEN, J. E., et al.: Cardiovascular responses to experimental pulmonary embolism. Am. J. Cardiol. **20,** 3–9 (1967).
4. ELLIOTT, F. M., REID, L.: Some new facts about the pulmonary artery and its branching pattern. Clin. Radiol. **16,** 193–198 (1965).
5. FLEISCHNER, F. G.: Roentgenology of the pulmonary infarct. Sem. Roentg. **2,** 61–76 (1967).
6. HYLAND, J. W., et al.: Effect of selective embolization of various sized pulmonary arteries in dogs. Am. J. Physiol. **204,** 619–625 (1963).
7. JAMES, A. E., Jr., et al.: An experimental model to study pulmonary microembolism. Radiology (Work in Progress Section) **92,** 924–927 (March 1969).
8. KRAHL, V. E.: The lung as a target organ in thromboembolism. In: Pulmonary Embolic Disease (SASAHARA, A. A., STEIN, M., Eds.). New York: Grune & Stratton 1965, pp. 13–22.
9. MADDISON, F. E., et al.: Pulmonary microembolism — radiologic findings. Radiology **90,** 1176 (June 1968).
10. PRITCHARD, M. M. L., et al.: Peripheral ischemia of the lung. Brit. J. Radiol. **27,** 93–96 (1954).
11. Pulmonary Microembolism, leading articles. Lancet **1,** 429 (1967).
12. SMITH, G. T., et al.: Postmortal arteriographic studies of the human lung in pulmonary embolization. JAMA **188,** 143–151 (1964).
13. SMITH, G. T., et al.: Human systemic-pulmonary arterial collateral circulation after pulmonary thrombo-embolism. JAMA **188,** 452–458 (1964).
14. WESSLER, S., et al.: Experimental pulmonary embolism with serum induced thrombi. Am. J. Path. **38,** 89–101 (1961).
15. WAGNER, H. N., Jr., et al.: Diagnosis of massive pulmonary embolism in man by radioisotope scanning. New Engl. J. Med. **271,** 377 (1964).
16. WEST, J. B., DOLLERY, C. T.: Distribution of blood flow and ventilation — perfusion ratio in the lung, measured with radioactive CO_2. J. Appl. Physiol. **15,** 405 (1960).
17. WIENER, S. N., et al.: Observations on pulmonary embolism and the pulmonary angiogram. Am J. Roentg. **98,** 859–873 (1966).

Radiation Doses to Patients from Radionuclide Investigations

A Survey Carried Out in New Zealand in 1967

A. C. McEwan

National Radiation Laboratory, Christchurch, New Zealand

Abstract

Data has been collected on all in vivo radionuclide tests and treatments carried out in New Zealand over a 3 month period in 1966. For each type of investigation an assessment has been made of the radiation doses received by organs of increased uptake and by the gonads. From the statistical data of the numbers and types of investigations and treatments and the calculated gonad doses, an annual genetic dose to the population of 0.062 mrad per person was determined, of which only 0.015 mrad arose from the use of radionuclides in diagnosis.

Introduction

Since the production of the first artifically-produced radionuclides of medical importance about 30 years ago there has been a continuing increase both in the total quantities of radionuclides used in medical practice and in the diversity of applications. It is evident that the internal administration of radioactive materials constitutes an addition to the various sources of radiation to which man is already subject. Calculations of the average radiation dose received by individuals to whom diagnostic amounts of radionuclides have been administered have shown that the doses are in general acceptably low and may be compared favourably with the doses received by patients undergoing radiological examinations. It has been considered that because of the magnitude of the radiation doses received by patients and the relatively few tests carried out compared with X-ray examinations, the contribution to the genetically significant dose from the medical use of radionuclides would be less than 1 mrem per year (Report of UNSCEAR, 1958), and this has been confirmed by calculated values in some countries. The annual genetic dose in the United Kingdom in 1957 arising from the use of P^{32} and I^{131} was estimated to be 0.181 milliroentgens (Adrian Committee Report, 1960). For these two radioisotopes the genetic dose contribution in Canada in 1956 was estimated to be 0.42 mR (Johns and Taylor, 1958). Since these estimates were made 10 years have elapsed, in which time much more extensive use has been made of radionuclide techniques. In New Zealand this increase is indicated by the fivefold increase in the numbers of orders for radioactive shipments over the last 10 years (National Radiation Laboratory Report AR/17). There does not, however, appear to be any recent or more precise calculation of the genetic dose arising from the use of radionuclides, perhaps partly because the dose is generally believed to be small, and partly because of the difficulties inherent in making a more precise estimate.

In this paper a description is given of a survey of the numbers and types of in-vivo radionuclide tests and treatments carried out in New Zealand in 1966, and of the method of calculation of gonad and organ radiation doses arising from these tests, and the genetic dose, $G = \frac{\Sigma D_i N_i P_i}{\Sigma N_i P_i}$ (I. C. R. U. 1957), to the population.

Statistical Survey

Collection of Data

Letters were sent to each establishment in New Zealand where in-vivo radionuclide tests were performed, and data was requested on the numbers and types of all tests and treatments carried out in the 3 month period September-November 1966. The information requested included the type of test or treatment, the radionuclide used and its chemical form, the mode of administration, the dose in microcuries, the sex and age of the patient, and any further information which could have had a bearing on the radiation doses received by the patient, e.g. the use of a blocking dose.

The numbers and types of in-vivo tests and treatments performed in each hospital during the 3 month period are shown in Table 1. In two cases data from medical units attached to hospitals, but operating independently of the radioisotope laboratories, have been included with the data from the respective hospitals. It may be noted that subsequent to the period sampled by the survey a further 5 establishments have commenced in-vivo radionuclide studies.

Table 1 indicates that the numbers and types of tests carried out in any one hospital were very dependent upon the local research or diagnostic interests. At hospital B only renograms and blood volume determinations were carried out and

Table 1. *Numbers and types of in-vivo tests performed in each establishment during Sept.-Nov. 1966*

	A	B	C	D	E	F	G	H	I	J	Total
Thyroid Uptake and Scans	548		39	13	117	64	17	122	28		948
Brain Scan	52				29			9			90
Liver Function and Scan	43				28			17	5		93
Kidney Scan	7				15						22
Schilling Test	22			1	84			4			111
Renogram		34		6				4			44
Blood Volume	2	126		1	8			1			138
Red Cell Survival	7				2	3		1	1		14
Bone Scan				2	14						16
Lung Scan					19						19
Exchangeable Sodium										28	28
Miscellaneous	10			3	1	1		2	1		18
I^{131} Therapy	22		3	1	2			13	9		50
P^{32} Therapy				5					3		8
Total	713	160	42	32	319	68	17	173	47	28	1599

almost all patients were pregnant. The smaller hospitals C and G performed mainly thyroid function tests. At hospital E a particular interest was shown in lung scanning, bone scanning, and in Vitamin B_{12} absorption studies, as is evidenced by the relatively large number of Schilling tests performed there. At establishment J a research programme required exchangeable sodium determinations to be carried out.

In addition to the tests listed in Table 1 a relatively large number of in-vitro tests or other tests carried out in conjunction with in-vivo tests, were also performed. From Department of Health statistics the numbers of protein bound iodine determinations and "Triosorb" or "Triomet" tests for the year ending 31st December 1966 were about 1085 and 760 respectively. Thyroid scans were normally performed in conjunction with an uptake measurement and have not therefore been listed separately. On the average about 1 scan was carried out for every 10 uptakes. The total number of administrations of radionuclides to patients in the 3 month period was 1599, or a yearly rate of 6396, about 60% of which were for thyroid uptake measurements. These figures indicate a rate of 2.4 tests or treatments per 1000 population per year, and a rate of 1.7 per 1000 for thyroid uptake measurements.

Age and Sex Distribution

The 1599 tests and treatments were analysed into sex and age groups. The distribution for thyroid uptake tests (Fig. 1) indicates a very marked preponderance of females, the female to male ratio being 5.8 to 1.

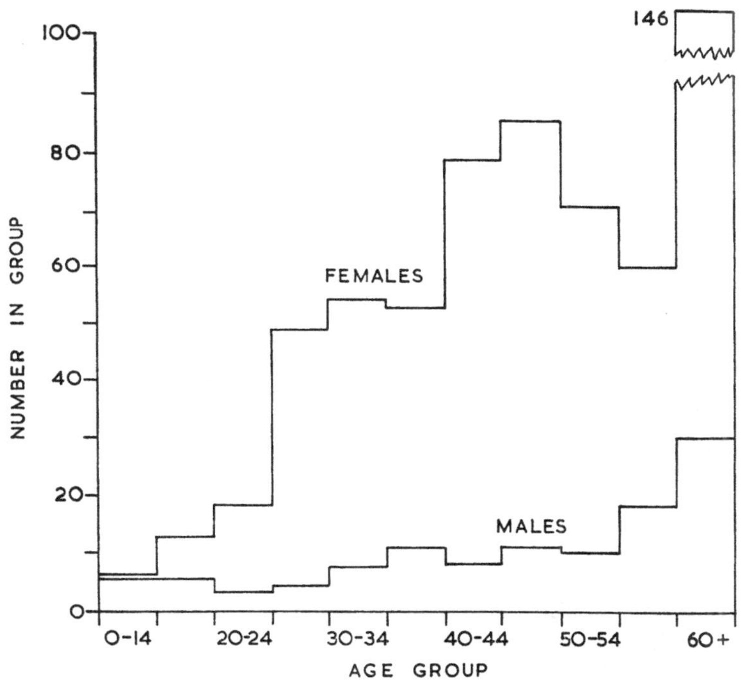

Fig. 1. Age and sex distributions of 736 thyroid uptake measurements

Apart from renograms and blood volume determinations, the distributions of which were strongly influenced by the studies of pregnant women at hospital B, all tests were carried out more frequently with increasing age. This feature would be even more marked if the numbers of tests in each age group were expressed as rates per 1000 population and could perhaps be interpreted as a reluctance on the part of referring practitioners to subject younger persons to in-vivo radionuclide studies. Only 1.3% of tests were carried out on persons in the 0—14 age group and only 17.8% on persons under 30 years of age, whereas the percentages of the total population in these groups were 32.7 and 54.9 respectively. This, however, can be partly accounted for by the fact that the incidence of some diseases is greater in the later age groups. The 40—50 year peak in the distribution of thyroid uptake measurements on females, for instance, appears to be directly related to menopausal disturbances.

Demographic Statistics

Data on the age and sex distribution of the total population of New Zealand on 31 December, 1965 (the latest date for which information was available) and on the numbers of women in each age group having confinements resulting in live births during 1965, were obtained from the Report "Vital Statistics, 1965", of the Department of Statistics. It was necessary to know what proportion of women were pregnant at any given time in order to assess the possible radiation exposure of foetuses.

Mean future child expectancies were also supplied by the Department of Statistics for quinquennial age groups of the total population averaged over the three years 1963—1965, and are based on all births for mothers and nuptial births only for fathers with no allowance for parental mortality. The mean future child expectancy for foetuses has been taken as the average of the male and female 0—14 group values.

Radiation Dose Calculations

Method of Calculation

In general a soluble radioactive material which has been administered to a patient will undergo a complex pattern of distribution throughout the body, with accumulation in different organs, and eventual elimination. While it is not possible to determine precisely the radiation dose received by a particular organ without a knowledge of the distribution and uptake of the administered radiopharmaceutical in the organ, it is possible to calculate an average radiation dose using the formula

$D = C\,T_e(73.8\,\bar{E}_\beta + 0.0346\,\Gamma \bar{g} f)$ (Hine and Brownell, 1956, pp. 827 and 961),

where the symbols have their usual meaning. The formula assumes that the biological retention in the organ may be described by an exponential function, that the tissue is of unit density, and that there is no contribution to the radiation dose received by the organ from radioactivity in surrounding organs.

Average doses resulting from the ingestion of inorganic solutions of radionuclides which have been used in medicine were calculated as an initial step in the determination of radiation doses received by patients from the administration of radiopharmaceuticals. The calculated average values are presented in Table 2.

Table 2. *Radiation doses to organs from ingestion of various radionuclides used in medicine*

Nuclide	Organ	T_e (days)	$\bar{E}_\beta$ (MeV)	Γ (R-cm/mCi-hr)	$73.8\,\bar{E}_\beta$	$0.0346\,\Gamma\bar{g}f$	Dose from 1 μCi in organ (mrad)	Fraction to organ from ingestion	Dose from 1 μCi ingested (mrad)
H^3	Whole Body	12	0.0055	—	0.406	—	0.070	1.0	0.070
C^{14}	Whole Body	10	0.049	—	3.62	—	0.52	1.0	0.52
Na^{22}	Whole Body	11	0.19	12	14.0	57.6	11.2	1.0	11.2
Na^{24}	Whole Body	0.6	0.55	19	40.6	112	1.30	1.0	1.30
P^{32}	Whole Body	13.5	0.69	—	51.0	—	9.7	0.75	7.2
P^{32}	Bone	14.1	0.69	—	51.0	—	72	0.375	27
S^{35}	Whole Body	44.3	0.049	—	3.62	—	2.3	1.0	2.3
S^{35}	Testes	76.4	0.049	—	3.62	—	6900	$1.3 \cdot 10^{-3}$	9.0
Cl^{38}	Whole Body	0.026	1.52	7.1	112	39.4	0.056	1.0	0.056
K^{42}	Whole Body	0.52	1.45	1.4	107	7.2	0.85	1.0	0.85
Ca^{45}	Whole Body	162	0.076	—	5.61	—	13.0	0.6	7.8
Ca^{45}	Bone	162	0.076	—	5.61	—	91	0.54	49
Ca^{47} Sc^{47}	Whole Body	4.9	0.342 0.161	5.5 0.5	25.2 11.9	27.4 2.8	4.7	0.6	2.8
Cr^{51}	Whole Body	26.0	0.0061	0.17	0.45	0.90	0.50	$5 \cdot 10^{-3}$	0.0025
Fe^{55}	Whole Body	463	0.0059	—	0.435	—	2.9	0.1	0.29
Fe^{55}	Spleen	665	0.0059	—	0.435	—	1900	$2 \cdot 10^{-3}$	3.9
Fe^{59}	Whole Body	42.7	0.120	6.2	8.85	31.1	24	0.1	2.4
Fe^{59}	Spleen	43.9	0.120	6.2	8.85	6.10	4400	$2 \cdot 10^{-3}$	9.0
Co^{57}	Whole Body	9.2	0.0229	0.54+0.31	1.69	3.07	0.63	0.3	0.19
Co^{58}	Whole Body	8.4	0.035	5.5	2.58	25.6	3.4	0.3	1.0
Co^{60}	Whole Body	9.5	0.094	13.1	6.94	65.9	9.9	0.3	3.0
Ni^{63}	Whole Body	652	0.018	—	1.33	—	12.4	0.3	3.7
Ni^{63}	Bone	492	0.018	—	1.33	—	65	0.15	9.8
Cu^{64}	Whole Body	0.53	0.13	1.1	9.58	5.17	0.11	0.28	0.031
Zn^{65}	Whole Body	194	0.01	2.7	0.74	13.3	39	0.1	3.9
Zn^{65}	Testes	128	0.01	2.7	0.74	1.67	7700	$9 \cdot 10^{-5}$	0.69
Zn^{65}	Ovaries	74	0.01	2.7	0.74	0.97	15800	$4 \cdot 10^{-5}$	0.63
Se^{75}	Whole Body	10.1	0.0192	2.0	1.42	8.68	1.46	0.9	1.3
Br^{82}	Whole Body	1.3	0.142	15.6	10.5	74.6	1.6	1.0	1.6

Table 2 (continued)

Nuclide	Organ	T_e (days)	$\bar{E}_\beta$ (MeV)	Γ (R-cm/mCi-hr)	73.8 $\bar{E}_\beta$	0.0346 $\Gamma \bar{g} f$	Dose from 1 μCi in organ (mrad)	Fraction to organ from ingestion	Dose from 1 μCi ingested (mrad)
Sr^{85}	Whole Body	64.7	0.0151	3.0	1.11	13.6	13.6	0.3	4.1
Sr^{89}	Whole Body	50.3	0.56	—	41.4	—	30	0.3	8.9
Sr^{89}	Bone	50.4	0.56	—	41.4	—	210	0.09	19
Sr^{90}	Whole Body	5700	0.20	—	14.8	—	6800	0.3	2000
Y^{90}			0.93	—	68.6	—			
Sr^{90}	Bone	6400	0.20	—	14.8	—	53000	0.09	4800
Y^{90}			0.93	—	68.6	—			
Tc^{99m}	Whole Body	0.20	0.014	0.58+0.12	1.03	3.21	0.0121	0.5	0.0060
I^{125}	Whole Body	42	0.028	1.3	1.53	2.43	2.4	1.0	2.4
I^{125}	Thyroid	42	0.028	1.3	1.53	0.24	7400	0.3	2200
I^{125}	Testes	6.3	0.028	1.3	1.53	0.31	290	$5 \cdot 10^{-3}$	1.45
I^{129}	Whole Body	138	0.056	0.05+0.78	4.20	1.49	11.2	1.0	11.2
I^{129}	Thyroid	138	0.056	0.05+0.78	4.20	0.15	30000	0.3	9000
I^{129}	Testes	7	0.056	0.05+0.78	4.20	0.19	770	$5 \cdot 10^{-3}$	3.8
I^{130}	Whole Body	0.52	0.285	12.4	21.0	57.4	0.58	1.0	0.58
I^{130}	Thyroid	0.52	0.285	12.4	21.0	5.7	690	0.3	210
I^{130}	Testes	0.48	0.285	12.4	21.0	7.2	340	$5 \cdot 10^{-3}$	1.7
I^{131}	Whole Body	7.6	0.188	2.2	13.9	10.3	2.6	1.0	2.6
I^{131}	Thyroid	7.6	0.188	2.2	13.9	1.0	5300	0.3	1600
I^{131}	Testes	3.73	0.188	2.2	13.9	1.3	1420	$5 \cdot 10^{-3}$	7.1
I^{132}	Whole Body	0.097	0.49	11.9	36.2	57.4	0.13	1.0	0.13
I^{132}	Thyroid	0.097	0.49	11.9	36.2	5.7	200	0.3	60*
I^{132}	Testes	0.097	0.49	11.9	36.2	7.2	105	$5 \cdot 10^{-3}$	0.53
Cs^{137}	Whole Body	70	0.23	3.1+0.08	17.0	14.5	32	1.0	32
Au^{198}	Whole Body	2.6	0.331	2.4	24.4	11.2	1.3	0.1	1.3
Au^{198}	Kidneys	2.7	0.331	2.4	24.4	2.2	240	$3 \cdot 10^{-3}$	0.72
Hg^{197}	Whole Body	2.1	0.0794	0.31	5.85	1.72	0.23	0.75	0.17
Hg^{197}	Kidneys	2.3	0.0794	0.31	5.85	0.34	47	0.26	12
Hg^{203}	Whole Body	8.2	0.099	1.30+0.04	7.30	6.39	1.6	0.75	1.2
Hg^{203}	Kidneys	11.0	0.099	1.30+0.04	7.30	1.25	370	0.26	97

* 20 when a factor of 0.32 is used to correct for uptake time (Hine and Brownell, p. 840).

The organs selected were the whole body, the critical organ, if this was not the whole body, and the gonads, if the dose to these was included in the organs listed in the Tables of the I. C. R. P. publication, Report of Committee II, 1959. The effective half-lives used were also taken from I. C. R. P. II.

Average beta energies and photon yields for the nuclides Cr^{51}, Co^{57}, Se^{75}, Sr^{85}, Tc^{99m}, Hg^{197} and Hg^{203} were obtained from SMITH (1966). Average beta energies for other radionuclides were obtained from HINE and BROWNELL (Tables 1 B and 2B, p. 899 and 903) or computed from Fig. 4, Ch. 16-I-A, p. 698, of the same reference. Low energy electromagnetic radiations and conversion electrons for radionuclides other than those listed by SMITH were treated according to the method of HINE and BROWNELL (Ch. 17-II-A, Sections 8 and 9, pp. 842—847) so that X-radiation below 11.3 keV was included as beta-like radiation.

Specific gamma ray constants, average geometric factors and roentgen-to-rad factors were found independently for each gamma energy emitted by a radionuclide in order to determine the gamma dose contribution. The contributions from each gamma energy, weighted according to the fraction emitted per disintegration, were then summed. Specific gamma ray constants were taken from HINE and BROWNELL (Fig. 36, Ch. 16-II-A, p. 757) and f factors from I. C. R. U. 62 data for muscle as tabulated by JOHNS, 1961 (Table A 10, p. 703).

Average Geometric Factors

Average geometric factors for both cylinders and bodies of different sizes have been determined by HINE and BROWNELL (Tables V and VI, p. 858), using the integral doserate calculations of BUSH, 1946. ELLETT, CALLAHAN and BROWNELL (1964) have pointed out, however, that the values in these tables under-estimate the true geometric factor by about 15%, and revised values have subsequently been prepared for cylinders of different sizes by FOCHT, QUIMBY and GERSHOWITZ (1965). By comparing the values given by HINE and BROWNELL for different cylinders and body sizes with those of FOCHT et al. an average geometric factor of 160 was selected as appropriate for a 70 kg standard man. This value applies to gamma ray energies corresponding to a true linear absorption coefficient of about 0.028/cm in water. At other gamma ray energies a correction to the geometric factor was made by multiplying by $0.028/\mu_w$, where μ_w is the absorption coefficient in water for the particular gamma energy. Values of μ_w were obtained from the Radiological Health Handbook, p. 141.

This correction to the average geometric factor could be expected to fail at very low or very high gamma energies, and to obtain some assessment of the accuracy of the gamma dose contributions ($0.0346\,\Gamma\bar{g}f$) at these energy extremes the calculated values were compared with values derived using the absorbed fraction method of ELLETT, CALLAHAN and BROWNELL (1964 and 1965). The gamma dose contribution using the absorbed fraction method is given by $73.8\,(AF)\,E_\gamma$, where (AF) is the absorbed fraction for the gamma energy E_γ. Absorbed fractions given by ELLETT et al. for a uniform distribution of activity in a 70 kg ellipsoid were used to derive the gamma contributions shown in Table 3. For energies below about 80 keV the values derived using average geometric factors began to fall significantly below those calculated using absorbed fractions, and at 20 keV the $\bar{g}$ factor derived value was only about 40% of the (AF) derived value. Since these low values could lead to significant errors in the radiation doses from low energy emitters, and in particular from I^{125}, $\bar{g}$ values for energies below 80 keV were adjusted so that the derived gamma contributions were equal to those found using absorbed fractions (Table 3).

Table 3. *Comparison of gamma dose contributions to whole body calculated using $\bar{g}$ values and absorbed fractions*

E_γ (MeV)	0.01	0.02	0.04	0.08	0.16	0.364	0.662	1.46	2.75
Γ	8.00	1.86	0.489	0.362	0.79	2.13	3.81	7.40	12.2
$\bar{g}$	(2.9)	(24)	(114)	171	157	137	139	154	181
f	0.932	0.923	0.927	0.947	0.967	0.964	0.965	0.965	0.964
$0.0346\, \Gamma \bar{g} f$	0.74	1.44	1.79	2.07	4.05	9.74	17.7	38.0	73.6
AF	1.00	0.98	0.606	0.375	0.344	0.347	0.348	0.312	0.263
$73.8\,(AF)\,E_\gamma$	0.74	1.44	1.79	2.21	4.06	9.31	17.0	33.6	53.4

At high energies the $\bar{g}$ factor derived dose contributions were greater than those calculated using absorbed fractions, although even for Na^{24}, which is the most energetic emitter included in Table 2, the gamma dose contribution did not deviate from the absorbed fraction dose by more than 25%, and the average whole body dose was greater by only 18%.

While the absorbed fraction method could have been used here for the calculation of radiation doses it was necessary to derive and use specific gamma ray constants for the calculation of gonad dose contributions from organs of increased

Table 4. *Average geometric factors for various organs*

Organ	Mass (g)	Radius (cm)	$\bar{g}$	a
Liver	1700	7.40	69.8	0.436
Thyroid	20	1.68	15.8	0.099
Spleen	150	3.30	31.1	0.194
Kidney	300/2	3.30	31.1	0.194
Testes	40	2.12	20.0	0.125
Ovary	8/2	0.98	9.3	0.058
Thyroid	25	1.81	17.1	0.107

uptake, and it was then convenient to make use of these values of Γ and proceed to the calculation of gamma dose contributions using $\bar{g}$ values. Average geometric factors for organs other than the whole body were calculated assuming the organs to be unit density spheres with masses equal to those of the organ of the standard man using $\bar{g} = 3\pi R$ (Hine and Brownell, Eq. 80, p. 857). Kidneys and ovaries were considered as 2 equal spheres. Factors obtained for various organs are shown in Table 4 together with the ratio a of $\bar{g}$ for the organ to $\bar{g}$ for the whole body. The gamma dose contribution to the organ was then calculated by multiplying the whole body gamma $\bar{g}$ by a.

Doses from Radiopharmaceuticals

It is evident that the average doses shown in Table 2 are applicable only to those radiopharmaceuticals which may be considered as being uniformly distributed throughout the whole body and which in general are in the form of inorganic compounds or salts. Colloidal preparations and some labelled compounds may have very different distributions and metabolic pathways in the body and consequently altered effective half-lives. To determine patient gonad doses in any particular diagnostic test or therapy treatment it was therefore necessary to take account not only of the particular radionuclide forming the radioactive label but also the nature of the test material. In certain cases it was in addition necessary to take account of contributions to the gonad doses from the concentration of the radioactive material in blood or in certain organs of increased uptake, or in the gut or bladder.

Gonad doses and some organ doses of interest for in-vivo radionuclide tests and treatments carried out in New Zealand in 1966 were derived from published data on biological turnover and distribution and the data of Table 2 taking account of dose contributions from tissue, blood, organs of high uptake, and bladder. Details of these calculations have been given in a National Radiation Laboratory Report (McEwan, 1967).

Dose to Tissue from Blood

In general when a labelled compound is introduced into the vascular space, some of the radioactive material is absorbed in various tissues and some may be taken up in high concentration in a few organs. While gamma radiation arising from radioactive material in blood may be considered as giving rise to an approximately uniform whole body dose distribution, the dose to tissue from a beta emitter in blood will depend both on the size of the blood vessels in the tissue and the energy of emission. For P^{32}, which emits beta particles with a maximum energy of 1.7 MeV, the maximum range of the particles in unit density material is 7.9 mm. There will therefore be some deposition of energy over a distance of almost 0.8 cm. The average distance of penetration of the energy carried by the beta particles (R), however, is only about 0.2 of the maximum range (Hine and Brownell, p. 740).

An estimate was made of the size of blood vessels which were sufficiently small so that tissue surrounding them was subject to beta irradiation. By taking account of the geometry and the directions of all particles emitted, half the energy of emission was found to be deposited outside a vessel when the radius was about

$^3/_4$ R. Vessels having diameters less than about 1.5 R could then be considered as providing contributions to the tissue radiation dose. Since in practice there are a large number of vessels and capillaries with diameters very much less than practical values of 1.5 R, and only a small percentage with diameters of a similar magnitude, the precise value chosen for the limiting vessel diameter is not critical for maximum energies below about 1.5 MeV. The volumes of blood contained in vessels of different sizes have been estimated by Green (1950) from measurements on dogs. Assuming that the dog data is applicable to man and that small blood vessels and capillaries are uniformly distributed throughout the body, the fraction of the beta radiation emitted in blood which gave rise to uniform tissue doses were calculated for different energies of emission and are shown in Fig. 2.

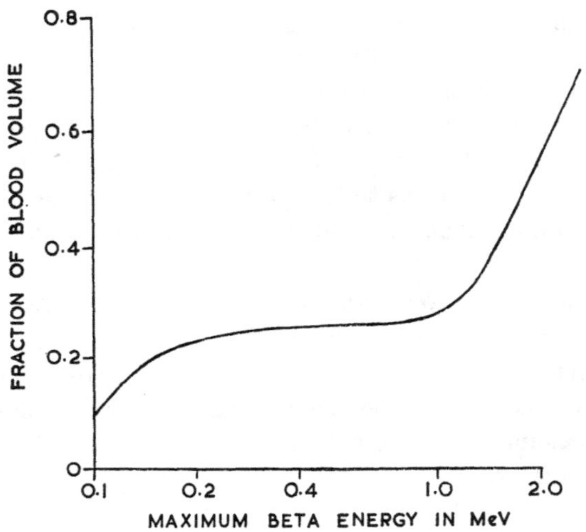

Fig. 2. Fraction of blood volume contributing to tissue dose from beta radiation

Doses to Gonads from Organs of Increased Uptake

A gamma emitting radionuclide which is selectively concentrated in an organ may give rise to a higher dose to the gonads than the simple calculation of the average whole body dose would indicate. If the organ of high concentration is considered as a point source, the doserate at a distance of r cm in an infinite absorbing medium is given by

$$R_\gamma = \frac{B\Gamma\, A f e^{-\mu_t r}}{r^2}$$

where R_γ is the doserate in rads per hour, A is the activity in the organ in mCi, μ_t is the total absorption coefficient, and B is a build-up factor appropriate to the gamma ray energy. Build-up factors for point sources of monoenergetic γ-ray emitters in infinite absorbing media have been calculated (Hine and Brownell,

p. 774), although when used to determine the dose at a point in the body will over-estimate the true dose. If the build-up factor is ignored and the true absorption coefficient is substituted for the total absorption coefficient, the reduction in doserate due to the neglect of the build-up factor is approximately compensated for by the increase resulting from the reduction in the absorption coefficient (HINE and BROWNELL, p. 776).

The expression $R_\gamma = \Gamma A f e^{-\mu r}/r^2$, where μ is the true absorption coefficient, has been used here to calculate the contribution to the gonad doses from organs of increased uptake. Approximate values of r for average adults were estimated after consulting standard texts on anatomy and making appropriate measurements on a skeleton, and are listed in Table 5.

Table 5. *Approximate average distances from the centres of various organs to the gonads (cm)*

Organ	Testicles	Ovaries
Lungs	37	27
Liver	32	22
Kidney	27.5	18
Intestines	22	10
Bladder	10	7.5

Dose to Gonads from Radioactive Material in Bladder

The contribution to the gonad dose from a gamma emitter in the bladder can be calculated in the same manner as for organs of increased uptake. For a radiopharmaceutical which is eliminated from the body entirely in urine, the activity excreted on day t is $(0.693 A_t/T_b)$ (assuming the body retention may be described by an exponential function), where T_b is the biological half-life of the test material, and A_t is the activity t days after administration of the dose. If the patient voids on the average 6 times per day and excretes a total of 1400 ml (I. C. R. P. II value for the standard man), the maximum bladder volume is 233 ml and the average volume may be taken as 117 ml. The average activity in the bladder on day t is then $(117/1400) \cdot (0.693 A_t/T_b)$, and the effective number of microcurie-hours in the bladder for complete decay is

$$\frac{117}{1400} \cdot \frac{0.693}{T_b} \cdot 1.44\, T_e \cdot 24 \cdot A_0 = \frac{2.0\, T_e A_0}{T_b}$$

where T_e is the effective half-life in days and A_0 is the dose administered in μCi.

The total dose to the gonads is then given by $D = (2.0\, T_e A_0 \Gamma f e^{-\mu r})/(T_b r^2)$ mrad/μCi, where the appropriate r is taken from Table 5. Average values of Γ were used for each radionuclide, in general excluding contributions from X-rays. If however, the X-ray contribution formed a significant fraction of the dose then this was calculated separately with an appropriate value for μ. The column of average Γ values in Table 2 indicates those radionuclides which have a significant low electromagnetic energy contribution.

Table 6. *Average doses to gonads and to organs of high uptake in mrad/µCi administered*

Test	Nuclide	Organ	Organ Dose	Male Gonad Dose	Female Gonad Dose
Thyroid Uptakes, Therapy	I^{131}	Thyroid	230—1900	0.87—7.3	0.16—1.3
		Blood	0.42—2.5		
Brain Scanning	Tc^{99m}	Whole Body	0.013	0.014	0.015
Brain and	Hg^{197}	Kidneys	19	0.011	0.026
Kidney Scanning	Hg^{203}	Kidneys	76	0.076	0.21
Brain Scanning, Placental	RISA I^{131}	Blood	6.4	1.7	1.7
Localization, Blood Volume Determination	RISA I^{125}	Blood	1.2	0.62	0.63
Liver Scanning	Colloidal Au^{198}	Liver	40	0.11	0.25
		Spleen	48		
Liver Scanning	Rose Bengal I^{131}	Liver	5.0	0.032	0.17
	Rose Bengal I^{125}	Liver	0.81	0.001	0.024
Liver Function	PVP I^{131}	Liver	36	0.20	0.38
	PVP I^{125}	Liver	41	0.029	0.099
Schilling Test	Co^{57}	Liver	160	0.78	2.1
	Co^{58}	Liver	210	2.6	8.3
	Co^{60}	Liver	8100	81	210
Renogram	Hippuran I^{131}	Kidneys	0.3	0.016	0.030
	Hippuran I^{125}	Kidneys	0.04	0.0025	0.0064
Red Cell Survival	Cr^{51}	Blood	1.4	0.22	0.22
Spleen Scanning	Cr^{51}			0.030	0.076
Bone Scanning	Sr^{85}	Bone	5	2.9	2.9
Calcium Metabolism	Ca^{45}	Bone	58	0.18	0.18
	Ca^{47}	Bone	17	3.1	3.1
Lung Perfusion Scanning	I^{131} M.A.A.	Lung	7.7	0.074	0.11
Lung Ventilation Scanning	I^{131} RISA	Lung	1.9	0.011	0.018
Exchangeable Sodium	Na^{24}	Whole Body	1.3	1.3	1.3
	Na^{22}	Whole Body	11	11	11
Lipoprotein Turnover	I^{131}	Whole Body	0.35	0.35	0.35
	I^{125}	Whole Body	0.045	0.045	0.045
Pancreas Scanning	Se^{75}	Whole Body	8.1	8.1	8.1
Iron Metabolism	Fe^{59}	Blood	97	22	22
Polycythaemia Therapy	P^{32}	Bone	27	3.5	3.5

Genetic Doses

Method of Calculation

Contributions to the total annual genetic dose to the population from the in-vivo use of radionuclides were assessed making an allowance for the irradiation of foetuses. Where a test was associated with pregnancy (as in placentograms, and most radionuclide tests carried out at hospital B), the doses to the foetal gonads in late pregnancy were calculated. For all other women of child bearing age it was assumed that a test could be carried out during the first 3 months of pregnancy, since the patient's practitioner might refer her for a diagnostic test without necessarily being aware that the patient was pregnant. It was assumed that no woman would undergo an in-vivo test when more than 3 months pregnant unless the test was related to the pregnancy.

The doses of I^{131} administered to patients varied between 5 and 30 μCi for uptake measurements, and from 30 to greater than 150 μCi for scans. The mean activity administered was 22.5 μCi, which led to mean gonad doses of 42.5 mrad and 7.14 mrad, for males and females respectively.

For brain scans and other tests where more than one radiopharmaceutical was used, the relative numbers of tests performed with each agent were assumed to be constant throughout all the age groups, and mean gonad doses were calculated by weighting the gonad doses calculated for each radiopharmaceutical, by the proportion of tests performed using it. For therapy treatments, and tests of which fewer than 30 were performed in the 3 month period, the quantity $\Sigma N_i P_i D_i$ was determined by summing the contributions calculated for each patient individually. Child expectancies for patients undergoing bone scans for metastases detection, and iodine therapy for carcinoma, were reduced by a factor of 10. It may be noted that the comparatively large numbers of pregnant women undergoing renograms and blood volume determinations at hospital B gave rise to the much larger genetic dose contributions from females and foetuses than males for these tests.

Errors in the Calculated Doses

Table 7 indicates that the therapeutic use of I^{131} contributes 75% of the total genetic dose. There is, however, considerable uncertainty in this value, particularly in the contribution from males, owing to the statistical uncertainty which arises from the fact that there were only a comparatively few male patients in the younger age groups. Some uncertainty in the result is also introduced through the choice of the arbitrary factor of one-tenth used to allow for the lower child expectancies of carcinoma patients.

The largest single source of error in the derivation of the gonad doses would appear to be in the published data on the metabolic pathways, body distribution, and organ uptakes of the administered radiopharmaceuticals by the gonads. In the calculation of gonad doses the concentration of I^{131} in testes was taken from I. C. R. P. II data obtained from studies on rabbits, and the extrapolation of rabbit data to man clearly introduced a source of error. The assumption was also made that iodine in the gonads was in equilibrium with that in plasma, and that the concentration of iodine in the ovaries was not significantly different from that in other tissues.

A further source of error in the radiation dosimetry arose from the fact that all gamma dose contributions were assessed for a 70 kg standard man and no allowance was made for individual or sex variations in weight and height. The average New Zealand woman may in fact weigh less than 60 kg, and since the $\bar{g}$ factor is only very slightly smaller for a 60 kg person than for a 70 kg man, the average whole body gamma dose will be greater for 60 kg persons owing to the higher body concentration. The error from this source implies that female and foetal genetic dose contributions arising from radionuclides having uniform distributions in the body should be increased by about 1.17.

Table 7. *Annual genetic doses in mrad/person · 10^{-3}*

Test	Male	Female	Foetus	Total	% of Total
Thyroid Uptake and Scan	2.4	1.8	0.16	4.4	7.1
Brain Scan	3.3	1.4	0.22	4.9	7.9
Liver Scan	0.23	0.76	0.040	1.03	1.7
Kidney Scan	0.040	0.056	—	0.096	0.15
Schilling Test	0.024	0.044	0.004	0.072	0.12
Renogram	0.003	0.88	0.25	1.13	1.8
Blood Volume	0.012	1.7	1.2	2.9	4.7
Red Cell Survival	0.032	—	—	0.032	0.051
Bone Scan	—	0.052	—	0.052	0.083
Lung Scan	0.008	0.024	—	0.032	0.051
Exchangeable Sodium	0.012	0.028	—	0.040	0.064
Miscellaneous	0.044	0.45	0.024	0.52	0.83
Sub-Total	6.1	7.2	1.9	15.2	24.4
I^{131} Therapy	23	20	4.0	47	75.5
P^{32} Therapy	0.024	0.052	—	0.076	0.12
Total	29.1	27.3	5.9	62.3	100.0

Diagnostic doses administered to children under 15 were in general smaller than those administered to adults and for the very few patients in this group it was assumed that the lowered dose and smaller body weight would result in about the same gonad doses as for adults.

Discussion

Comparison with Earlier Estimates

The annual genetic dose contribution from the in-vivo use of I^{131} in New Zealand in 1966 of 0.05 mrad per person is considerably less than the value of 0.11 mR found in the United Kingdom in 1957 (Adrian Committee Report, 1960) and the 0.31 mR assessed for Canada in 1956 (Johns and Taylor, 1958), in spite of the more widespread use of this radioisotope in this country in 1966 than in either the

United Kingdom or Canada at the time the earlier estimates were made. A total of 9.9 Curies were shipped to hospitals and medical research laboratories in New Zealand in 1966 (National Radiation Laboratory Report AR/17) which corresponds to 3.7 Curies per million of population, and from the data collected for this survey an estimated 5.2 Curies were administered to patients, corresponding to 2.0 Curies per million of population. These values may be compared with the 1.0 Curies per million administered to patients in the United Kingdom in 1957 and the 1.9 Curies shipped to medical users in Canada in 1956 (SOWBY, 1961). Although the rate of use of I^{131} in New Zealand in 1966 was twice the rate of use in Canada and the United Kingdom in 1956–57, the genetic dose contribution from this source in New Zealand is less than half the values obtained in these countries. The reason for this appears to be largely attributable to the age distributions in New Zealand of the total population, and of persons receiving I^{131} tests or treatments. From Table 2, the percentages of persons undergoing I^{131} tests or therapy in the under 15 and under 30 age groups were only 0.5 and 13.1, whereas it was noted in the section "Demographic Statistics", that for these two groups the contributions to the denominator of the expression for the genetic dose were 59.2% and 91.9% respectively. Differences in the genetic dose estimates could also be partly accounted for by differences in the gonad dose calculations.

The genetic dose contributions from the use of P^{32} were estimated to be 0.11 mR in Canada in 1956, and 0.71 mR in the United Kingdom in 1957. The value shown in Table 10 for P^{32} use in New Zealand in 1966 is negligible by comparison. This appears to be due very largely to a shift in the treatment of polycythemia vera over the last few years from radiation to chemotherapy.

The Significance of the Calculated Doses

The value for the annual genetic dose from the diagnostic use of radionuclides of 0.015 mrad per person, indicates that these procedures at present contribute a negligible fraction of the total average genetic dose received by the population of New Zealand. The value is, in fact, only about one-thousandth of the annual genetic dose of 14 milliroentgens per person arising from all types of diagnostic radiology (WILLIAMSON and MCEWAN, 1965). It is evident that any restriction on the numbers of diagnostic radionuclide tests and on the activities administered to patients should continue to be imposed by consideration of the radiation doses received by patients, rather than by the genetic dose burden on the population. The radiation doses received in diagnostic tests by organs of high uptake (Table 6) may be very much greater than the average dose to the whole body, and this indicates the importance of administering a blocking dose to an organ when some other organ or function is being studied using a radionuclide which may be specifically concentrated in an organ of a high uptake. Calculation showed that a normal thyroid gland weighing 25 g receives an average radiation dose of about 180 rads from a scanning dose of I^{131} as large as 150 μCi. A radiation dose of this magnitude cannot be viewed as other than very large for a diagnostic procedure. It is evident that doses of the order of several rads would be received by the unblocked thyroid gland when only a small percentage of the I^{131} of a labelled compound was free or was liberated in the body.

References

Adrian Committee Report: Radiological Hazards to Patients, Second Report of the Committee. Ministry of Health, H.M.S.O., London 1960.

Bush, F.: Brit. J. Radiol. **19**, 14 (1946).

Department of Statistics: Publication "Vital Statistics, 1965", Wellington.

Ellett, W. H., Callahan, A. B., Brownell, G. L.: Brit. J. Radiol. **37**, 45 (1964). Ibid. **38**, 541 (1965).

Focht, E. F., Quimby, E. H., Gershowitz, M.: Radiology **85**, 151 (1965).

Green, H. D.: In: Medical Physics (Glasser, O., Ed.), Vol. 2. Chicago: Year Book Publishers Inc. 1950, p. 228.

Hine, G. J., Brownell, G. L. (Eds.): Radiation Dosimetry. New York: Academic Press 1956.

I.C.R.U. Phys. Med. Biol. **2**, 107 (1957).

I.C.R.P. II. Report of Committee II on permissible dose for internal radiation. London: Pergamon Press 1959.

Johns, H. E., Taylor, R. M.: J. Can. Ass. Rad. **9**, 4 (1958).

McEwan, A. C.: Publication NRL/PDS/1967. National Radiation Laboratory, Christchurch, New Zealand.

National Radiation Laboratory: Annual Report 1966, NRL.AR/17.

Radiological Health Handbook. U.S. Department of Health, Education and Welfare, Washington, D.C. 1960.

Report of the United Nations Scientific Committee on the Effects of Atomic Radiation (UNSCEAR), General Assembly, Official Records: Thirteenth Session, Supplement No. 17 (A/38, 38), New York 1958.

Smith, E. M.: Nucleonics **24**, 33 (1966).

Sowby, F. D.: Radiation Protection Division Publication RPD-15, Department of National Health and Welfare, Ottawa 1961.

Williamson, B. D. P., McEwan, A. C.: Publication NRL/PDS/1965. National Radiation Laboratory, Christchurch, New Zealand.

Visualization of Radioactivity in the Dog Following Administration of Various ^{11}C-Carboxylates*

H. S. WINCHELL and M. B. WINSTEAD

Donner Laboratory, University of California, Berkeley, California, U.S.A.

Introduction

Prior to the general availability of ^{14}C, the positron-emitting radioisotope, ^{11}C, with a half-life of 20.4 minutes, was the principal radioisotopic tracer of carbon in chemical and biological studies [1]. Curie quantities of ^{11}C were produced and a large number of ^{11}C compounds were synthesized [1]. Unfortunately, devices for imaging the *in vivo* distribution of radioisotopes in the body were not available at that time, and the virtues of the annihilation gamma rays arising from the decay of ^{11}C were insufficient to offset the inconvenience of its short half-life in comparison to ^{14}C. The present availability of devices for imaging the distribution of positron-emitting radioisotopes within the body has led to renewed interest in ^{11}C-labeled compounds [2]. For example, we anticipate that ^{11}C-dopamine would be useful in visualization of the adrenal medulla and for the localization of metastases from neuroblastoma and melanoma [3], while similarly labeled asparagine might be useful in visualizing asparagine dependent tumors [4].

In an initial effort to develop techniques for the production and rapid incorporation of ^{11}C into organic compounds, we have synthesized a series of ^{11}C-labeled carboxylic acids and evaluated the relationship between their chemical structure and their distribution within the body. The *in vivo* distribution of these ^{11}C-carboxylates was visualized, as a function of time, utilizing a rapid imaging whole body scanner and a positron camera equipped with a tomographic attachment allowing for simultaneous visualization of the distribution of radioisotopes at various depths within the body. The results of these studies are presented in the present communication and discussed with regard to the potential clinical usefulness of these compounds.

Materials and Methods

Fig. 1 is a diagram showing the method of preparation of the ^{11}C-carboxylic acids investigated in this study. The B_2O_3 was fused on a corrugated surface set at 30° with regard to the incident beam of charged particles. When deuterons were used as the bombarding particle, the B_2O_3 that was used was 90% enriched in the ^{10}B isotope (20th Century Electronics, Ltd., New Addington, Surrey, England), and the reaction which occurred was $^{10}B(d, n)^{11}C$. When protons were used as the bombarding particle, natural B_2O_3 was used containing 80.4% of the ^{11}B isotope, and the reaction was $^{11}B(p, n)^{11}C$. When deuteron beam currents

* This work was supported under AEC contract No. W-7405-ENG-48.

greater than 10 microamps were used, rupture of the isolation or target foils (1 mil Al or 2 mil Pt) usually occurred, and, thus, deuteron beam currents were kept below 10 microamps. However, up to 35 microamperes of proton beam current could be used without rupture of the isolation or target foils. In each case, the deuteron or proton beam was maximally defocused prior to bombardment.

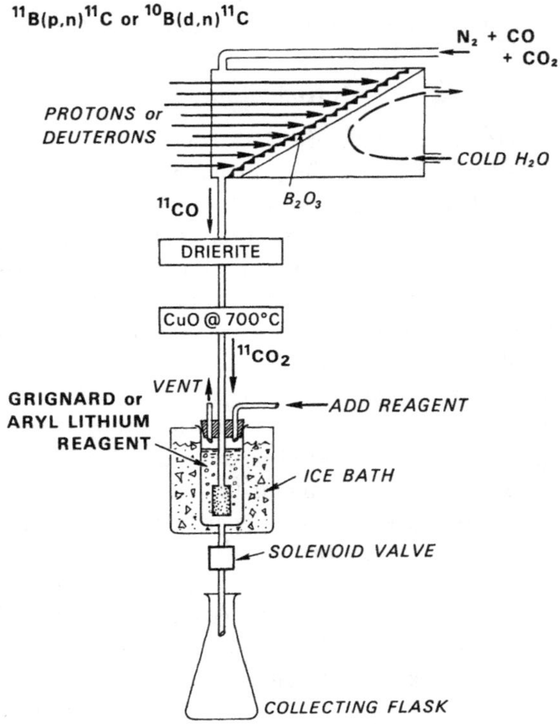

Fig. 1. Method of production of ^{11}C and its incorporation into carboxylic acids

The ^{11}C was liberated from the target by recoil primarily as ^{11}CO and $^{11}CO_2$ and was carried from the target by a gas stream composed of dry nitrogen containing 0.5% carbon monoxide and 0.5% carbon dioxide. The gas stream was passed over a column of Drierite to remove water vapor and the carbon monoxide was oxidized to carbon dioxide by passage through a 33 cm long column packed with CuO and heated at 700 °C. The gas was then cooled by passage through copper coils immersed in an iced bath. The gas flow rates were held at approximately 0.5—1 liter of gas per minute.

The stream of gas containing $^{11}CO_2$ and carrier $^{12}CO_2$ was passed for 20—30 minutes (or until a maximum activity reading was obtained with a radiation detection device mounted against the outside of the iced-water bath, see Fig. 1) into a cold solution of 5—10 mmoles of either the appropriate Grignard reagent (R(or Ar) — MgX, X = Br or Cl) or aryl lithium reagent (ArLi) dissolved in 20—25 ml of anhydrous ether. Following the carbonation of the Grignard or aryl lithium reagent, the reaction mixture was hydrolyzed by the addition of

5 ml of 6N hydrochloric acid. A stream of nitrogen gas was passed through the reaction mixture during both the acid hydrolysis and subsequent bicarbonate extraction in order to obtain adequate mixing of the reagents.

After removal of the lower hydrochloric acid layer by use of the externally operated solenoid valve, the upper ether layer containing the ^{11}C-carboxylic acid was extracted with 20—25 ml of 6 percent aqueous sodium bicarbonate. The aqueous bicarbonate layer containing the sodium ^{11}C-carboxylate was removed and heated to boiling to remove traces of ether. The aqueous solution was then transferred to a serum bottle and sterilized prior to its administration by passing through a Millipore filter.

Representative reactions for the preparation of the sodium ^{11}C-carboxylates listed in Table 1 are illustrated as follows for sodium ^{11}C-octanoate (reaction 1) and sodium 1-^{11}C-naphthoate (reaction 2):

$$\begin{aligned} &CH_3(CH_2)_5CH_2MgBr + CO_2 \longrightarrow CH_3(CH_2)_6COOMgBr \xrightarrow{HCl} \\ &CH_3(CH_2)_6COOH \xrightarrow{NaHCO_3} CH_3(CH_2)_6COONa \end{aligned} \qquad (1)$$

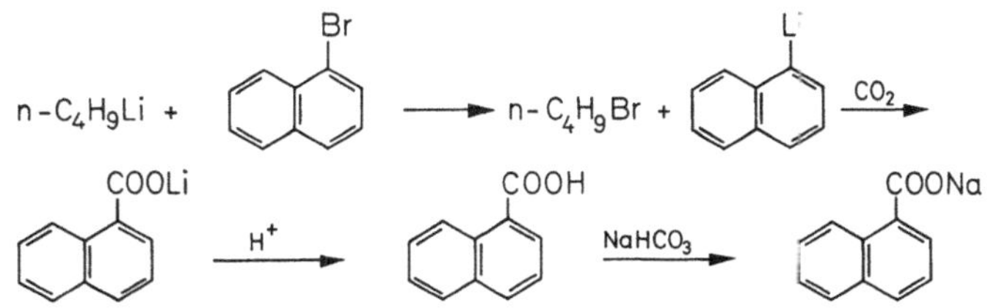

(2)

The total preparation time from the addition of hydrochloric acid to the carbonated Grignard or aryl lithium reagent to preparation of the material for administration is 15—20 minutes. Yields of up to 184 mCi of ^{11}C-carboxylates have been obtained.

Prior to the preparation of each of the ^{11}C-carboxylates listed in Table 1, experiments using ^{14}C-labeled carbon dioxide with carrier carbon dioxide demonstrated that from 50 to over 90 percent of the ^{14}C introduced into the Grignard or aryl lithium reagent was converted into the respective sodium ^{14}C-carboxylate. Essentially all of the $^{14}CO_2$ that was introduced into the system was absorbed in the solution containing the Grignard or aryl lithium reagent, as negligible quantities of radioactivity were recovered in a 2-methoxyethanol-ethanolamine trap placed in series with the system.

Prior to the ^{11}C and ^{14}C carbonation reactions, the carboxylic acids listed in Table 1 were prepared by carbonating the appropriate Grignard or aryl lithium reagent with $^{12}CO_2$ using experimental conditions similar to those described for the preparation of the radioactive carboxylic acid. The resulting carboxylic acids were isolated and purified, and their yields and melting points of the crystalline acids were ascertained as a confirmation of their respective values reported in the literature.

The distribution of ^{11}C-radioisotope within the entire body of the dog following the administration of the ^{11}C-carboxylates was obtained using the rapid imaging whole body scanner previously described by H. O. ANGER [5]. Localization of the

Table 1. *Localization of radioactivity in the dog following administration of various* ^{11}*C-Carboxylates*

Compound		Localisation of radioactivity
Name	Structure	
Acetate[a]	$CH_3—^{11}COONa$	Homogeneous whole body
9-Phenanthrenecarboxylate[b]	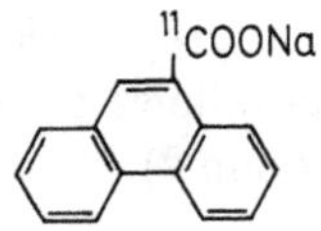	Homogeneous whole body[c]
Butyrate[a]	$CH_3(CH_2)_2\,^{11}COONa$	Entire abdomen
Octanoate[a]	$CH_3(CH_2)_6\,^{11}COONa$	Entire abdomen
Benzoate[a]	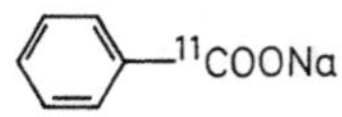	Kidneys, Bladder
p-Chlorobenzoate[a]	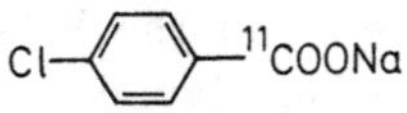	Kidneys, Bladder
2-Thiophenecarboxylate[d]	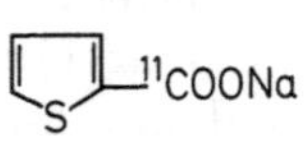	Kidneys, Bladder
Phenylacetate[a]	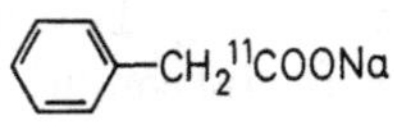	Liver, Kidneys, Bladder
3, 4-Dimethoxybenzoate[e] (Veratrate)	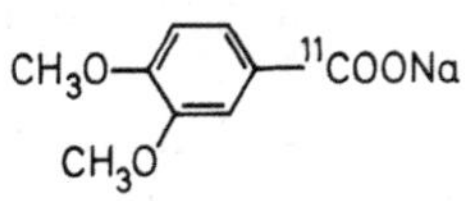	Liver, Kidneys, Bladder

Table 1 (continued)

Compound		Localisation of radioactivity
Name	Structure	
o-Hydroxybenzoate[f] (Salicylate)	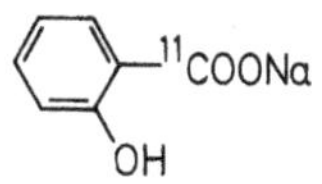	Liver, Kidneys, Bladder (Heart ?)
Cyclohexanecarboxylate[a]	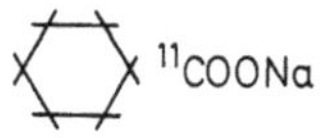	Liver, Kidneys, Bladder (Heart ?)
1-Naphthoate[g]	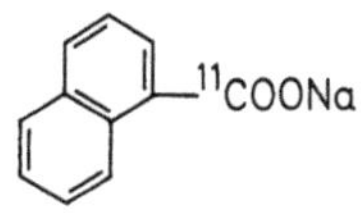	Liver, Gall bladder, Kidneys, Bladder
Pentanoate[a]	$CH_3(CH_2)_3{}^{11}COONa$	Liver, Gall bladder, Kidneys, Bladder

[a] This compound was prepared by carbonating the appropriate Grignard reagent which was obtained from either Arapahoe Chemicals, Boulder, Colorado or Alfa Inorganics, Inc., Beverly, Massachusetts.

[b] This compound was prepared by carbonating 9-phenanthryl lithium, which was prepared from 9-bromophenanthrene and n-butyl lithium [GILMAN, H., COOK, T. H.: J. Am. Chem. Soc. **62**, 2813 (1940)].

[c] During the initial 10 minutes, whole body scans were similar to those obtained with ^{131}I-albumin. After 2 hours whole body scans demonstrated an accumulation of radioactivity in the right upper quadrant of the abdomen.

[d] This compound was prepared by carbonating 2-thienylmagnesium bromide, which was prepared from 2-bromothiophene and magnesium [SHIRLEY, D. A.: Preparation of Organic Intermediates. New York: John Wiley and Sons 1951, p. 282].

[e] This compound was prepared by carbonating, at — 60 °C, 3,4-dimethoxyphenyl lithium, which was prepared from 4-bromoveratrole and n-butyl lithium [CALVIN, M., HEIDELBERGER, C., REID, J. C., TOLBERT, B. M., YANKWICH, P. F.: Isotopic Carbon. New York: John Wiley and Sons 1949, p. 183—184].

[f] This compound was prepared by carbonating the aryl lithium intermediate which was obtained from the reaction of o-bromophenol with n-butyl lithium [GILMAN, H., ARNITZER, C. E.: J. Am. Chem. Soc. **69**, 1537 (1938)].

[g] This compound was prepared by carbonating 1-naphthyl lithium, which was prepared from 1-bromonaphthalene and n-butyl lithium [GILMAN, H., MOORE, F. W.: J. Am. Chem. Soc. **62**, 1843 (1940)].

positron emitting ^{11}C within specific regions of the animal was determined utilizing the positron camera previously described [2] and modified by H. O. Anger (Donner Laboratory, University of California, Berkeley, California) to obtain tomographic visualization along six different focal planes while simultaneously obtaining standard positron scintiphotos at varying exposures.

Results

Fig. 2 presents whole body scans obtained following intravenous administration of representative compounds synthesized and studied in the present work. In the left hand portion of the figure a whole body scan obtained during the 10 to 16

(Scan following ^{131}I-albumin administration shown for comparison)

Compound	^{131}I-albumin	Acetate	p-Chloro-benzoate	Cyclohexane-carboxylate		Butyrate
Time of scan following admin. (min)	10-16	$32\frac{3}{4}$-$37\frac{1}{2}$	4-6	$5\frac{1}{4}$-6	$36\frac{1}{2}$-$42\frac{1}{2}$	$2\frac{1}{2}$-$3\frac{1}{2}$

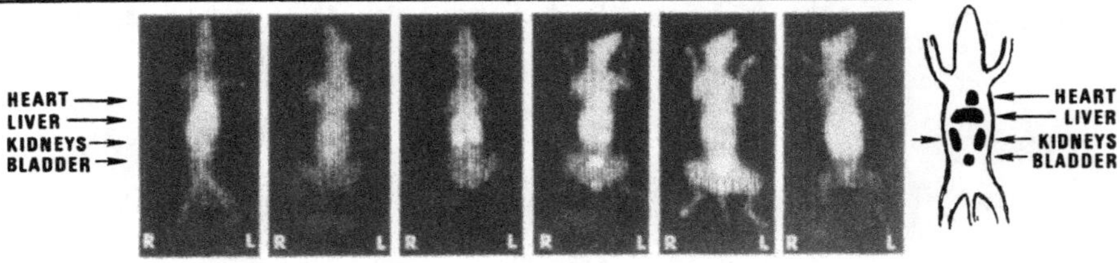

Fig. 2. Whole body scans following i. v. administration of representative ^{11}C-carboxyl labeled carboxylic acids to dogs

minute time interval after the administration of ^{131}I-albumin is presented to demonstrate the pattern of distribution of materials limited to the intravascular space. Location of the heart, liver, kidneys and bladder are indicated on the figure. On these and subsequent figures the scans represent P—A views. The scans were performed during the time interval indicated above each scan. The right and left hand sides of the body are indicated by R and L, respectively.

Homogeneous whole body distribution of ^{11}C is demonstrated by the scan obtained $32\frac{3}{4}$ to $37\frac{1}{2}$ minutes subsequent to the intravenous administration of ^{11}C-acetate. Rapid concentration and excretion by the kidney is shown for ^{11}C-labeled p-chlorobenzoate. Early accumulation of ^{11}C activity is present in the heart, liver, kidneys and bladder $5\frac{1}{4}$ to 6 minutes after intravenous administration of ^{11}C-cyclohexanecarboxylate. A much larger quantity of this material is present in the bladder $36\frac{1}{2}$ to $42\frac{1}{2}$ minutes after its administration, but there is significantly less activity seen in the region of the heart. Generalized distribution of activity in the entire abdomen without specific organ visualization is shown $2\frac{1}{2}$ to $3\frac{1}{2}$ minutes following intravenous administration of ^{11}C-butyrate.

Fig. 3 shows serial whole body scans following intravenous administration of ^{11}C-pentanoate to a dog. It can be seen that there is a rapid initial accumulation of activity in the liver with subsequent excretion of ^{11}C activity in the gall bladder. An additional small amount has been cleared apparently by the kidneys and is found in the urinary bladder.

Time of scan following admin. (min)	6½-7½	36-37½	49-60	Positron scintiphoto of ^{11}C activity in gall bladder 123-131½

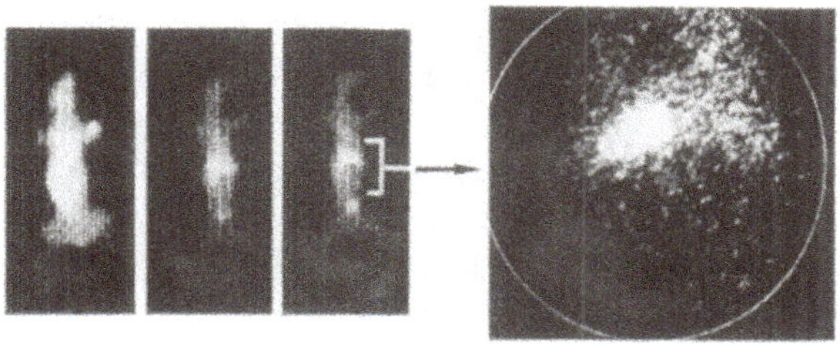

Fig. 3. In vivo temporal distribution of ^{11}C following i. v. administration of ^{11}C-Pentanoate to dogs

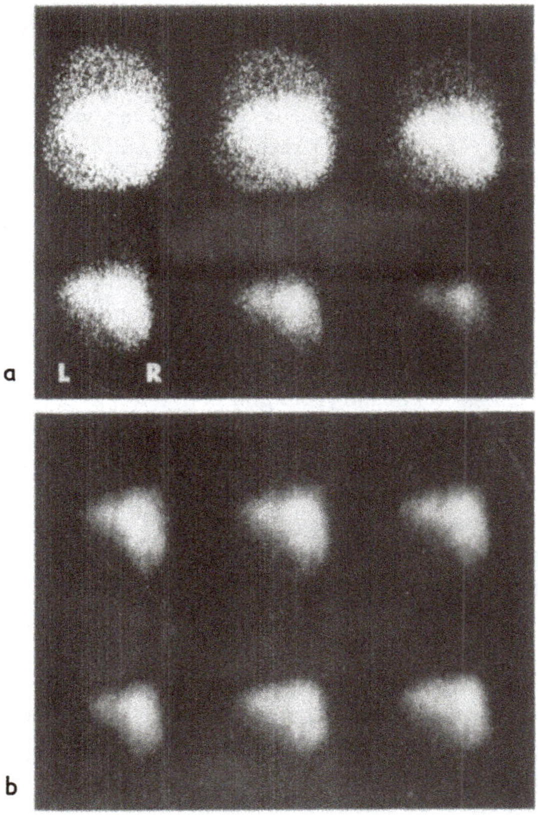

Fig. 4. Tomographic positron scintiphoto of liver 74—84$^3/_4$ min following i. v. administration of 1-^{11}C-Naphthoate to a dog.

a 6 lens representation with varying intensity. Focal plane fixed approx. 2 inches below detector head.

b Intensity fixed. 6 serial focal planes at 1, 2, 3, 4, 5 and 6 inches below detector head

The upper scan of Fig. 4 shows a six-lens representation of ^{11}C activity in the liver at varying intensities following the intravenous administration of 1-^{11}C-naphthoate to a dog. The focal plane was fixed at approximately 2 inches below the detector head and from the scintiphotos the localization of activity in the gall bladder is not obvious. The lower portion of this figure illustrates scintiphotos at a fixed intensity with 6 serial focal plane settings at 1, 2, 3, 4, 5 and 6 inches below the detector head. It can be seen from the lower portion of this figure that localization of activity in the region of the gall bladder is obvious at focal planes 4, 5 and 6 inches below the detector head.

Table 1 summarizes the localization of radioactivity within dogs subsequent to the intravenous administration of various ^{11}C-carboxylic acids. ^{11}C activity is found homogeneously throughout the whole body subsequent to the administration of ^{11}C-acetate and 9-phenanthrenecarboxylate. Subsequent to the administration of butyrate and octanoate most of the activity is found to be distributed homogeneously throughout the entire abdomen, while subsequent to the administration of benzoate, p-chlorobenzoate and 2-thiophenecarboxylate, localization of ^{11}C activity is confined primarily to the kidneys and urinary bladder. When phenylacetate or 3,4-dimethoxybenzoate are administered, significant activity is seen in the liver as well as the kidneys and bladder. After administration of o-hydroxybenzoate (salicylate) and cyclohexanecarboxylate, there appears to be a significant accumulation of activity in the region of the heart, in addition to the liver, kidneys and bladder. ^{11}C-labeled 1-naphthoate and pentanoate show some accumulation of activity in the gall bladder in addition to the liver, kidneys and urinary bladder.

Discussion

Knowledge of the spatial distribution of ^{11}C-labeled compounds within the body adds an additional dimension to our understanding of metabolism of organic compounds. The localization of butyrate and octanoate homogeneously throughout the abdomen suggest that plasma fatty acids are in more rapid equilibrium with fat located within mesentery and omentum than other fat stores throughout the body. Benzoate, p-chlorobenzoate and 2-thiophenecarboxylate are very rapidly cleared from the body by the kidney. Thus, if any significant conjugation within the liver occurs it must proceed quite rapidly, possibly during a single pass of the material through the liver, since insufficient ^{11}C accumulates in the liver at any time to allow for its visualization. The potential usefulness of ^{11}C-benzoate in delineation of renal lesions has been discussed in a previous communication [6]. On the other hand phenylacetate and 3,4-dimethoxybenzoate are concentrated by the liver as well as the kidneys suggesting that their hepatic metabolism requires a significant time. The results with o-hydroxybenzoate (salicylate) and cyclohexanecarboxylate are similar to those for the compounds just mentioned. However, they appear to have some cardiac localization soon after their administration. Whether this is indeed the case or whether the scans simply represent retention of these materials within the blood remains to be determined. If this latter situation is what is actually being observed, then the radioisotopic cardiac silhouette is merely a reflection of activity in the blood pool. In addition to being concentrated in the liver and kidneys, 1-naphthoate and pentanoate are excreted into the bile allowing for visualization of the gall bladder. The homogeneous whole body distribution of activity following administration of ^{11}C-labeled acetate or 9-phenanthrenecarboxylate suggests that either the carboxyl group of these compounds is rapidly degraded to bicarbonate or these materials rapidly distribute themselves homogeneously throughout body spaces. Simultaneous kinetic analysis of plasma ^{11}C clearance curves could have elucidated these points, but, unfortunately, were not performed in the present study.

It is noteworthy that chlorination of benzoic acid at the para position (p-chlorobenzoate) does not alter the spatial distribution of the resulting compound in comparison to benzoate. However, the addition of a carbon atom between the carboxyl group and the benzene ring (phenylacetate) results in marked accumulation of the resulting material in the liver relative to that observed with benzoate. Similar results are obtained by the addition of methoxy groups at the 3 and 4 positions of benzoate.

It is also noteworthy that the naturally occuring butyrate and octanoate are not significantly excreted by the liver or kidneys, but the odd-number carbon atom aliphatic fatty acid pentanoate, which is normally not found within the body, is concentrated in the liver and kidneys and excreted in the bile and urine.

Positron scintiphotographic tomography shows promise of being clinically useful as can be appreciated from Fig. 4, and discussions of the technique will be published in the future by Mr. H. O. ANGER. The apparent exclusion of these ^{11}C-carboxylates from the brain and their failure to concentrate in salivary glands or nasal sinuses suggests their possible usefulness in delineation of brain lesions.

Summary

Low radiation dose, high resolution tomographic positron scintiphotography and potential for isologous labeling of a large variety of organic compounds attendant to its use makes ^{11}C a potentially extremely important radioisotope in nuclear medicine. The present work with thirteen ^{11}C-carboxylates demonstrates rapid renal concentration and excretion of benzoate, p-chlorobenzoate and 2-thiophenecarboxylate suggesting their usefulness in morphological as well as functional studies of this organ. Hepatic as well as renal concentration of phenylacetate, 3,4-dimethoxybenzoate, o-hydroxybenzoate (salicylate) and cyclohexanecarboxylate suggests their use in morphological and functional studies of both these organs. (The latter two compounds appear to show some concentration in the heart as well.) In addition to accumulation and excretion of 1-naphthoate and pentanoate in the liver and kidneys, these compounds localized in the gall bladder suggesting their use in morphological and functional studies of this organ. The apparent exclusion of all of these compounds from the brain and their lack of concentration in salivary glands and nasal sinuses suggest their possible usefulness in visualization of brain lesions. The diffuse abdominal distribution of butyrate and octanoate suggest accumulation in mesenteric and omental fat, and if such is the case further studies with these materials might be useful in visualizing the kinetics of fatty acid deposition.

Acknowledgements

The authors wish to thank Dr. W. G. Myers of the Ohio State University, Columbus, Ohio for his encouragement and suggestions concerning cyclotron production of ^{11}C. They are grateful also for the assistance of Dr. R. Fawwaz and G. Armaly of the Lawrence Radiation Laboratory, University of California, Berkeley, California.

References

1. Kamen, M. D.: Short-lived Radioactive Carbon (^{11}C). In: Radioactive Tracers in Biology. Second Edition, New York: Academic Press 1951, Ch. 8, p. 224—243.
2. Anger, H. O.: Gamma-ray and Positron Scintillation Camera. Nucleonics **21,** 56 (1963).
3. Lieberman, L. M., Beierwaltes, W. H., Varma, V. M., Weinhold, P., Ling, R.: Labeled Dopomanie Concentration in Human Adrenal Medulla and in Neuroblastoma. J. Nucl. Med. **10,** 93 (1969).
4. Old, L. J., Boyse, E. A., Campbell, H. A.: L-Asparagine and Leukemia. Scient. Amer. **219,** 34 (1968).
5. Anger, H. O.: Whole Body Scanner Mark II. J. Nucl. Med. **7,** 331 (1966).
6. Winstead, M. B., Winchell, H. S., Fawwaz, R.: The Use of Sodium ^{11}C-Benzoate in Renal Visualization. Int. J. Appl. Radiat. and Isotopes (in press).

Die Bedeutung des Zeitfaktors für die Beurteilung von Strahlendosen in der Nuklearmedizin

Significance of the Time Factor in the Assessment of Radiation Doses in Nuclear Medicine

P. Pfannenstiel

Abteilung für klinische Nuklearmedizin der Medizinischen Universitätsklinik Freiburg i. Br.

Summary

From assessments of radiation doses after incorporation of radionuclides, internal radiation at low doserates appears to be significantly less effective than fractionated external irradiation at high doses of equal physical magnitude. The difference is related to the "time factor", i.e. there is more opportunity for tissue recovery during the longer irradiation period.

An evaluation of 53 patients treated with ^{131}I for toxic and nontoxic goiter, thyroid cancer and malignant lymphoma revealed that thyrotoxic patients received some three to six times greater total-body irradiation doses than patients with nontoxic goiters or thyroid cancer. When the thyroid is overactive, the greater portion of the dosage to the blood comes from the slow-release protein-bound ^{131}I phase. When an initial high dose is given in thyroid cancer therapy, an increase of reticulocytes has definitely been established. In contrast, patients receiving a low total-body ^{131}I irradiation delivered over a longer period have shown no hematological response. In accordance with these observations, the author emphasizes the important biological aspects for dosimetry in diagnostic nuclear medicine procedures.

Jede nuklearmedizinische Untersuchung ist mit einer Strahlenbelastung verbunden. In zahlreichen Publikationen wurde die Dosis für das kritische Organ bzw. den Ganzkörper bei den heute üblichen diagnostischen Radionuklidanwendungen zusammengestellt. Dabei wird die Strahlenbelastung immer wieder mit den bei der Röntgendiagnostik in Kauf genommenen Strahlenbelastungen verglichen.

Anhand einer Analyse der bei der ^{131}J-Therapie von Schilddrüsen- und Lymphknotenerkrankungen am Herd bzw. im Ganzkörper zur Wirkung kommenden Strahlendosen ist uns aufgefallen, daß die Herddosen nach interner Strahlentherapie um das 2–3fache über denen liegen, die bei der äußeren Bestrahlung des Menschen den gleichen biologischen Effekt haben.

So ist bekannt, daß zur Normalisierung des Stoffwechsels einer hyperthyreoten Schilddrüse eine Herddosis von etwa 3000–4000 rad genügt (Zimmer 1944) und daß bei der malignen Lymphogranulomatose als Herdvernichtungsdosen ebenfalls etwa 4000 rad ausreichen (Musshoff, Stamm, Lummel u. Gössel 1964).

Bei der ^{131}J-Therapie müssen dagegen in einer hyperthyreoten Schilddrüse 6000 bis 8000 rad und in malignen Lymphomen bis zu 400 000 rad zur Wirkung kommen, um den gleichen strahlentherapeutischen Effekt zu erzielen.

Für diesen Unterschied zwischen äußerer und innerer Bestrahlung wurde in der Literatur immer wieder die inhomogene Dosisverteilung nach der Inkorporation β-Strahlen aussendender Radionuklide diskutiert, deren Reichweite im Gewebe nur wenige mm, für ^{131}J z.B. maximal nur 2,2 mm beträgt (ODDIE 1951, MYANT 1964).

Nachdem ANSPAUGH (1965) überzeugend nachweisen konnte, daß dieser Faktor infolge der Reichweite der β-Partikel bei den kleinen Gewebsstrukturen nur eine untergeordnete Rolle spielt, stießen wir bei weiterer Analyse des unterschiedlichen Dosiseffektes auf ein Phänomen, das seit den Anfängen der perkutanen Strahlentherapie diskutiert wird (LIECHTY 1929), aber unseres Wissens in der klinischen Nuklearmedizin bisher keine Beachtung fand, das Phänomen des „Zeitfaktors".

Die Bedeutung des Zeitfaktors soll an einem Schema veranschaulicht werden.

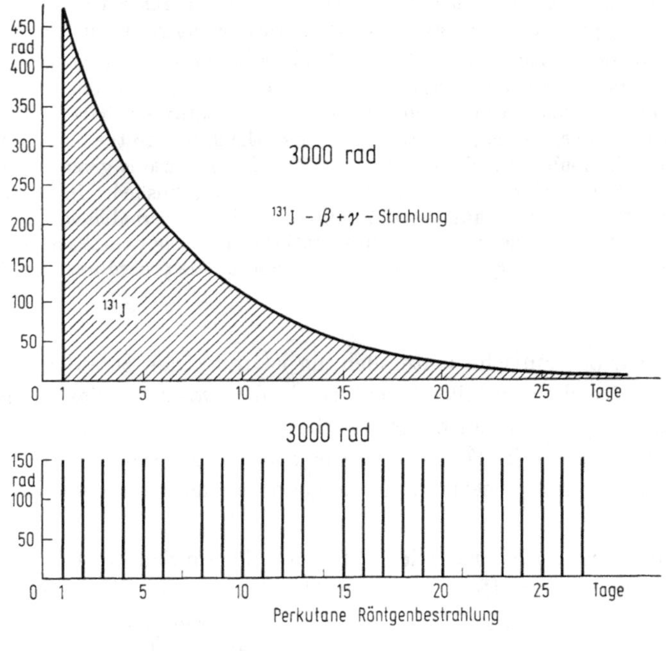

Abb. 1. Schema der Dosisleistung bei interner und externer Strahlentherapie der Schilddrüse

In Abb. 1 ist die Dosisleistung bei interner ^{131}J-Bestrahlung derjenigen bei externer Röntgenbestrahlung gegenübergestellt. Es wird angenommen, daß eine hyperthyreote Struma, deren Gewicht 100 g betragen soll, sowohl bei der inneren, als auch bei der äußeren Bestrahlung eine Herddosis von 3 000 rad akkumuliert.

Nach der für die ^{131}J-Dosimetrie u. E. gültigen Formel[1]

$$D_{\beta+\gamma} = 15 \cdot \frac{\text{max. thyr. } ^{131}\text{J-Aufnahme } (\mu\text{Ci})}{\text{Schilddrüsengewicht (g)}} \cdot T_{\text{eff}}$$

wäre für die ^{131}J-Therapie bei einer maximalen thyreoidalen Radiojodaufnahme von etwa 70% der ^{131}J-Dosis und einer angenommenen effektiven Halbwertszeit des ^{131}J in der Schilddrüse von etwa 4 Tagen eine ^{131}J-Dosis von 7 mCi erforderlich.

Die ^{131}J-Aktivität in der Schilddrüse nimmt exponentiell ab. Die Dosisleistung ist kurz nach der ^{131}J-Gabe hoch, liegt jedoch bereits in der zweiten Woche unter 150 rad. Im unteren Teil der Abbildung sind die Verhältnisse bei einer über 4 Wochen fraktioniert verabreichten perkutanen Strahlenbehandlung schematisch dargestellt. Die Dosisleistung ist während des gesamten Beobachtungszeitraumes mit 50 R/min gleich anzunehmen, d.h. die tägliche Herddosis von 150 rad wird an je 5 Behandlungstagen der Woche in jeweils 3 Minuten akkumuliert.

Nach HUG (1966) kann der Dosiseffekt zwischen einer stark protrahierten Bestrahlung und einer Kurzzeitbestrahlung mit hoher Dosisleistung um den Faktor 2 bis 10 variieren. Würde man diese Beobachtung aus der Strahlenbiologie auf das hier gewählte Beispiel übertragen, dann müßte die Herddosis nach ^{131}J-Therapie mindestens doppelt so hoch sein, um den gleichen Dosiseffekt wie nach perkutaner Bestrahlung zu erreichen.

In diese Überlegung läßt sich die von BERMAN, RALL u. HESLIN (1957) sowie von GREIG, CROOKS u. MCGREGOR (1966) gemachte Beobachtung eines Wirkungsverhältnisses von etwa 2 : 1 zwischen perkutaner Strahlentherapie und interner ^{131}J-Bestrahlung der Schilddrüse einordnen. Es erklärt sich auch, warum bei der ^{131}J-Therapie die empirisch ermittelte Energiedosis etwa doppelt so hoch liegt wie die in der externen Röntgentherapie benötigte.

Die Beobachtungen, daß Strahlendosen physikalisch gleicher Größenordnung einen biologisch unterschiedlichen Effekt haben, wenn sie vom Gewebe in unterschiedlichen Zeiträumen akkumuliert werden, lassen u. E. die unvermeidliche Strahlenbelastung in der Nuklearmedizin in einem anderen Licht erscheinen.

Bei Untersuchungen über die Strahlenbelastung des Ganzkörpers nach therapeutischer ^{131}J-Inkorporation beobachteten wir, daß die vom Ganzkörper aufgrund der ^{131}J-Blutradioaktivität akkumulierte Ganzkörperstrahlenbelastung entscheidend von der Schilddrüsenfunktion abhängt. Die Ganzkörperstrahlenbelastung wurde aufgrund wiederholter Blutradioaktivitätsmessungen nach dem Verfahren von SEIDLIN, YALOW u. SIEGEL (1952) aus der zwischen zwei Meßpunkten liegenden jeweiligen Dosis bestimmt nach der Formel[2]

$$D_{\text{Ganzkörper}} = 0{,}33 \cdot \frac{\text{mittlere } ^{131}\text{J-Konzentration}}{\text{ml Vollblut}} \text{ rad/h.}$$

[1] Der Umrechnungsfaktor 15 ergibt sich aus der Multiplikation des Faktors 73,8, der die physikalischen Konstanten zusammenfaßt, mit der mittleren pro Zerfall absorbierten Energie von $\bar{E}_{\text{abs}} = 0{,}203$ MeV. $\bar{E}_{\text{abs}}$ setzt sich zusammen aus der vollständig absorbierten mittleren β-Energie von $\bar{E}_\beta = 0{,}187$ MeV und einem Anteil für die Absorption der γ-Strahlung, der mit 0,016 MeV etwa 10% der mittleren β-Energie entspricht (LOEVINGER, HOLT u. HINE 1956).

[2] Der Umrechnungsfaktor 0,33 ergibt sich aus der Zusammenfassung der physikalischen Konstanten (2,13) und der mittleren im Ganzkörper von der Blutradioaktivität absorbierten β- und γ-Energie des ^{131}J von 0,154 MeV (QUIMBY u. FEITELBERG 1963).

Die bei 53 Patienten bestimmten mittleren Ganzkörperstrahlendosen sind in Tab. 1 zusammengefaßt.

Tabelle 1. *Mittlere Ganzkörperstrahlenbelastung*

	Hyperthyreote Strumen ($n = 8$)	Euthyreote Strumen ($n = 13$)	Maligne Strumen ($n = 10$)	Maligne Lymphome ($n = 22$)
appl. mCi ^{131}J	6,9	16,1	100	33
rad Ganzkörper	6,29	4,33	11,45	0,42
rad Ganzkörper pro mCi ^{131}J	0,84	0,28	0,15	0,013

Die Ergebnisse zeigen, daß die Ganzkörperstrahlenbelastung bei malignen Strumen mit 11,5 rad am höchsten, jedoch pro verabreichtes mCi am niedrigsten liegt. Die vom Patienten mit hyperthyreoter Struma mit 0,84 rad pro verabreichtes mCi ^{131}J während 5 Wochen akkumulierte Ganzkörperstrahlendosis liegt aufgrund einer Varianzanalyse[1] hochsignifikant sowohl über derjenigen bei euthyreoten als auch über derjenigen bei malignen Strumen.

Analysiert man ähnlich wie bei der Herddosis die mittlere Dosisleistung für die verschiedenen Patientengruppen in Abhängigkeit von der Zeit nach der ^{131}J-Therapie, so ergeben sich die in Abb. 2 dargestellten Verhältnisse:

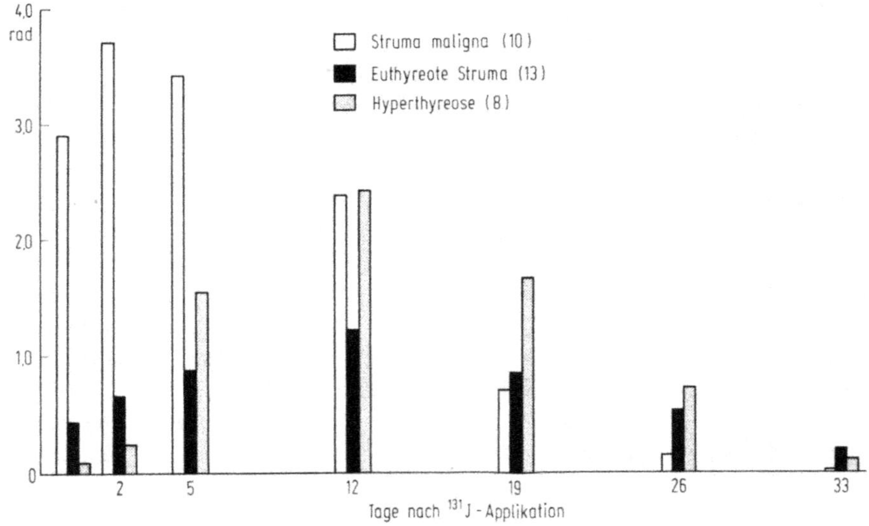

Abb. 2. Mittlere Dosis bei der Ganzkörperstrahlenbelastung nach ^{131}J-Therapie

---

[1] Lineares Modell der Varianzanalyse:

$$Y_{ij} = \mu + \alpha_i + \varepsilon_{ij}.$$

Hierbei bedeutet μ = allgemeiner Mittelwert, α_i = Patientengruppe (fester Faktor), ε_{ij} = Versuchsfehler. Es ergibt sich $MQ_\alpha = 1{,}15^{***}$ und $MQ_\varepsilon = 0{,}062$. Der lineare Vergleich der Gruppen untereinander ergibt nach dem Prüfmaß von SCHEFFÉ bei 28 Freiheitsgraden beim Vergleich hyperthyreoter Strumen mit euthyreoten Strumen 26,6*** und beim Vergleich hyperthyreoter Strumen mit malignen Strumen 33,6***.

Während Patienten mit ^{131}J-anreichernder Struma maligna infolge geringer oder ganz fehlender ^{131}J-Hormonproduktion die gesamte Ganzkörperstrahlendosis in der ersten Woche während der anorganischen Jodstoffwechselphase akkumulieren, findet sich bei den hyperthyreoten Strumen der Gipfel der Dosisleistung in der zweiten Woche nach ^{131}J-Therapie infolge der Abgabe größerer Mengen ^{131}J-markierter Schilddrüsenhormone in die Blutbahn. Patienten mit euthyreoter Struma akkumulieren wegen des langsamen intrathyreoidalen ^{131}J-Umsatzes die Ganzkörperstrahlendosis nur allmählich bei einer fast gleichbleibenden Dosisleistung.

Bei hämatologischen Verlaufsbeobachtungen konnten wir lediglich bei malignen Strumen, die die Ganzkörperstrahlenbelastung innerhalb einer Woche fast vollständig akkumulierten, den in Abb. 3 dargestellten, aufgrund einer Varianzanalyse signifikanten Anstieg der Retikulozyten feststellen.

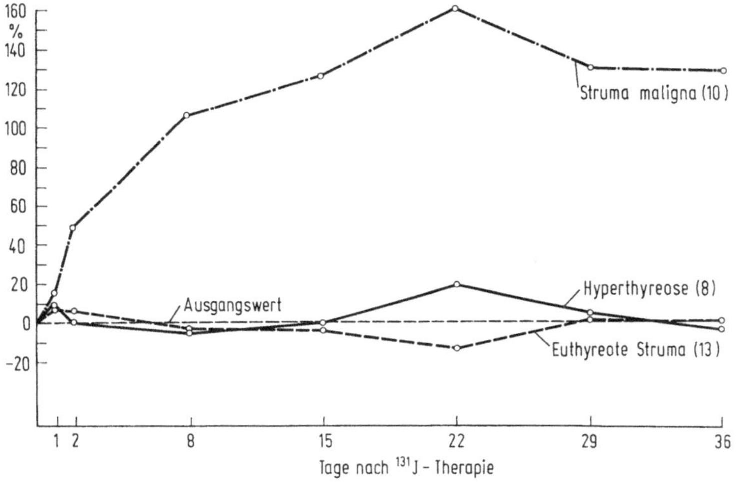

Abb. 3. Verhalten der mittleren prozentualen Retikulozytenzahlen nach ^{131}J-Therapie

Da Dienstbier, Arient u. Pospisil (1967) eine ähnliche Beobachtung im Tierexperiment machten, kann es sich hierbei um eine möglicherweise strahleninduzierte Verminderung der zentralen Erythrozytenpopulation handeln.

Daß wir bei geringeren und über einen längeren Zeitraum einwirkenden Ganzkörperstrahlendosen keine sicheren hämatologischen Veränderungen im peripheren Blut fanden (Simon, Pfannenstiel, Heimpel u. Hoffmann 1968; Klemm, Pfannenstiel, Weissleder u. Hoffmann 1968) kann damit erklärt werden, daß eine kontinuierliche Ganzkörperbestrahlung mit einer geringen Dosisleistung vom Knochenmark durch Umstellung auf ein neues Gleichgewicht rasch kompensiert wird (Lamerton, Pontifex, Blackett u. Adams 1960; Lajtha u. Oliver 1962).

Hieraus läßt sich ableiten, daß trotz aller Bemühungen, eine physikalisch exakte Strahlendosimetrie für die Nuklearmedizin zu entwickeln, biologische Faktoren einen bedeutenden Einfluß auf das Ausmaß des Strahlenschadens haben können. Das Phänomen der unterschiedlichen Dosisleistung sollte in der Strahlenschutzgesetzgebung und in der Abschätzung der bei diagnostischen Radionuklidanwendungen entstehenden lokalen bzw. generellen Strahlenbelastung des Patienten u. E. mehr Beachtung finden. Eine nach innerer Bestrahlung ermittelte rad-Zahl ist nicht unmittelbar mit den rad-Zahlen nach äußerer Bestrahlung vergleichbar.

Literatur

Anspaugh, L. R.: Special problems of thyroid dosimetry — considerations of ^{131}I dose as a function on gross size and inhomogenous distribution. USAEC—UCRL—12492, Lawrence Radiation Laboratory, 25. 3. 1965.

Berman, M., Rall, J. E., Heslin, J.: Some physical considerations governing the choice of internally administered radioisotopes for therapy. Phys. Med. Biol. **1**, 243 (1957).

Dienstbier, Z., Arient, M., Pospisil, J.: Post-irradiation changes in the bone marrow and peripheral blood. In: Effects of ionizing radiations on the hemapoetic tissue. Proc. IAEA, Symp., Wien 1967, S. 60.

Greig, W. R., Crooks, J., McGregor, A. G.: Clinical and radiobiological consequences of therapeutic irradiation. Proc. Roy. Soc. Med. **59**, 599 (1966).

Hug, O.: Zur Analyse der Dosiswirkungsbeziehungen in der Strahlenbiologie. Strahlenschutz in Forschung und Praxis **6**, 23 (1966).

Klemm, D., Pfannenstiel, P., Weissleder, H., Hoffmann, G.: Auswirkungen der endolymphatischen ^{131}J-Therapie auf Proteine und zelluläre Bestandteile des menschlichen Blutes. Fortschr. Röntgenstr. **109**, 588 (1968).

Lajtha, L. G., Oliver, R.: Cell population kinetics following different regimes of irradiation. Brit. J. Radiol. **35**, 131 (1962).

Lamerton, L. F., Pontifex, A. H., Blackett, N. M., Adams, K.: Effect of protracted irradiation on the bloodforming organs of the rat. Part I: Continuous exposure. Brit. J. Radiol. **33**, 287 (1960).

Liechty, A.: Über den Zeitfaktor der biologischen Strahlenwirkung. Strahlentherapie **33**, 1 (1929).

Loevinger, R., Holt, J. G., Hine, G. J.: Radiation dosimetry of internally administered isotopes. In Hine, G. J., Brownell, G. L.: Radiation dosimetry. New York: Academic Press 1956, S. 801.

Musshoff, K., Stamm, K., Lummel, G., Gössel, K.: Zur Prognose der Lymphogranulomatose. Klinisches Bild und Strahlentherapie. Freiburger Krankengut 1938—1958. In Keiderling, W.: Beiträge zur inneren Medizin. Stuttgart: F. K. Schattauer 1964, S. 549.

Myant, N. B.: The problem of dosage in the treatment of thyrotoxicosis. Minerva nucl. **8**, 87 (1964).

Oddie, T. H.: Dosage from radioisotopes uniformly distributed within a sphere. Brit. J. Radiol. **24**, 333 (1951).

Quimby, E. H., Feitelberg, S.: Radioactive isotopes in medicine and biology. Philadelphia: Lea and Febiger 1963.

Seidlin, S. M., Yalow, A. A., Siegel, A.: Blood radioiodine concentration and blood radiation dosage during ^{131}I therapy for metastatic thyroid cancer. J. Clin. Endocr. **12**, 1197 (1952).

Simon, A., Pfannenstiel, P., Heimpel, H., Hoffmann, G.: Hämatologische Verlaufskontrollen in Beziehung zur Ganzkörperdosis nach ^{131}J-Inkorporation. Strahlenschutz in Klinik und Forschung **8**, 209 (1968).

Zimmer, E. A.: Zur Strahlentherapie der Hyperthyreosen. Strahlentherapie **133**, 481 (1947).

Die Bestimmung des Erythrozyteneisenumsatzes mit Hilfe ^{51}Cr-markierter Erythrozyten

Calculation of Iron Turnover in Erythrocytes by Means of ^{51}Cr Labelling

U. Wellner

Nuklearmedizinische Abteilung der Universitätskliniken Köln

Summary

The normal lifespan of human erythrocytes is 100—120 days. A "hemolytic index" is defined, having a value of zero when all erythrocytes in the blood attain the age of 110 days. All earlier losses of potential erythrocyte populations are treated as hemolytic losses. Labelling red cells with ^{51}Cr enables the hemolytic index γ to be calculated from the apparent erythrocyte half-life by an equation which includes allowance for "early loss" and ^{51}Cr elution after application. The overall exchange rate of red cells can be calculated from γ and the blood volume, also determined by chromium labelling. It is then only necessary to determine the hemoglobin content of the blood in order to calculate the iron turnover in erythrocytes in mg/day.

Die Alterspyramide einer normalen Erythrozytenpopulation P_0 im Blut des Menschen ist zu jedem Zeitpunkt näherungsweise eine Kastenfunktion.

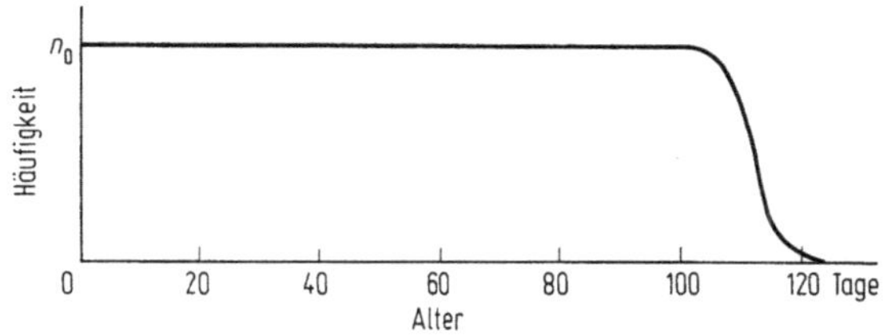

Abb. 1. Alterspyramide von Erythrozyten, die zur Population P_0 gehören

Werden Mitglieder einer solchen Population zum Zeitpunkt $t = 0$ mit einem Indikator versehen, dann läßt sich deren Verschwinden aus dem Blut verfolgen.

Nach der Agglutinationsmethode von Ashby wurde ein zeitproportionaler Abbau von Erythrozyten gemäß dem Gesetz

$$N = N_0 \left(1 - \frac{t}{L}\right) \tag{1}$$

bei Normalpersonen nachgewiesen [1, 2].

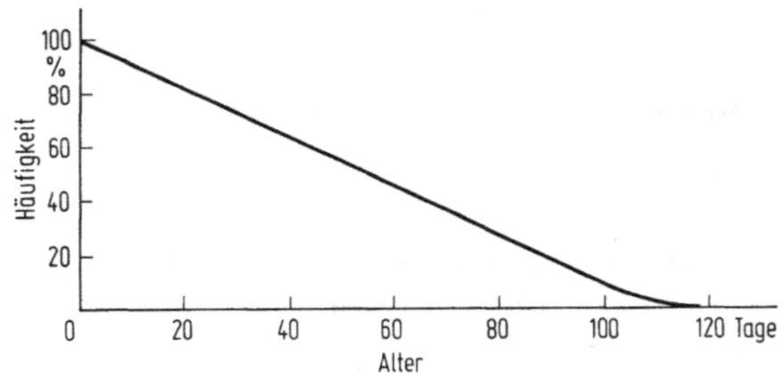

Abb. 2. Anteil der nach der Zeit t überlebenden Mitglieder der zur Zeit $t = 0$ markierten Population P_0

Dabei ergeben sich Zahlenwerte für L, die zwischen 100 und 120 Tagen schwanken.

Erythrozytenpopulationen, die von P_0 abweichen, werden in der Folge P genannt.

P kann durch zwei Parameter beschrieben werden:

1. durch die Lebensdauer L, die verkürzt sein kann,

2. durch eine Hämolysekennzahl, γ Tag^{-1}, die die tägliche relative Eliminationsrate von Erythrozyten aus der Gesamtpopulation P und aus jeder zu P gehörenden Teilpopulation p angibt.

Die Zahl der im Blut vorhandenen Erythrozyten ändert sich nur sehr langsam. Sie soll für die hier vorgenommenen Untersuchungen als $N = \text{const}$ angesetzt werden. Unter dieser Voraussetzung ist die Erythrozytenerzeugungsrate gleich der Erythrozytenvernichtungsrate n_0.

Nach diesen Vorbemerkungen kann die Alterspyramide der Population P, die aus N Erythrozyten besteht, angegeben werden:

$$n(t) = n_0 e^{-\gamma t} \text{ für } 0 \leq t \leq L. \tag{2}$$

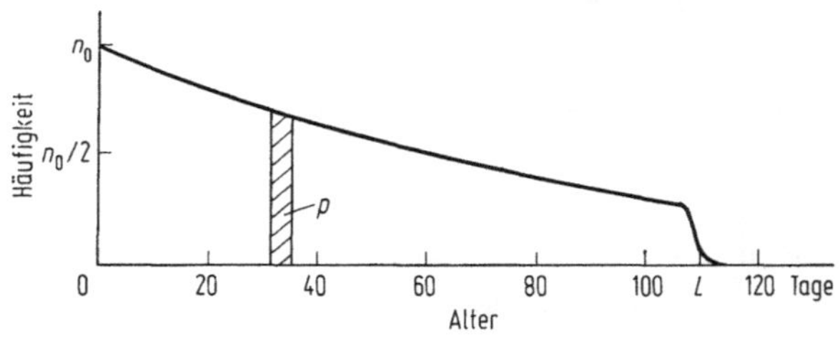

Abb. 3. Population P mit Teilpopulation p und Erythrozytenerzeugungsrate n_0

Die Zahl N der zu P gehörenden Erythrozyten ist dann:

$$N = \int_0^L n_0 e^{-\gamma t} dt = \frac{n_0}{\gamma}(1 - e^{-\gamma L}). \tag{3}$$

Das Verschwinden der Population P aus dem Blut kann durch Zählung beobachtet werden, wenn die zugehörigen Erythrozyten mit einer geeigneten Indikatorsubstanz markiert werden. Bei idealer Markierung wird die Änderung der Zahl der markierten Erythrozyten im Blut durch die folgende Beziehung beschrieben:

$$\frac{dN}{dt} = -\gamma N - \gamma N_0 \frac{e^{-\gamma L}}{1 - e^{-\gamma L}} \tag{4}$$

mit N_0 = Anzahl der ursprünglich markierten Erythrozyten.

Als Ergebnis des Zählversuchs wird die Lösung der obigen Differentialgleichung, die auch von DORNHORST [3] gefunden wurde, erwartet. —

$$N(t) = N_0 \frac{e^{-\gamma t} - e^{-\gamma L}}{1 - e^{-\gamma L}}. \tag{5}$$

Für den Fall $\gamma = 0$ findet man:

$$\lim_{\gamma \to 0} N(t) = N_0 \left(1 - \frac{t}{L}\right). \tag{1}$$

Diese Lösung enthält als Sonderfall das Verhalten der normalen Erythrozytenpopulation P_0.

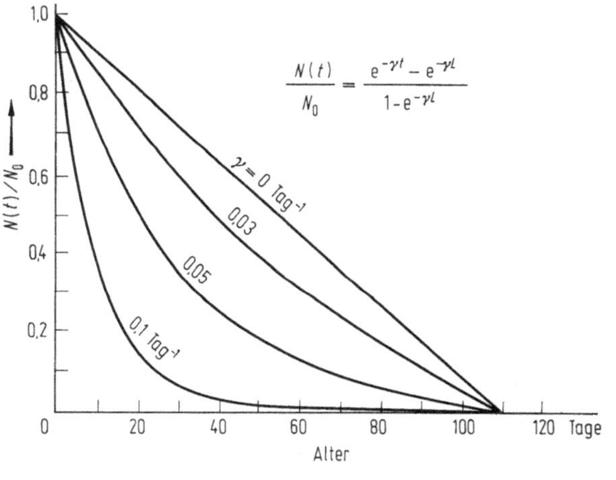

Abb. 4. Ergebnis des Zählversuchs an Populationen von markierten Erythrozyten mit einer Lebenserwartung $L = 110$ Tage und der $100\,\gamma$-prozentigen täglichen Eliminationsrate als Scharparameter

Das Radioisotop ^{51}Cr ist ein geeigneter Indikator, mit dem die Erythrozyten markiert werden können [4, 5]. Es ist kein idealer Indikator, da der „Frühe Verlust" eintreten kann und da wegen der geringen Chromkonzentration im menschlichen Körper eine Auswaschung von ^{51}Cr aus den Erythrozyten stattfindet [6].

Beide Effekte bewirken, daß das ^{51}Cr schneller aus dem Blut verschwindet als die Erythrozyten der markierten Population P. Der „Frühe Verlust" der ^{51}Cr-

Markierung von Erythrozyten ist durchweg nach zwei bis fünf Tagen abgeschlossen. Die ^{51}Cr-Auswaschung wird durch eine Exponentialfunktion mit einer Halbwertszeit von 64 Tagen beschrieben.

Durchführung und Auswertung des Zählversuchs

Die Erythrozyten aus 12 ml einer patienteneigenen Blutprobe werden mit 50 μCi $Na_2{}^{51}CrO_4$ versetzt. Nach einer Inkubationszeit von 30 Minuten werden die Erythrozyten in einer Waschflüssigkeit (3 Teile Serum und 97 Teile phys. NaCl) gewaschen, zentrifugiert und wieder mit frischer Waschflüssigkeit versetzt. Die markierten Erythrozyten befinden sind dann in einem Volumen von 15 bis 20 ml. Ein ml dieser Erythrozytenaufschwemmung wird als Ursprung für ein Standardpräparat verwendet. Das übrige Volumen wird dem Patienten reinjiziert. Der Zeitpunkt der Injektion gilt für den Zählversuch als Zeitnullpunkt $t = 0$.

10 Minuten sowie 1, 3, 6, 10 und 15 Tage nach der Injektion werden dem Patienten Blutproben entnommen, die alle zusammen mit dem Standardpräparat in einem Strahlungsmeßplatz mit Probenwechsler auf ^{51}Cr-Aktivität untersucht werden.

Die Probe, die 10 Minuten nach der Injektion dem Patienten entnommen wurde, dient der Bestimmung des Blut-, Plasma- und Erythrozytenvolumens.

Das Verschwinden der Aktivität aus dem Blut wird an den Proben vom 1. bis 15. Tag nach der Injektion beobachtet.

Zur Grundlage der Auswertung wird die Bestimmung der „Scheinbaren Erythrozytenhalbwertszeit" $T_{1/2}$ gemacht. $T_{1/2}$ hat keine reale physiologische Bedeutung, da das Verschwinden des ^{51}Cr nicht durch eine einfache Exponentialfunktion beherrscht wird, wie schon die Gl. (1) und (5) zeigten. Die Abweichung der Zählergebnisse von einem exponentiellen Verlauf würde offensichtlich, wenn länger als 15 Tage gemessen werden würde.

Als Normalwert findet man bei diesen Untersuchungen $T_{1/2} = 29$ Tage.

Diese effektive Halbwertszeit beruht auf den drei gleichzeitig wirksamen Vorgängen:

1. Erythrozytenabbau durch Überschreiten der natürlichen Altersgrenze $L = 110$ Tage und die Einwirkung der Hämolyse,
2. Früher Verlust der ^{51}Cr-Markierung an den Erythrozyten,
3. ^{51}Cr-Auswaschung aus den Erythrozyten.

Die Vorgänge 2. und 3. verlaufen unabhängig von dem Zustand der markierten Erythrozytenpopulation P. Sie bewirken einen Chromabbau, der mit einer schädlichen Halbwertszeit $T_{1/2\,S}$ von 49,4 Tagen vor sich geht.

Die von den erwarteten Meßergebnissen bei der Population P in der Zeit zwischen dem ersten und fünfzehnten Tag abgeleitete Halbwertszeit hängt von γ ab und ist für $\gamma = 0$: 70,3 Tage.

Tab. 1 zeigt den Zusammenhang zwischen der extrapolierten Halbwertszeit $T_{1/2\,E}$ als Funktion von γ. Die hier angegebenen Halbwertszeiten wurden aus den zu den Entnahmezeiten berechneten Erwartungswerten (Gl. (5)) aus der logarithmischen Regression berechnet, bei der die Erythrozytenlebensdauer $L = 110$ Tage gesetzt und γ variiert wurde.

Tabelle 1. *Zusammenhang zwischen der extrapolierten Halbwertszeit bei ideal markierten Erythrozythen der Population P* (L = 110 *Tage*; γ)

γ Tag^{-1}	$T_{1/2\,E}$ Tage
0,00	70,3
0,025	21,3
0,05	12,05
0,075	8,8

Da bei gleichzeitig verlaufenden Vorgängen mit den Einzelhalbwertszeiten $T_{1/2\,E}$ und $T_{1/2\,S}$ die effektive Halbwertszeit $T_{1/2}$ nach der Formel

$$\frac{1}{T_{1/2}} = \frac{1}{T_{1/2\,E}} + \frac{1}{T_{1/2\,S}} \tag{6}$$

berechnet wird, ist der Zusammenhang zwischen der scheinbaren Erythrozytenhalbwertszeit $T_{1/2}$ und der Hämolysekennzahl γ durch die Gleichung

$$\gamma = \frac{0{,}754}{T_{1/2}} - 0{,}026 \text{ Tag}^{-1} \tag{7}$$

gegeben.

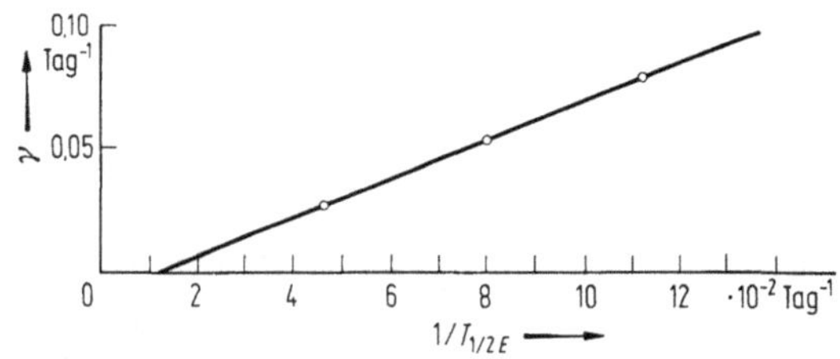

Abb. 5. Zusammenhang zwischen dem Erwartungswert der Erythrozytenhalbwertzeit $T_{1/2E}$ und der Hämolysekennzahl γ

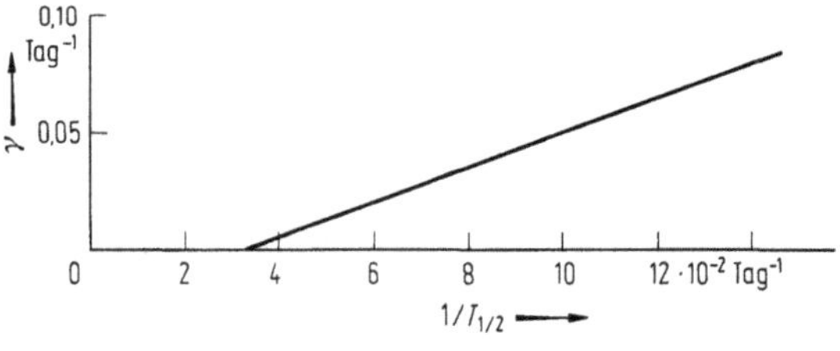

Abb. 6. Zusammenhang zwischen der scheinbaren Erythrozytenhalbwertzeit $T_{1/2}$ und der Hämolysekennzahl γ

Mit Hilfe von γ läßt sich nun leicht die Zeit berechnen, nach der die Gesamtzahl N der Erythrozyten aus dem Blut ausgeschieden ist.

Die Gesamtzahl der im Blut vorhandenen Erythrozyten ist durch Gl. (4) gegeben:

$$N = \frac{n_0}{\gamma}\,(1 - e^{-\gamma L}). \tag{4}$$

Wegen des Kontinuitätsprinzips war die Erzeugungsrate gleich der Erythrozytenvernichtungsrate n_0. Die Erythrozytenaustauschzeit T_E wird derart definiert, daß

$$n_0 T_E = N \tag{8}$$

gilt. Damit ergibt sich sofort

$$T_E = \frac{1 - e^{-\gamma L}}{\gamma}. \tag{9}$$

Da mit der Bestimmung der scheinbaren Erythrozytenhalbwertszeit zugleich das Blutvolumen V_B bestimmt werden kann, ist das Gesamterythrozyteneisen entsprechend der Gleichung

$$\text{GEE} = \text{Hgb} \cdot 3{,}4 \cdot V_B \cdot 10^{-2}\ \text{mg}, \tag{10}$$

mit dem Meßwert für den Hämoglobingehalt des Blutes Hgb g/100 ml, berechenbar.

Zusammenfassung

In der vorgelegten Arbeit wird eine Hämolysekennzahl definiert, die den Wert $\gamma = 0$ hat, wenn die Erythrozyten im Blut des Menschen alle das Lebensalter $L = 110$ Tage erreichen.

Die Verluste aller möglichen Erythrozytenpopulationen, die vor Erreichen dieser Altersgrenze eintreten, werden wie hämolytische Verluste behandelt.

Die ^{51}Cr-Markierung von Erythrozyten bietet die Möglichkeit, die Hämolysekennzahl γ aus der scheinbaren Erythrozytenhalbwertszeit $T_{1/2}$ mit Hilfe der Gleichung

$$\gamma = \frac{0{,}754}{T_{1/2}} - 0{,}026 \quad \text{Tag}^{-1}$$

zu bestimmen. Die Einflüsse des „Frühen Verlustes" und der ^{51}Cr-Auswaschung sind dabei berücksichtigt.

Aus γ kann die Gesamterythrozytenaustauschzeit nach der Formel

$$T_E = \frac{1 - e^{-110\gamma}}{\gamma}\ \text{Tage}$$

berechnet werden. Für $\gamma = 0$ ist $T_E = 110$ Tage.

Damit ergibt sich aus dem Blutvolumen V_B, das ebenfalls mit Hilfe der Erythrozytenchromierung bestimmt werden kann, und dem Hämoglobingehalt des Blutes in g/100 ml der tägliche Erythrozyteneisenumsatz zu

$$\text{EEU} = \frac{V_B \cdot 3{,}4 \cdot \text{Hgb} \cdot 10^{-2}}{T_E}\ \text{mg/Tag}.$$

Literatur

1. ASHBY, W.: The determination of the length of life of transfused blood corpuscles in man. J. exp. Med. **29**, 267 (1919).
2. BÖTTNER, H., SCHLEGEL, B.: Ashby-Technik zur Messung der Überlebensdauer von Erythrozyten. Klin. Wschr. **29**, 498 (1951).
3. DORNHORST, A. C.: The interpretation of red cell survival curves. Blood **6**, 1284 (1951).
4. GRAY, S. J., STERLING, K.: The tagging of red cells and plasma proteins with radioactive chromium. J. clin. Invest. **29**, 1604 (1950).
5. STERLING, K., GRAY, S. J.: Determination of the circulating red cell volume in man by radioactive chromium. J. clin. Invest. **29**, 1614 (1950).
6. HUGHES, N. C., MOLLISON, P. L.: The interpretation of measurements with ^{51}Cr-labelled red cells. Clin. Sci. **15**, 207 (1956).

Interpretation of Radioisotope Renogram Changes During Acute Toxic Renal Insufficiency Due to Oxalic Acid

An Experimental Study

D. Pavel

Nuclear Medicine Laboratory of the Vth Medical Clinic, Bucharest, Roumania

In previous clinical works we emphasized the fact that isotope renogram might point to a certain alteration of the proximal tubular cell which could not be revealed by usual explorations [7, 12, 13]. We considered this alteration to be a disfunction of the cell which, though able to normally take up radio-Hippuran from peritubular blood, could not secrete it normally in the tubular lumen. We suggested the hypothesis that the cell passes through this stage both at the onset of insufficiency and during recovery.

The discussion was based at that time on three types of diseases: acute toxic tubular nephropathies in the polyuric stage, advanced arterial hypertensions and the advanced stage of ascitogenic cirrhosis. The "princeps" and most instructive case then presented was that of an acute renal insufficiency in the polyuric stage, secondary to a voluntary intoxication by means of a mixture containing oxalic acid. The possibilities of investigating homogeneous groups of such patients being, however, reduced, we decided to try an experimental model which might permit the detailed study of renogram changes in such cases under satisfactory statistical conditions.

Material and Method

At least 40 days prior to the first exploration, left side nephrectomy was performed on rats. We used 123 male white Wistar-type rats, of mixed stock, weighing 180–270 g; 304 renograms were performed under anaesthesia with amital-Na (9–10 mg/100 g b.w.) i. peritoneal. The substance used for renogram was ^{131}I Hippuran, administered i. v. in a dosis of 4–6 μCi in a tail vein. Tracings were obtained by means of a NaI crystal detector of 2/1 inches and of a Friesecke-Hoepfner single-channel potentiometric rectilinear strip chart recorder having a speed of 3.33 mm/min. The cylindrical collimator (6.3 cm long and 2.2 cm diameter) was located in the immediate vicinity of the rat loin facing the kidney localized by palpation.

Normal parameters were established on 59 rats hydrated "ad libitum". Two classical parameters were used (t_{max} and $t_{3/4}$) as well as a third one U.I. ("uptake index") which characterizes the uptake capacity of the tubular cell and was lately introduced by us in assying renograms in man [13] (see Fig. 1). As the exact measurement of all three parameters was not possible in all cases, t_{max} was determined in 54 rats, $t_{3/4}$ in 53 and U. I. in 43. Reproducibility was tested by repeating the renogram in several cases, at 2—3 days interval. At the same time, the effect of 4—5 days water restriction on the three parameters was studied on 14 rats maintained on dry food. Intoxication was carried out by stomach tube with an

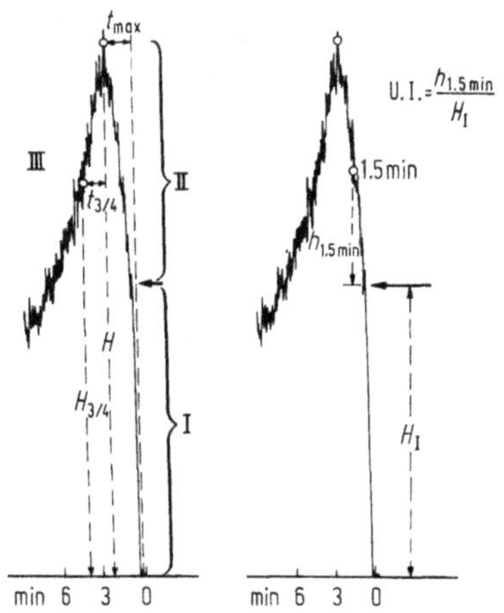

Fig. 1. The three parameters used: t_{max}, $t_{3/4}$ and "uptake index" (U.I.)

oxalic acid solution, ranging between 60—80 mg/100 g b. w. A first registration was carried out after 20 to 50 hours, then after 5—9 days and in some cases it was repeated as late as 14 or 21 days after intoxication. 101 rats were intoxicated but only 53 could be used for the experiment as the others died within the first 24 hours or did not present renographic changes at all.

Control renograms before intoxication could be carried out in 37 animals. The time difference also did not allow to intoxicate all the rats tested for normal values.

Urine output was studied on 33 animals by collecting urine over a period of 20 hours separately from the faeces. In 26 of them measurements were done before and after intoxication.

Results

The normal aspect of the renogram is shown in Fig. 2. The normal range of the parameters, with a confidence limit calculated at 95%, was found, for t_{max}: 0.9—2.56 min, for $t_{3/4}$: 0.51—2.15 min and for the U.I.: 0.26—0.86. For the interpretation of t_{max} and $t_{3/4}$ only the maximum limits are significant. For more certainty we considered as maximum limit for normal values the 99% confidence limit that is 2.83 for t_{max} and 2.48 min for $t_{3/4}$. These values had not been exceeded by any of the normal tracings studied. For the "uptake index", on the contrary, it is the lower limit which is essential for interpretion. Although the 95% confidence limit gives a lower limit of 0.26, in our series of normal rats the lowest U.I. value was 0.33.

Reproducibility was very good in all the cases where renograms were repeated at several days interval under the same conditions (Fig. 2).

Water restriction for 4—5 days led to a urine output lower than 1 ml/20 h, and resulted in a small but significant increase of t_{max} and $t_{3/4}$. As concerns U.I. it is important that it does not decrease under those conditions (it even tends to slightly increase average 0.76; $p < 0.05$). As a whole renograms although modified do not change their shape in a striking manner after water restriction.

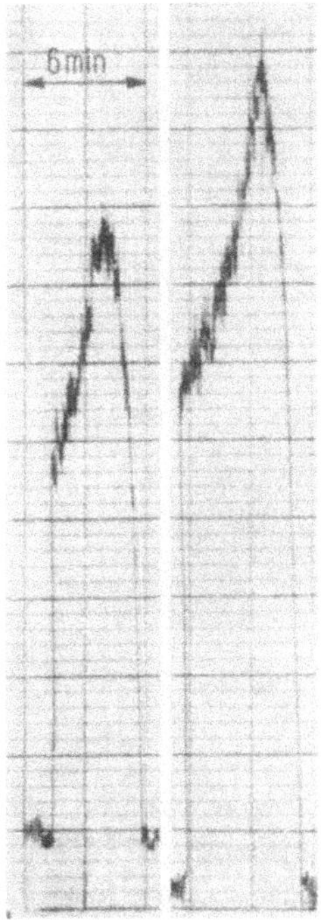

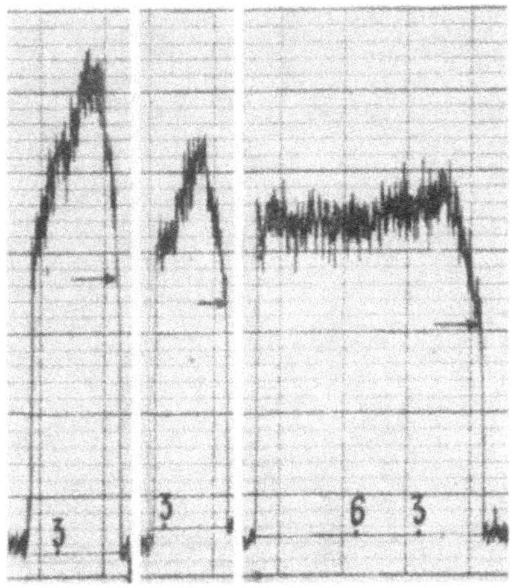

Fig. 3. First degree renographic changes. Right: 26 hours after intoxication. Middle and left: 4 and 8 days later, complete recovery

Fig. 2. Normal right rat renogram repeated under same conditions at four days interval

Rat intoxication produced renographic changes in 53 rats. On the others there were no changes at all (32 rats), or they died during the first day without being examined (16 rats). One could observe 3 types of renogram modifications:

a) Changes that we called "first degree", in 9 cases (Fig. 3, Tables 1 and 3), characterized by slightly prolonged t_{max} (average 2.74 ± 0.6) but significantly different from the normal value ($p < 0.01$); $t_{3/4}$ obviously prolonged (average > 10.06*) and an unmodified mean of U.I. (average 0.63 ± 0.21, not significantly different from normal: $p > 0.05$).

b) Second degree changes (Figs. 4 and 5, Tables 1 and 3) occurred most frequently (32 animals) and they actually form the main object of our discussion. The general aspect of the curve took, as we have called it, a "dome-like" shape. These curves are characterized by an extremely evident prolongation of t_{max} (average > 14.1) and also by a great delay of $t_{3/4}$ (which practically can no longer be measured). Neither in this group did U.I. tend to decrease; it remained practically unchanged (average 0.598 ± 0.26; not significantly different from the normal: $p > 0.05$). If the average value of t_{max} is considered to be 14.4 min (although it is higher), it will obviously very significantly differ from the normal

* The average is marked as "greater than" when the respective parameter could not be measured exactly as a very long-lasting registration should have been necessary.

but it will also differ very significantly from the average obtained during water restriction.

c) Third degree changes (Fig. 6 and Tables 1 and 3) occurred in 12 cases. t_{max} was much prolonged or in extreme cases it no longer existed (complete absence of segment II) and $t_{3/4}$ was also very prolonged so that it could practically no longer be measured. However, these tracings were mainly characterized by an U.I. always under the value of 0.26, when segment II existed (average 0.11 ± 0.09 is significantly different from the normal: $p < 0.001$).

Table 1. *Values of parameters in the three degrees of renographic changes. P: plateau curve; ≫: very long, could not be measured*

Group with first degree renographic changes

Rat No.	t_{max}	$t_{3/4}$	U.I.
17	2.13	>12.3	0.46
20	3.4	≫	0.72
54	2	≫	0.5
61	1.5	11.3	0.39
80	3.33	10.1	0.76
86	3.63	>10	0.73
91	3.15	6.6	0.43
102	3	≫	0.97
111	2.55	≫	0.69
Total: 9	Mean: 2.74		Mean: 0.63

Group with second degree renographic changes

Rat No.	t_{max}	$t_{3/4}$	U.I.
14	20.2	≫	0.42
18	>10.5	≫	—
21	20.1	≫	0.56
22	4.68	8.2	0.58
26	>17	≫	0.37
27	12.8	≫	0.51
29	>24	≫	0.52
30	8.28	≫	0.34
31	8.25	16	0.5
32	17.9	≫	0.77
38	6.4	12	0.97
42	>14	≫	0.53
43	18.8	≫	0.63
48	>18	≫	0.48
51	8	>14	0.44
55	4.7	13.1	0.64
65	4.2	>12.7	0.44
68	6.42	8	0.52
77	>14.4	≫	1.15
79	19.7	≫	0.88
93	15.3	≫	0.315
95	16	≫	0.87
96	17.5	≫	0.6
97	13.7	≫	0.59
98	16.8	≫	0.53
100	18.4	≫	1.59
101	15.9	≫	0.42
106	38.6	≫	0.43
107	> 6.4	≫	0.465
110	7.5	22.6	0.88
113	29.7	≫	0.39
114	>17.4	≫	0.56
Total: 32	Mean: >14.4		Mean: 0.598

Group with third degree renographic changes

Rat No.	t_{max}	$t_{3/4}$	U.I.
23	22	≫	0.17
28	*P*	≫	0.00
35	15.6	≫	0.23
52	*P*	≫	0.00
57	18.3	≫	0.21
58	19.6	≫	0.175
60	18	≫	0.095
64	*P*	≫	0.00
70	11.2	≫	0.196
78	>10	≫	0.02
90	10.8	≫	0.2
108	>10.1	≫	0.04
Total: 12			Mean: 0.11

Urine output studied in 33 normals presented an average of 7.3 ± 2.1 ml/20 h. In no case did it fall under 3 ml/20 h. After intoxication, the general tendency observed was not towards decrease in any of the categories mentioned above but

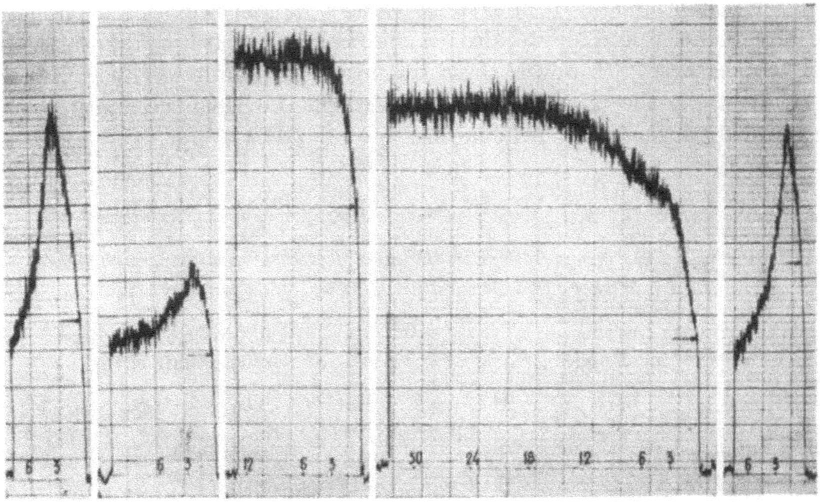

Fig. 4. Second degree renographic changes. From right to left: normal control renogram, and renograms 1, 5, 8 and 25 days after intoxication. Typical "dome-like" curve 24 hours after intoxication, followed by progressive improvement of the renogram which passes through a first degree stage before complete recovery

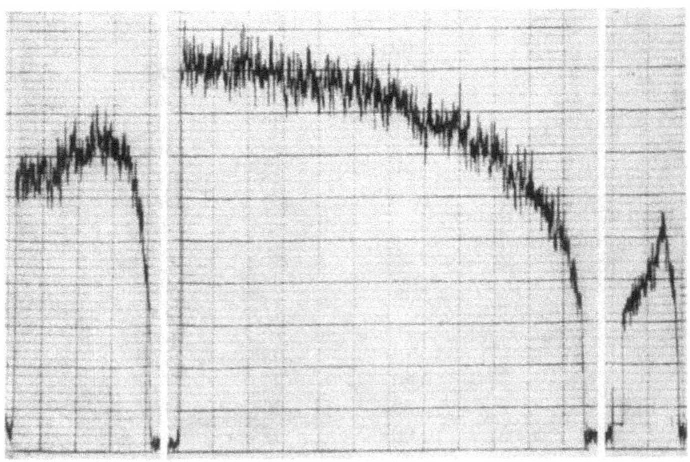

Fig. 5. Second degree renographic changes. From right to left: normal control renogram (urine output 5 ml/20 h), "dome-like" curve 20 hours after intoxication (urine output 7 ml/20 h), improvement of renogram 9 days later (urine output 12 ml/20 h)

rather towards increase. In order to realise a statistical analysis of this tendency by means of the couple method, we grouped in Table 2 the 17 cases (with first to third degree changes) where urine output was estimated both before and after intoxication. No significant changes are observed in the first three days while by the fourth day there is a significant trend towards urine output increase as compared to control values ($p < 0.05$). For the rats that had no renographic changes after oxalic acid administration, diuresis behaved in the same manner. In none of the

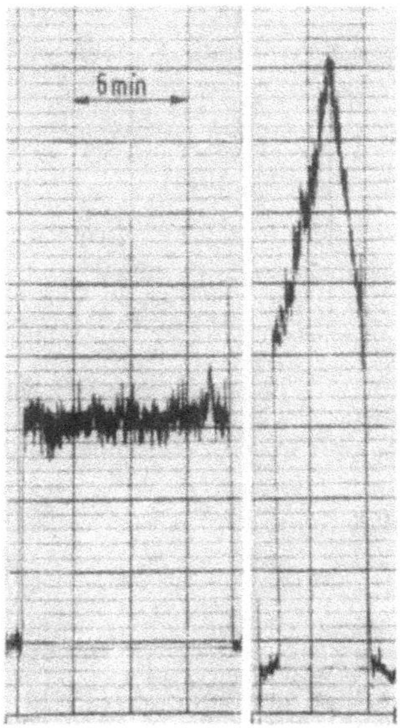

Fig. 6. Extreme third degree renographic changes. Right: normal control renogram. Left: renogram 26 hours after intoxication

Table 2. *Urine output evolution before and after intoxication*

Renographic changes	Rat No.	Control urine output before intoxication (ml/20 h)	Urine output after intoxication (ml/20 h)				
			0—1st day	1—2nd day	2—3rd day	3—4th day	mean after intoxication
	111	7	—	12.5	—	10.5	11.5
First	102	6	7	9	—	5	7
degree	91	9	3	5	—	—	4
	86	9	—	8.75	7.5	—	8.2
	114	6.5	—	11.5	—	12.5	12
	113	5	7	—	—	10	8.5
	110	10.25	—	9	—	—	9
	107	5.5	—	4	—	—	4
Second	106	5	5.5	5	9.5	—	6.6
degree	101	7	8	8	—	10	8.6
	100	5	6	6	—	10	7.3
	96	10	7.5	8.5	—	16	10.6
	95	11	12	—	—	9.5	10.75
	93	11	11.5	—	—	12	11.75
	79	3.75	6.25	—	—	5.5	5.8
Third	108	3.25	—	3.5	—	—	3.5
degree	90	3.5	3	—	—	3	3
Mean		6.92 ± 2.65	7.68	7.44		9.45	7.76 ± 2.98

intoxicated animals did urine output fall under 3 ml/20 h, and the lower values occurred precisely in the two rats with third degree changes that had also had low values previously (Table 2).

Table 3. *Evolution of parameters and urine output before and after intoxication, each rat serving as its own control. P: plateau curve;* $\gg$ *: very long, could not be measured*

Renographic changes	Rat No.	t_{max} (min)		$t_{3/4}$ (min)		U.I.		Urine output (ml/20 h)	
		before	after	before	after	before	after	before	after (mean)
	14	1.5	20.2	1.53	$\gg$	0.52	0.42		
	18	1.83	10.5	0.96	$\gg$	0.63	—		
	21	2.1	20.1	0.9	$\gg$	0.44	0.56		
	22	1.92	4.68	2.1	8.2	0.66	0.58		
	26	1.8	>17	≈ 2.8	$\gg$	0.48	0.3		
	27	2	12.8	≈ 2.63	$\gg$	0.59	0.51		
Second degree	32	1.72	17.9	1.69	$\gg$	0.64	0.77		
	68	1.5	6.42	1.9	8	0.54	0.52		
	77	1.52	>14.4	1.23	$\gg$	0.96	1.15		
	96	1.8	17.5	1.5	$\gg$	0.7	0.6	10	10.6
	100	1.5	18.4	1.55	$\gg$	1.18	1.59	5	7.3
	107	1.53	>6.4	1.14	$\gg$	0.93	0.46	5.5	4
	110	1.2	7.5	0.6	22.6	>0.56	0.88	10.25	9
	113	1.26	29.7	1.44	$\gg$	>0.50	0.4	5	8.5
	17	1.8	2.13	1.34	12.3	0.63	0.46		
First degree	61	1.32	1.5	0.9	11.3	>0.46	0.39		
	102	2	3	0.81	$\gg$	0.54	0.97	6	7
	111	1.74	2.55	0.9	$\gg$	0.51	0.69	7	11.5
	23	2.7	22	1.71	$\gg$	0.38	0.17		
Third degree	28	2.1	*P*	1.23	$\gg$	0.55	0.00		
	78	1.5	$\gg 10$	1.67	$\gg$	0.59	0.02		
	90	1.8	10.8	1	$\gg$	0.61	0.2	3.5	3
	108	1.8	>10.1	1.2	$\gg$	0.63	0.04	3.25	3.5

Table 3 is a synthesis of the data from animals with renographic changes where complete examination was possible (renogram before and after intoxication and in some of them diuresis also). The above described changes for the three degrees appear even more striking as each animal is its own control. The significance of this changes appears obviously, except for the paired U.I. values in the first and second degree categories. Therefore we calculated it for this later categories and, as expected, the difference was not significant ($p > 0.05$).

Discussion

On unilateral nephrectomized rats and under amital-Na anaesthesia the reproducibility of renogram was very good (Fig. 2) as we also demonstrated previously [11].

For the correct and rapid estimation of a renogram the three parameters indicated above may be sufficient. The uptake index (U.I.) we used permitted to estimate in figures the normal or modified aspect of the first part of segment II. The advantage as compared to other similar indices, in which total height is used [6, 21], is that our ratio is made between the height proper, on the segment II, of the point that interests us and segment I (Fig. 1). The index is therefore more evidently influenced by slope modifications of segment II. It should be briefly recalled that it is unanimously agreed that two main factors are involved in producing the first portion of segment II [1, 8, 9, 10, 15, 17, 18, 21]: 1. the functions of the proximal tubular cell which removes Hippuran from the blood (in the same way as PAH) and 2. renal blood flow. Less steep slopes of segment II (and hence a decrease of the U.I. value) indicate a diminution either of irrigation or of the functional capacity of the cell or of both.

Water restriction produces the same effect on renogram as in man [13, 20]. It may be observed that the general aspect of the curve is not modified although the urinary flow decreases remarkably (under 1 ml/20 h as compared to the average of normally hydrated animals: 7.3 ml/20 h).

If the effect of intoxication is considered, it may be observed first that irrespective of the degree or absence of change in the renogram, urine output does not decrease and even tends to increase by the fourth day. Although urine output was not measured in all the rats, the results obtained are homogeneous enough to be extrapolated to the whole group. Moreover, average urine output is very far from the values below 1 ml/20 h obtained in rats explored after water restriction. Even if the extrapolation of results to the whole group is not accepted, it may, however, be asserted at least that anuria was never present at the moment of exploration. Indeed all the rats urinated during manipulation previous to anaesthesia.

As regards the effect on the renogram, it is non-homogeneous. This non-homogeneity of the response is well known wherever toxic substances are used in experiments. Anyhow the present work has not been affected in any way by this fact, as its aim was to obtain the greatest possible number of rats with "dome-like" tracings, without oliguria.

It can be noticed that the first degree of renogram alteration is the marked change of segment III, expressed by the prolongation of $t_{3/4}$ (Fig. 3 and Tables 1 and 3). In the next stage the "dome-like" aspect shows besides the change of segment III, also a great delay in t_{max}. These two changes occur, however, in the presence of a segment II with no modification in its initial portion: U.I. does not differ significantly from the normal ($p > 0.05$) (Figs. 4 and 5 and Tables 1 and 3). Third degree changes occur in a more advanced stage: they are characterized by the diminution down to disappearance of the slope of segment II ($0 \leq$ U.I. < 0.26) (Fig. 6 and Tables 1 and 3). In extreme cases of this stage the curve has a plateau aspect or presents a continuous decrease without any segment II.

Successive registrations showed that the three degrees of renographic changes are proportional to the severity of the renal lesions and may pass from one to the other in both directions, depending on the evolution of the lesion. The most frequently found was the succession shown in Fig. 4 and 5, where, in time, between the "dome-like" aspect and the normalization of the tracing one could observe the first degree aspect. The reverse evolution could also be observed from second degree to third degree changes, followed by the death of the animal. Moreover, in almost all cases with third degree renograms, animals showed symptoms of severe intoxication and died in short time.

The possible physiopathological interpretation of these renographic changes has now to be discussed. The "dome-like" tracings which occur in the second degree renographic changes in the presence of normal or increased urine output are the most striking and characteristic. Three explanations may be forwarded:

1. Theoretically, oligo-anuria might produce such tracings, by slowing down Hippuran flow in the tubular lumen and in the renal pelvis. However, in practice even prolonged water restriction with oliguria below 1 ml/20 h cannot produce such evident renographic changes. Anuria, on the other hand, might produce them. However, neither anuria nor oliguria can be taken into account since, as we have shown, rats were far from being oligo-anuric and even tended to an increased urine output.

2. Recent ureteral obstruction could also yield such changes. However, ureteral obstruction occurring in some scores of rats, 1—2 days after intoxication, under normal diuresis seems quite unlikely.

3. Hippuran storage within the proximal tubular cell seems the only possible explanation of the "dome-like" curve in the presence of normal or increased diuresis. Indeed, the prolongation of t_{max} and $t_{3/4}$ points out that labelled substance persists for a longer time in front of the probe. As in the tubular lumen and in the renal pelvis the substance cannot persist because of sufficient urine output, it means that it can lag only in the blood or in the tubular cell. If, however, lagging occurred in the blood, because of complete damage of tubular cell function, there would be no reason for the normal shape of the first portion of segment II on the renogram. On the contrary one should expect to find the classical type of curve noticed during chronic renal insufficiency in man and rat, when the proximal tubular cell is severely altered. The characteristic feature of the renogram in such cases is a much lower or even absent slope of segment II. However, in our experiment this type of curve is found only in the group of rats with above-described third degree changes and it is completely different from the "dome-like" curve.

Intracellular storage appears thus as the only possible explanation. This implies a dissociated alteration of the cell function, which is still able to take up blood Hippuran but becomes unable (or has a lower capacity) to eliminate it by the apical pole in the tubular lumen. The theoretic existence of such a possibility is practically emphasized by the normal aspect of the first part of segment II (normal U.I.) combined to a "dome-like" shape of the renogram. It has been already mentioned that segment II of the renogram is directly related to the uptake function of the tubular cell and to renal blood flow. A normal "uptake index" implies, therefore, that these two components should be at least normal.

We had already reached the conclusion of intracellular storage also in previous studies on renogram in man [7, 12, 13]. At that time we had no sufficient statistical confirmation, especially for cases of acute tubular insufficiency. The present experiment corroborates that conclusion. Moreover, the process could be now studied more thoroughly by establishing the three evolution stages. Thus, it seems that the first sign of tubular disfunction is the modification of the secretion process alone, towards tubular lumen (prolongation of $t_{3/4}$ alone, in first degree changes). In the next stage this secretory deficiency is more marked, subsequently leading to "dome-like" tracings, but uptake is still unaltered (normal U.I.). The uptake process at the level of the basal membrane seems to be disturbed (third degree, with U.I. decrease) only when the action of the toxic drug is very strong and then, most likely the lesion is irreversible. The recovery of the tubular function after intoxication occurs in the opposite direction, passing through the same stages.

Such alterations seem to appear not only during intoxications but also in some other acute or chronic disease [2, 13, 16, 19, 22]. Thus it may be possible to consider that each aggression on the proximal tubular cell results in a more or less transient disfunction of this type.

If we assume the existence of an intracellular Hippuran storage at certain moments during evolution of acute kidney insufficiency, PAH clearance interpretation should be done very carefully. Indeed it seems very likely that urine PAH concentrations may be low owing to intracellular trapping of PAH. Thus clearance value will be low, according to a cause other than plasma flow decrease.

Renogram therefore has not only an important diagnostic role but seems even to be the only method for diagnosis of this particular stage of tubular disfunction.

Oxalic acid has proved to be a very adequate substance for this study, especially owing to the fact that diuresis was not diminished by intoxication. The action of other toxic drugs tested experimentally—e.g. the sublimate used by several authors [3, 21] on renogram did not yet lead to such conclusions, firstly on account of anuria occurring in the first days and maybe also owing to its much more brutal action on the tubular cell (the tracings of these authors show generally a clear-cut alteration of segment II).

Renogram may also permit a step forward in the understanding of the physiology of tubular secretion. It points out that in mammals the secretory transport of Hippuran has two steps—whose existence is sometimes still argued about [4, 5, 14]: uptake and secretion proper. Moreover it pointed to the evidently higher sensitivity of the latter to intoxication with oxalic acid. This strengthens the belief that an active transport occurs in this last secretion step, which fact is also sometimes controverted [5].

In conclusion, we feel that more and further attention should be granted to this particular stage of tubular failure, emphasized by isotope renogram.

Summary

The effect of acute intoxication with oxalic acid on isotope renogram and diuresis in rats is studied in view of trying the reproduction of certain renographic changes previously observed in man. Thus it was noticed that the first days after intoxication diuresis remained unchanged or had even a tendency to increase, while the renogram was more or less altered according to the severity of the intoxication. In most of these cases there occurred a "dome-like" curve in which the first part of segment II remained unaltered.

The possible interpretation of such curves in the presence of normal diuresis is discussed reaching the conclusion that they can only be due to an intracellular storage in the proximal cells. The cell maintains its capacity of achieving blood Hippuran uptake but it loses its capacity to secrete it into the tubular lumen. This tubular lesion seems to be an intermediary stage in the setting-in of proximal tubular insufficiency and seems to appear under the influence of various causes. Renogram has proved to be of particular importance in revealing this stage.

References

1. Becker, G.: Nierenfunktionsanalyse mit ^{131}I-Hippuran (Theoretische und klinische Medizin, Bd. 40). Heidelberg: Alfred Hüthig 1969.
2. Burke, G., Halko, H.: Scintillation camera renography in the study of prolonged renal transit time. Radiology **88**, 704—712 (1967).
3. Dore, E. K., Taplin, G. V., Johnson, D. E.: Current interpretation of the sodium iodohippurate I^{131} renocystogram. J.A.M.A. **185**, 925—932 (1963).
4. Forster, R. P., Copenhaver, J. H.: Intracellular accumulation as an active process in a mammalian renal transport system "in vitro". Am. J. Physiol. **186**, 167 (1956).
5. Foulkes, E. C., Miller, B. J.: Steps in PAH transport by kidney slices. Am. J. Physiol. **196**, 86 (1959).
6. Frey, K. W., Hör, G., Heinze, G. H.: Radioisotopennephrographie bei der Nierentuberkulose. Nuclear Medizin **4**, 249—262 (1965).
7. Gheorghescu, B., Pavel, D., Jovin, Gh.: Studiul modificărilor funcţiei renale în ciroze cu ajutorul renogramei izotopice. Medicina Int. (Bucharest) **19**, 215—220 (1967).
8. Hundeshagen, H., Heimsoth, V., Graul, E. H., Hartmann, F.: Neue Aspekte in der Radionephrographie. Atompraxis **11** (August 1965).
9. Magnusson, G.: Kidney function studies with ^{131}I-tagged sodium-o-iodohippurate. Acta med. scand., Suppl. **378** (1962).
10. Pabst, H. W.: Isotopendiagnostik von Nierenkrankheiten. Münch. med. Wschr. **34**, 1581—1590 (1965).
11. Pavel, D., Chisiu, N., Oproiu, A.: Model experimental de studiu al funcţiei renale cu ajutorul nefrocistogramei izotopice la şobolan. Simpozion: Modelarea în biologie, Braşov, Romania, 1964.
12. — Gheorghescu, B., Herscovici, H., Cavulea, G., Lupu, Irène, Jovin, Gh.: Objectivation à l'aide du néphrogramme isotopique d'une certaine étape de l'insuffisance tubulaire rénale. Acta urol. belg. **36**, 59—67 (1968).
13. — — Jovin, G., Rotarescu, G.: Le néphrogramme isotopique au cours de certaines formes d'insuffisance tubulaire rénale. Essai d'interprétation physiopathologique. J. Urol. Nephrol. **74**, 645—654 (1968).
14. Pitts, R. F.: Physiology of the kidneys and body fluids. Chicago: Year Book Medical Publishers 1963.
15. Semprebene, L., Scuncio, G., Benedetti, V. J. Jr., Fiorani, P.: Studio comparativo della renografia in condizioni basali ed in corso di diuresi provocata. Importanza diagnostica della ipertensione nefrovascolare. Nuntius Radiol. **11**, 1197—1215 (1965).
16. Swyngedauw, J.: Néphrogramme de l'hypertendu. Ann. Biol. clin. **23**, 1015—1049 (1965).
17. Taplin, G. V., Dore, E. K., Johnson, D. G.: The hippuran I^{131} renocystogram. V. Internationales Symposium, Bad Gastein 1962. München: Urban & Schwarzenberg 1963.
18. Tauxe, W. N., Hunt, J. C.: Evaluation of renal function by isotope techniques. Med. clin. N. Amer. **50**, 937—945 (1966).
19. Tiegerman, T., Pompilian, V., Pavel, D., Grimberg, L., Petrescu, C., Erena, R.: Cercetări asupra dinamicii inbunătăţirii funcţiilor renale după insuficienţa renală acută. Medicina Int. (Bucharest) **21**, 1449—1462 (1969).
20. Wedeen, R. P., Goldstein, M. H., Levitt, M. F.: The radioisotope renogram in normal subjects. Amer. J. Med. **34**, 765—774 (1963).
21. Winkel, K. zum: Nierendiagnostik mit Radioisotopen. Stuttgart: Georg Thieme 1964.
22. — Hallwachs, O., Steinhausen, M.: Kameraszintigraphie und Isotopennephrographie an der Hundeniere und deren Überprüfung durch die Intravitalmikroskopie. Fortschr. Geb. Röntgenstr. **108**, 382—393 (1968).

Die renale Ganzkörperclearance

Renal Whole-Body Clearance

H. U. Pixberg

Institut für Nuklearmedizin und spezielle Biophysik des Radiologischen Zentrums
der Medizinischen Hochschule Hannover

Summary

With renal whole-body clearance without catheter, the removal of clearance substance is measured in the body. The clearance is calculated from the differential quotient and the plasma concentration 20 and 25 minutes p. i., applying the Fick principle. The whole-body clearance and the steady-state clearance are in good agreement. Two scintillation detectors, each consisting of a 2″ sodium iodide crystal without collimator and a well-type scintillation counter, are used for measurement.

Ein völlig neuer Weg, die renale Clearance mit radioaktiven Substanzen zu bestimmen, wurde erstmals von Oberhausen [1, 2] mit der Ganzkörperclearance beschritten. Bei den bisher bekannten katheterlosen Verfahren wurde nur der Ablauf der Plasmakonzentrationskurve (Dost [3]) oder die infundierte Menge der Testsubstanz (Earle und Berliner [4]) der Berechnung der renalen Clearance zugrunde gelegt. Alle Verfahren beruhen darauf, den „Gefäßraum" solitär und als repräsentatives Kompartiment für das Verteilungsvolumen der Clearancesubstanz zu betrachten. — Wie wir [5, 6] zeigen konnten, treten dabei unüberwindliche Schwierigkeiten auf, vor allem bei der Bestimmung des effektiven Nierenplasmastromes mit J^{131}-Hippuran.

Die Methode der Ganzkörperclearance umgeht diese Schwierigkeiten, indem sie die Mengenabnahme der radioaktiv markierten Clearancesubstanz im Körper mißt. Sie ist somit weitgehend unabhängig von den Konzentrationen in den einzelnen Kompartimenten des gesamten Verteilungsraumes, da diese gleichermaßen in die Messung eingehen. Oberhausen führte die skizzierte Methode mit einem Ganzkörperzähler unter Abdeckung der Nieren, Ureteren und der Blase durch.

Wir haben uns unter zwei Gesichtspunkten mit der renalen Ganzkörperclearance beschäftigt:

1. Läßt sich die Methode von einem Ganzkörperzähler, der nur an wenigen Forschungsstellen vorhanden ist, auf einen einfachen Funktionsmeßplatz mit zwei Szintillationssonden von 2″ Durchmesser übertragen?

2. Ist die renale Ganzkörperclearance in ihrer Aussage der Standardclearance gleichzusetzen?

Methodik

Die Patienten wurden wie zur Standardclearance vorbereitet. Zur Abdeckung der Nieren, Ureteren und Blase schoben wir einen Tisch mit einer Bleiplatte von 10 Halbwertsdicken (30 mm) über den Patienten. Die Bleiplatte schloß kranial mit dem Processus styloides, kaudal handbreit unterhalb der Symphyse ab.

Zur Ganzkörpermessung verwandten wir einen Funktionsmeßstand mit zwei Szintillationssonden, die jeweils mit einem 2″-NaJ(Th)-Kristall bestückt waren ohne Kollimator. Beide Sonden wurden vor der Untersuchung genau aufeinander abgestimmt und dann senkrecht über dem Patienten eingestellt. Rechnet man mit einem durchschnittlichen Abstand von 55 cm zwischen Kristall und Körpermitte, so wird die im Körper vorhandene Radioaktivität in einem Durchmesser von 38 cm annähernd homogen aufgelöst.

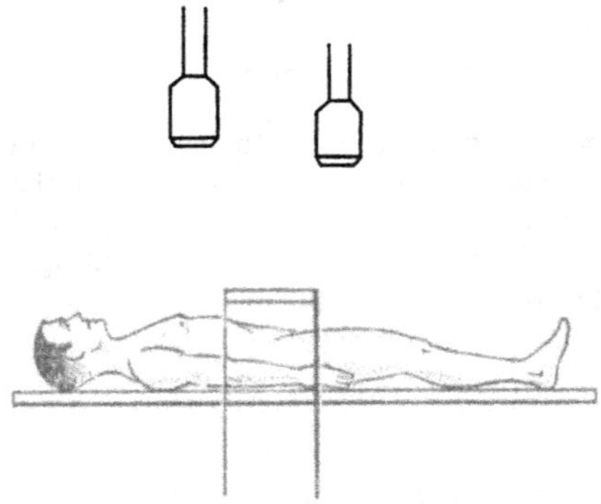

Abb. 1. Szintillationssondenanordnung und Abdeckung der Nieren-Blasenregion am Patienten zur Ganzkörperclearance

Abb. 1 zeigt die beschriebene Anordnung des Funktionsmeßplatzes. Nach den angegebenen Meßwerten entspricht das homogene Sichtfeld der Sonden bei einem normalgroßen Menschen dem Körperabschnitt von der kranialen Bleiabdeckung bis zur Orbita und von der kaudalen Bleiabdeckung bis zu den Kniegelenken. Das inhomogene Sichtfeld ist selbstverständlich größer und geht über die gesamte Körperlänge des Patienten hinaus.

Zur Bestimmung des effektiven Nierenplasmastromes und des Glomerulusfiltrates wurde J^{131}-Hippuran in einer Dosis von 50—60 μCi und Cr^{51}-EDTA in einer Dosis von 100 μCi intravenös injiziert. Vor der Injektion wurden die Dosen in geometrischer Abschwächung vom Szintillationskristall im automatischen Probenwechsler FHT 770D gemessen. Die Dosis wurde im Vergleich zu einem Bariumchlorid-Standard in μCi angegeben. — Sofort nach der intravenösen Injektion wird mit der Messung begonnen. Die Meßzeit beträgt jeweils 2 Minuten und erstreckt sich insgesamt über 40 Minuten. Die Impulse pro 2 Minuten werden im halblogarithmischen System gegen die Zeit (arithmetisch) aufgetragen. Anschließend wird die Kurve nach t_0 rückextrapoliert, um die Anfangskonzentration y_0 zu bestimmen. Die so ermittelte Kurve wird nach der Formel für die kubische Ausgleichsparabel der numerischen Differentiation unterzogen, um das Absinken des Indikators im Körper (dm/dt) über die Zeit zu bestimmen. Die Formel lautet (ZURMÜHL [7]):

$$\bar{y}_0 = \left[(y_{+1} - y_{-1})\, 8 + (y_{-2} - y_{+2})\right] \frac{1}{12 \cdot N}\,; \quad N = 5.$$

Blutproben zur Bestimmung der Serumkonzentration werden zwischen 10 und 40 Minuten p.i. aus der Kubitalvene entnommen. Die Aktivität der Serumproben wird wieder im automatischen Probenwechsler FHT 770D im Vergleich zu einem Ba-Standard gemessen und in nCi angegeben.

Die Ganzkörperclearance errechnet sich nach der Formel:

$$C_{GK}\ (\text{ml/min}) = \frac{-dm/dt}{c};$$

$-dm/dt$ = Indikatorabnahme im Körper, c = Plasmakonzentration.

In die Formel werden die zur Zeit der Blutentnahmen ermittelten Werte für dm/dt eingesetzt und die renale Clearance in ml/min berechnet. Das Ergebnis wird dann noch mit einem Faktor (Nomogramm Geigy-Tabellen) auf ml/min/1,73 m² normiert. — Im Anschluß an die Ganzkörperclearance wurde bei jedem Patienten die Standardclearance durchgeführt und beide Werte miteinander verglichen.

Ergebnisse der Ganzkörperclearance

In Tab. 1 sind die Ganzkörperclearances von Vorversuchen, die mit einem Magna Scanner V und Armac gemessen wurden, der Standardclearance mit J^{131}-Hippuran gegenübergestellt. Dazu wurde die prozentuale Abweichung von der Standardclearance angegeben.

Tabelle 1. *Ganzkörperclearance mit J^{131}-Hippuran*

Nr.	Apparat	C_{GK} J^{131}-Hipp. (ml/min/1,73 m²)	C J^{131}-Hipp. (ml/min/1,73 m²)	Abweichung von der Standard-clearance (%)
1	Magna Sc. V	366	493,1	— 26
2	Magna Sc. V	354	464,8	— 24
3	Magna Sc. V	279	410,9	— 32
5	Magna Sc. V	316	404,5	— 28
6	Magna Sc. V	47	50,5	— 7
7	Magna Sc. V	50	64,8	— 23
10	Magna Sc. V	432	487,8	— 11
12	Magna Sc. V	152	205,9	— 26
15	Armac	577	463,8	+ 24
16	Armac	464	393,7	+ 18

Unsere Ergebnisse der Ganzkörperclearance, gemessen mit einem serienmäßigen Funktionsmeßplatz, geben Abb. 2a und 2b in graphischer Darstellung wieder. In Abb. 2a sind die Ganzkörperclearances (Ordinate) mit Blutentnahmen nach 20, 25 und 30 Minuten denen der Standardclearance (Abszisse) gegenübergestellt. In Abb. 2b werden die Ganzkörperclearances mit Blutentnahmen nach 20 und 25 Minuten mit der Standardclearance verglichen. Die Regressionsgeraden betragen für 2a: $y = 9{,}17 + 0{,}948\,x$ und für 2b: $y = 9{,}83 + 0{,}967\,x$. Für beide Vergleiche fanden wir eine sehr enge Korrelation von $r_a = 0{,}948$ und $r_b = 0{,}967$. Zwischen den Mittelwerten der Ganzkörperclearances und der Standardclearance konnte kein signifikanter Unterschied festgestellt werden. Die Streuung von S_{xy} beträgt 30 bzw. 33,8 ml.

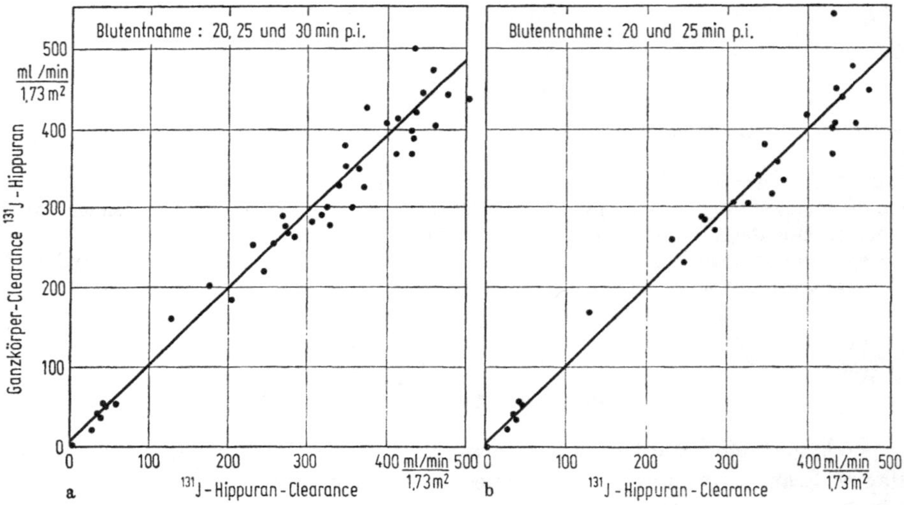

Abb. 2. Vergleich der J^{131}-Hippuranclearance im steady state mit der Ganzkörperclearance. a Blutentnahme nach 20, 25 und 30 Minuten, b nach 20 und 25 Minuten

Die Ergebnisse von sieben orientierenden Untersuchungen über die Ganzkörperclearance mit Cr^{51}-EDTA zur Bestimmung des Glomerulusfiltrates sind in Tab. 2 zusammengefaßt. Die Ergebnisse wurden wegen der geringen Zahl nicht statistisch ausgewertet.

Tabelle 2. *Ganzkörperclearance mit Cr^{51}-EDTA (Blutentnahmen nach 20, 25 und 30 Minuten)*

Nr.	C_{GK} Cr^{51}-EDTA (ml/min/1,73 m²)	C Cr^{51}-EDTA (ml/min/1,73 m²)	Abweichung von der Standard-clearance (%)
1	18	17	+ 5
2	100	94,3	+ 6
3	40	34,5	+ 16
4	101	98,6	+ 2
5	111	92,4	+ 20
6	49	50,3	— 3
7	47	42,6	+ 11

Diskussion

Wie wir bereits in der Einleitung herausstellten, liegt der besondere Vorteil der Ganzkörperclearance in der Unabhängigkeit von den Kompartimenten, in denen sich die Clearancesubstanz verteilt.

Voraussetzung für die Messung des Indikatorabfalls im Körper ist die Markierung der Testsubstanz mit einem Nuklid, dessen γ-Strahlung genügend Energie aufweist, um in bestimmter Entfernung vom Körper proportional registriert werden zu können. J^{131}-Hippuran und Cr^{51}-EDTA haben sich für die Untersuchung als brauchbar erwiesen. Ihre Energielinien liegen bei 360 bzw. 324 keV.

Bei der Verwendung von J^{131}-Hippuran und Cr^{51}-EDTA zur simultanen Bestimmung des effektiven Nierenplasmastromes und des Glomerulusfiltrates scheitert die Doppelnuklidmessung am Auflösungsvermögen der NaJ(Th)-Kristalle. Sollen beide Partialfunktionen mit der Ganzkörperclearance bestimmt werden, muß dies im zeitlichen Abstand von mehreren Stunden erfolgen. Die zeitliche Trennung ist zweifellos ein Nachteil. Er könnte überwunden werden, falls die Markierung einer Substanz zur Bestimmung des effektiven Nierenplasmastromes oder des Glomerulusfiltrates mit einem Nuklid gelingt, das sich von J^{131} oder Cr^{51} gut trennen läßt und genügend Energie aufweist, um in 50 cm Entfernung von der Strahlenquelle registriert zu werden.

Soll die Indikatorabnahme der Clearancesubstanz im Körper fortlaufend registriert werden, muß das System Niere — ableitende Harnwege — Blase mit Blei abgedeckt werden, da hier ja die Ansammlung der Substanz erfolgt. Vorausgehende Messungen mit einer Bleischicht von 30 mm Dicke ergaben eine genügende Absorption der γ-Strahlung, so daß aus der abgedeckten Körperzone keine das Ergebnis beeinflussende Strahlung auf die Meßsonden treffen kann.

Mit unserem Bleitisch wird der Körper vom Processus styloides des Sternums bis handbreit kaudal der Symphyse abgedeckt. Damit ist etwa $1/3$ des Körpers für die Meßsonden „nicht sichtbar“. Die Frage ist nun, sind Oberkörper und Beine repräsentativ für das Verteilungsmuster der Clearancesubstanzen oder, anders ausgedrückt, wird mit der Registrierung der Zeitaktivitätskurve von Oberkörper und Beinen, die zur Ganzkörperkurve addiert werden, der Konzentrationsabfall im Ganzkörper richtig wiedergegeben?

Um diese Frage zu klären, haben wir Vorversuche gemacht, bei denen einmal der Konzentrationsverlauf im Oberkörper (s. Tab. 1 unter Magna Scanner V), ein anderes Mal im Unterarm mit dem Armac aufgezeichnet wurde. Beide Kurven wurden nach dem gleichen Verfahren rechnerisch ausgewertet. Dabei zeigte sich, daß die Auswertung der Oberkörperkurve regelmäßig eine zu niedrige Ganzkörperclearance ergab. Im Vergleich zur Standardclearance lag sie ungefähr 20% unter dieser. Die Ganzkörperclearance, berechnet nach der Unterarmkurve (Armac), lag um fast den gleichen Prozentsatz über der Standardclearance. Wie kann man die Differenz erklären? Im Sichtfeld der Oberkörpersonde nimmt der vasale Anteil (Herz, große Gefäße, Lungengefäße) einen wesentlich größeren Raum am Gesamtvolumen ein als am Unterarm. Hier überwiegt die Muskelmasse den vasalen Raum. Das Verhältnis Gefäßraum—Extrazellulärraum ist also in beiden Meßbereichen sehr unterschiedlich. Parallel zu diesem Verhältnis verlaufen beide Kurven. Die Oberkörperkurve verläuft steiler als die Unterarmkurve. Sie zeigt damit den schnelleren Konzentrationsabfall des vasalen gegenüber dem extrazellulären Raum an. Gleichzeitig wird damit demonstriert, daß aus der Zeitaktivitätskurve eines Teilgebietes des Körpers die Ganzkörperclearance nicht berechnet werden kann. Diese Feststellung trifft auch für die Totalclearance zu.

Verwendet man zur Messung der Ganzkörperclearance zwei Sonden, die über Oberkörper und Beinen lokalisiert werden, treten diese Abweichungen nicht mehr auf. Bei einem Abstand von ungefähr 55 cm zwischen Kristall und Körpermitte werden von den Sonden genügend große Bereiche erfaßt, die in ihrer Addition offenbar der Verteilung im gesamten Körper entsprechen.

Abb. 3 zeigt im unteren Teil Bein- und Oberkörperkurve getrennt aufgezeichnet, im oberen Teil die Ganzkörperkurve, die sich durch Addition beider Kurven ergibt. Die Ganzkörperclearance ist mit der Standardclearance identisch, während

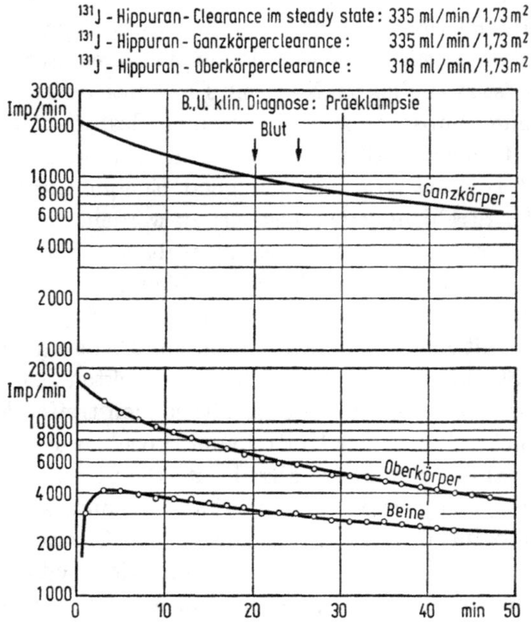

Abb. 3. Der untere Teil der Abbildung zeigt den Konzentrationsverlauf von J^{131}-Hippuran nach intravenöser Injektion im Oberkörper und in den Beinen. — Im oberen Teil ist die Addition beider Kurven als Ganzkörperkurve wiedergegeben

die Ganzkörperclearance, berechnet aus der Oberkörperkurve, unter der Standardclearance liegt. Genaue Werte im Vergleich zur Standardclearance sind bei unserer Meßanordnung nur durch Addition beider Kurven zu erreichen. — Der untere Teil des Bildes sagt außerdem noch etwas über die Messungsgenauigkeit beider Sonden aus. Die Einzelmeßwerte von Oberkörper und Beinen weichen kaum von der Kurve ab! Trotz der relativ kleinen NaJ(Th)-Kristalle wird also die γ-Strahlung mit ausreichender Genauigkeit registriert.

Nach den ersten Untersuchungen fiel uns auf, daß bei adipösen Patienten die Ganzkörperclearance gut mit der Standardclearance übereinstimmt, während bei normal- oder untergewichtigen Patienten die Ganzkörperclearance meistens 10% unter der Standardclearance lag. Bei diesen ersten Untersuchungen hatten wir beide Sonden wie bei einem Ganzkörperzähler in einer Ebene angebracht, so daß die Beinsonde durchschnittlich 5 cm weiter vom Volumenmittelpunkt entfernt war als die Oberkörpersonde. Nach dem Quadratabstandsgesetz wurde also die Radioaktivität der Beine im Verhältnis zum Oberkörper zu gering registriert. Modellmessungen bestätigten diesen unterschiedlichen Sondenabstand als Ursache der Ganzkörperclearanceabweichung von der Standardclearance, der bei einem

Ganzkörperzähler nicht in Erscheinung tritt. Die Modellmessungen ergaben, daß bei adipösen Patienten der Abstand zwischen Sonde und Beinoberfläche gleich dem Abstand zwischen Sonde und Oberkörperfläche sein muß. Bei Normal- und Untergewichtigen muß die Beinsonde 2 cm tiefer eingestellt werden, um das fehlende Volumen zwischen beiden Beinen auszugleichen. Verständlicher wird diese Forderung, wenn man sich beide Beine als ein Volumen vorstellt, das unter gleichen Bedingungen von der Sonde gemessen werden soll wie der kompakte Oberkörper. Werden diese Bedingungen beachtet, ist mit einem serienmäßigen Funktionsmeßplatz die Bestimmung der Ganzkörperclearance nach dem OBERHAUSENschen Prinzip durchaus möglich.

Zur Beantwortung der zweiten Frage, unter die wir diese Arbeit gestellt haben, ist der Vergleich mit der Standard-J^{131}-Hippuranclearance erforderlich. Wie die graphischen Darstellungen und die statistischen Auswertungen zeigen, ist die Übereinstimmung beider Methoden sehr gut. Voraussetzung hierfür ist die Einhaltung bestimmter Zeiten für die Blutentnahmen. Betrachtet man folgende Clearances als Beispiel (Tab. 3), wird der Grund für die Einhaltung bestimmter Entnahmezeiten klar.

Tabelle 3

J^{131}-Hippuran-Standardclearance (ml/min/1,73 m^2)	GK-J^{131}-Hippuranclearance		
	20 min	30 min	40 min
244,1	224	200	168
270,2	292	269	224
46,0	46	44	36

Mit Absicht wurden Beispiele gewählt, bei denen die Differenz zwischen den Blutentnahmen deutlich wird. Im allgemeinen sind die Unterschiede nicht so stark ausgeprägt. Aber die Tendenz zur abnehmenden Clearance mit größer werdendem Zeitabstand zur Injektion ist fast immer nachweisbar, und zwar unabhängig von der Höhe der Clearance. Der Sprung zwischen der Ganzkörperclearance 10 Minuten und 20 Minuten nach der Injektion ist am größten. Er wurde nicht in der Tabelle aufgeführt, da man auch bei dieser Methode eine bestimmte Verteilungszeit abwarten muß, die nach 10 Minuten noch nicht abgeschlossen ist.

Aus den genannten Beispielen wird klar, daß die Formel $C_{GK} = \frac{-dm/dt}{c}$ nicht unabhängig von der Zeit gültig ist, wie man zunächst annehmen möchte. Aber worin liegt die Ursache für dieses Verhalten, daß auch von DUTZ [8] und NÜSSGENS [9] bei der Totalclearanceuntersuchung mit PAH beobachtet wurde?

Zunächst haben wir angenommen, daß J^{131}-Hippuran ähnlich wie PAH so schnell durch die Nieren eliminiert wird, daß die Substanzmenge im Organismus nicht mehr zur genauen Berechnung ausreicht. Dagegen spricht der kontinuierliche Kurvenverlauf. Wenn unsere erste Annahme richtig gewesen wäre, mußte bei Cr^{51}-EDTA, das nur mit einem Fünftel der Geschwindigkeit von J^{131}-Hippuran über die Nieren ausgeschieden wird, die mit der

Zeit abnehmende Clearance nicht mehr zu beobachten sein. Der Effekt war aber bei der Cr^{51}-EDTA-Ganzkörperclearance im gleichen Maße vorhanden. Auch dies spricht gegen unsere erste Annahme. — Anschließend haben wir bei 30 Patienten das absolute und relative Verteilungsvolumen berechnet und zur Ganzkörperclearance in Beziehung gesetzt. Aber auch damit ließ sich das Problem nicht erklären. Ebenso brachten gleichzeitige venöse und arterielle Blutentnahmen keinen Hinweis für die Ursache.

Zwischen Körpergewicht und zeitlicher Abnahme der Ganzkörperclearance ließ sich eindeutig eine Abhängigkeit sichern. Benutzt man zur Abschätzung des Normalgewichtes die Broccasche Formel, war bei fast allen Übergewichtigen die Abnahme der Ganzkörperclearance nur gering ausgeprägt, bei den normal-, vorwiegend aber bei den untergewichtigen Patienten nahm die Ganzkörperclearance mit größer werdendem Abstand zur Injektion ab. Nach unserem Modellversuch kann man Meßfehler für diese Beobachtung nicht verantwortlich machen.

Das unterschiedliche Verhalten läßt aber vermuten, daß die Verschiebung der Clearancesubstanz zwischen Körper und vasalem Raum, in unserer Formel $\frac{-dm/dt}{c}$, bei übergewichtigen Patienten langsamer abläuft. Tatsächlich fanden wir unsere Vermutung bestätigt, als wir bei einigen Patienten zunächst den Quotienten von $-dm/dt$ der 20—40 Minuten Ganzkörperclearance berechneten und dann den Abfall für die Plasmakonzentration c ermittelten. Anschließend wurde der Quotient $-dm/dt$ durch den Wert der Plasmakonzentration c dividiert. Die erstbestimmte Ganzkörperclearance ist gleich 1. Je kleiner der Quotient wird, je ausgeprägter ist die Abnahme der Clearance. Bei fast gleichbleibender Clearance sinkt der Quotient kaum unter 1 ab.

Tabelle 4

		Gewicht (kg)	Größe (cm)	J^{131}-Hippuran-Ganzkörperclearance nach			
				20 min	25 min	30 min	40 min
GK-Cl.	a	61	156	426	403	398	402
	b	48	158	544	545	426	392
$-dm/dt$	a			1	0,74	0,58	0,41
	b			1	0,75	0,48	0,32
c	a			1	0,78	0,62	0,43
	b			1	0,74	0,62	0,44
$\frac{-dm/dt}{c}$	a			1	0,95	0,94	0,95
	b			1	1,01	0,78	0,73

Tab. 4 zeigt die Veränderung der Plasmakonzentration c unabhängig vom Gewicht, während der Quotient von $-dm/dt$ bei b stärker abfällt. Damit muß der Quotient von $\frac{-dm/dt}{c}$ bei b ebenfalls kleiner werden als bei a. Dieses Verhalten kann nur so interpretiert werden, daß bei vielen Normal- und Untergewichtigen der Abstrom der Clearancesubstanz aus dem Extrazellulärraum in den vasalen Raum über ein Konzentrationsgefälle relativ rasch erfolgt. Damit ist die Proportionalität von $\frac{-dm/dt}{c}$ jenseits der 25-, spätestens aber der 30-Minuten-Grenze nicht mehr gegeben. Bei größeren Fettgewebsmassen, die geringer als die Muskulatur durchblutet sind, findet der Rückstrom in den vasalen Raum langsamer statt. Deshalb tritt hier die zeitliche Beschränkung der Formel nicht so stark in Erscheinung. Wir sehen in den beschriebenen Veränderungen die Ursache für den unterschiedlich ausgeprägten Abfall der Ganzkörperclearance. Gleichzeitig ist nach unserer Ansicht damit die zunächst empirisch gefundene zeitliche Begrenzung der Ganzkörperclearance-Formel theoretisch begründet.

Die Stellung der Ganzkörperclearance in der nuklearmedizinischen Nierendiagnostik

Wie läßt sich nun die Ganzkörperclearance in die nuklearmedizinische Nierenfunktionsdiagnostik einbauen? Bisher führen wir Radionephrogramm und Kernspeicher-Nierenszintigraphie [10] nacheinander durch, da beide Untersuchungen kombiniert angewendet die besten diagnostischen Aussagen erlauben. Nach Möglichkeit sollte der Patient nicht durch eine dritte Untersuchung mit Radioisotopen belastet werden. Andererseits bietet die Ganzkörperclearance eine große Chance, beide Untersuchungen durch quantitative Meßdaten in erheblichem Maße zu ergänzen, so daß man auf diese einfach durchzuführende Untersuchung nicht verzichten sollte. Für simultane Untersuchungen bieten sich zwei Kombinationen an:

a) Ganzkörperclearance und Nephrogramm in einem Arbeitsgang mit nachfolgender Szintigraphie.

b) Computer-Kamera-Szintigraphie, kombiniert mit Nephrographie und Segment-Nephrographie mit nachfolgender Ganzkörperclearance.

Beide Verfahren haben ihre Vorteile. Welchem Weg der Vorzug gegeben werden soll, muß nach der klinischen Fragestellung entschieden werden.

Liegt die Ganzkörperclearance vor, läßt sich mit Hilfe des Nephrogramms die seitengetrennte Clearance abschätzen. Auf die Problematik der Clearancebestimmung über das Nephrogramm allein kann hier nicht eingegangen werden. Die Untersuchungen von Knudsen [11] und Taplin [12] lassen jedoch vermuten, daß die Steigung der zweiten Phase als Maß der Nierendurchblutung angesehen werden kann. Damit besteht die Möglichkeit, über das Nephrogramm die prozentuale Beteiligung der Einzelniere an der Gesamtclearance zu berechnen. Die Steigung der zweiten Phase läßt sich als Tangens rechts bzw. Tangens links leicht berechnen, wenn man die Ankathete gleich 1 setzt. Dann ist der Tangens sofort in mm aus dem Nephrogramm ablesbar. Wir gehen dabei so vor, daß wir durch die Nephrogramm-Kurve vom Erreichen der zweiten Phase bis zum Zeitpunkt zwei Minuten nach der Injektion die Tangente oder Sekante legen. Der Zeitpunkt von 2 Minuten p.i. ist nicht willkürlich gewählt. Knudsen [11] und Motzkus [13] konnten durch direkte Injektion von J^{131}-Hippuran in die Art. renalis zeigen, daß die Passagezeit durch die Nieren 2 bis 5 Minuten beträgt. Begrenzt man die Auswertung des Nephrogramms für die Clearance bis zum Zeitpunkt 2 Minuten nach der Injektion, liegt man mit Sicherheit in der „Blutflußphase", in der noch kein Abtransport von J^{131}-Hippuran stattgefunden hat.

Mit zwei Beispielen soll unser Vorgehen demonstriert werden. An einem 15 kg schweren Hund wurden in beide Ureteren Polyäthylenschläuche eingebunden. Im gleichen operativen Eingriff wurde die linke Art. renalis gedrosselt. Dann wurden Ganzkörperclearance und Nephrogramm registriert (Abb. 4). Anschließend bestimmten wir die seitengetrennte Nierenfunktion mit der Standardclearance.

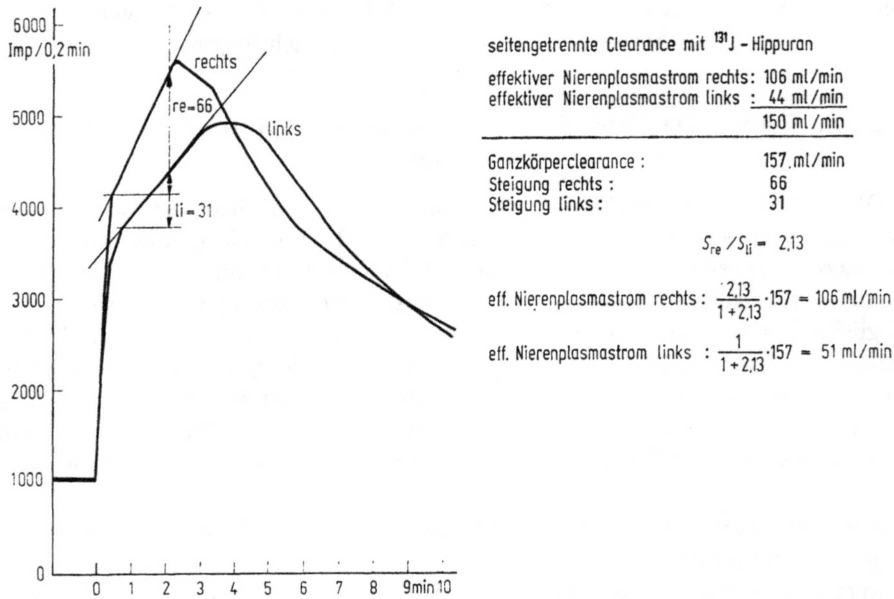

Abb. 4. Quantitative Auswertung des Nephrogramms zur seitengetrennten Nierenfunktion bei einem 15 kg schweren Hund. Die linke Art. renalis wurde gedrosselt

Im zweiten Versuch wurde zunächst das Nephrogramm geschrieben (Abb. 5) und danach die prozentuale Nierenleistung berechnet. Anschließend wurde im Rahmen der Hochdruckdiagnostik die seitengetrennte Nierenfunktion mit der Standardclearance bestimmt.

In beiden Fällen ließ sich die seitengetrennte Clearance nach dem Nephrogramm recht gut berechnen. Voraussetzung für die quantitative Auswertung des Nephrogramms ist eine optimale Lokalisation der Kollimatoren und symmetrischer Abstand der Nieren von der Meßsonde. Nachdem wir zeigen konnten, daß die Ganzkörperclearance vergleichbare Werte mit der Standardclearance ergibt, sollten weitere Anstrengungen unternommen werden, um die kombinierte Anwendung von Ganzkörperclearance und Nephrogramm mit der Fragestellung der seitengetrennten Clearance zu erarbeiten. Wenn sich die guten Ergebnisse, die wir in unseren Beispielen erhielten, an einem größeren Patientengut sichern lassen, wäre damit eine Methode gewonnen, die seitengetrennte Funktionsdiagnostik einer breiten Anwendung zuzuführen.

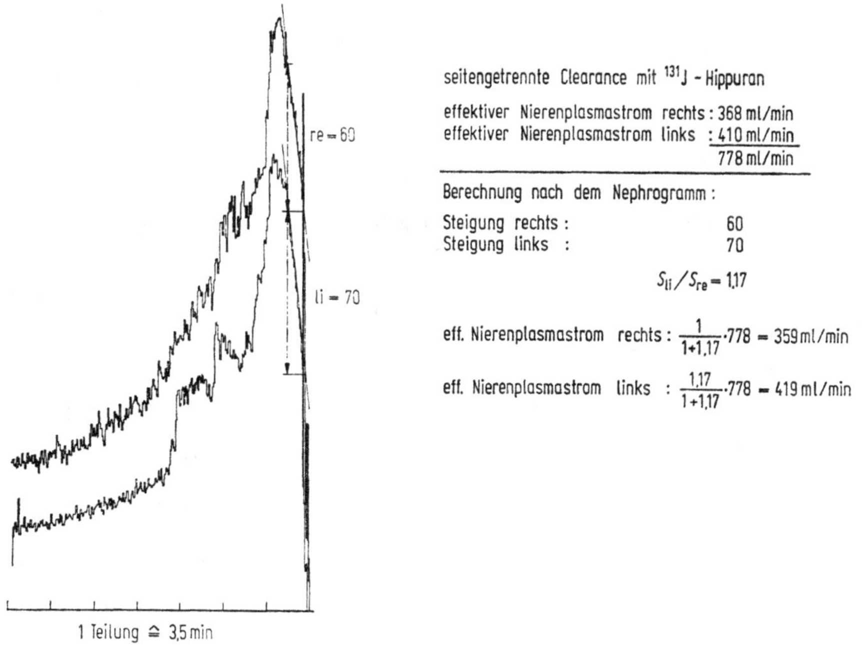

Abb. 5. Quantitative Auswertung eines Patientennephrogramms. Die seitengetrennte Nierenfunktion wurde durch Katheterisierung beider Ureteren bestimmt. Beide Bestimmungen ergeben fast gleiche Werte

Zusammenfassung

Das von OBERHAUSEN angegebene Verfahren der Ganzkörperclearance wurde im Gegensatz zur ursprünglichen Methode mit einer 5″-Meßsonde (Magna Scanner V mit Digitalzähler), einem Flüssigkeitsszintillationszähler (Armac) und einem normalen Funktionsmeßplatz (2 Sonden mit 2″NaJ(Th)-Kristall) durchgeführt. Dabei fanden wir, daß die Zeitaktivitätsmessung allein über dem Oberkörper oder Unterarm keine repräsentativen Ergebnisse für die Mengenabnahme der Clearancesubstanz im Körper ergab. Erst unter Anwendung des Funktionsmeßplatzes, mit dem die Indikatorabnahme über dem Oberkörper und den Beinen gleichzeitig registriert werden konnte, fanden wir eine sehr gute Übereinstimmung zwischen der Ganzkörperclearance und Standardclearance. Wird die Ganzkörperclearance mit den relativ kleinen Sonden eines serienmäßigen Funktionsmeßplatzes bestimmt, muß die beschriebene, in der Höhe unterschiedliche Sondeneinstellung beachtet werden. Weiterhin muß das angegebene Zeitintervall zur Blutentnahme eingehalten werden, da die Clearanceformel nur in diesem Bereich Gültigkeit hat. — Als Ursache für die Abnahme der Clearance mit zunehmendem zeitlichen Abstand zur Injektion konnte die unterschiedlich schnelle Verschiebung der Clearancesubstanz zwischen Extrazellulärraum und vasalem Raum ermittelt werden.

Zum Schluß wird an zwei Beispielen gezeigt, wie sich die Ganzkörperclearance in die nuklearmedizinische Nierendiagnostik sinnvoll einbauen läßt und wie in Kombination mit der Nephrographie eine seitengetrennte Clearance angegeben werden kann.

Literatur

1. Oberhausen, E., Romahn, A.: Bestimmung der Nierenclearance durch externe Gammastrahlenmessung. 5. Jahrestagung der Ges. f. Nuclearmedizin: Radionuklide in Kreislaufforschung und Kreislaufdiagnostik, Wien 1967.
2. — Krück, F., Romahn, A., Kirsch, W.: Vergleich zwischen der Cr-51 EDTA Clearance nach einmaliger Injektion und der klassischen Inulinclearance. Deutscher Röntgenkongreß, Hamburg 1968.
3. Dost, F. H.: Grundlagen der Pharmakokinetik. Stuttgart: Thieme 1968.
4. Earle, D. P., Berliner, R. W.: A simplified clinical procedure for measurement of glomerular filtration rate and renal plasma flow. Proc. Soc. Exp. Biol. Med. **62**, 262 (1964).
5. Pixberg, H. U., Hilger, P., Zielske, F.: Die Bestimmung des Glomerulusfiltrates mit J-125-Jothalamat. Fortschr. Röntgenstr. **111/1**, 148 (1969).
6. — — Bahlmann, J. O., Golnisch, J., Zielske, F.: Vergleich der PAH- und J-131-Hippuranclearance im steady state (im Druck).
7. Zurmühl, R.: Praktische Mathematik für Ingenieure und Physiker. Berlin-Göttingen-Heidelberg: Springer 1961.
8. Dutz, H.: Die Beziehung zwischen Verteilungsvolumen, Halbwertszeit und renaler Clearance. Z. Ges. Inn. Med. **7**, 14 (1952).
9. Nüssgens, H.: Zur Frage der Brauchbarkeit vereinfachter Clearancemethoden als Nierenfunktionsprüfung. Klin. Wschr. **32**, 491 (1954).
10. Hundeshagen, H., Henskes, D. Th., Geisler, S., Gettner, U., Creutzig, H.: Anschluß eines datenverarbeitenden Systems an einen modifizierten Dynapix-Scanner. Picker Nuclear Bulletin **2**/68.
11. Knudsen, E., Højberg, K. S.: Simulation of the radiohippuran renogram test on an analogue computer. Int. J. Appl. Radiat. **18**, 639 (1967).
12. Taplin, G. V.: The sodium iodohippurate J-131 renocystogram: Revised interpretation, terminology, and technique. In Croll, Brady (ed.): Recent Advances in Nucl. Med., Symposion 1966, New York: Appleton-Century Crofts, S. 20.
13. Motzkus, F., Bachmann, D.: Entstehung und Deutung des Radio-Renogramms. Nuclear-Medizin **7**, 46 (1968).

Der Einfluß von pH auf die Thyroxin-bindenden Proteine des Serums

Influence of pH on Thyroxine-binding Proteins in Blood Serum

A. Bekier

Isotopenlabor des Zentralen Strahleninstitutes des Kantonsspitals St. Gallen

> "... Such reports and the results of this study indicate the need for critical evaluation of the role of the various experimental techniques employed in attempting any physiologic assessment of this parameter of thyroid function ..."
>
> Hamolsky [16]

Summary

Artificial changes in the pH of the blood serum have been shown in vitro to affect the thyroxine-binding capacity of proteins; this is reduced above pH 8.5 and below pH 7.5. Intensive inhalation of O_2 and CO_2 by healthy subjects has produced discrete deviations in the thyroxine-binding proteins which may be observed even in vivo. Paper electrophoresis of ^{125}I-labelled T4 incubated for one hour in serum buffered at pH 7.0 and 8.6 and labelled at two levels of radioactivity, showed that under more acid conditions more thyroxine binds preferentially to the albumin fraction. Thus, false positive tests of thyroid function may be obtained in patients whose disease produces raised pCO_2, renal or metabolic acidosis and supraventricular arrhythmia.

Die Schilddrüsenhormone Thyroxin (T4) und Trijodthyronin (T3) gelangen nach der enzymatisch gesteuerten Proteolyse des Thyreoglobulins ins Blut [9].

Während der Transportphase im Blut werden die jodhaltigen Thyronine an spezifische Trägerproteine, das Thyroxin-bindende Präalbumin (TBPA), Thyroxin-bindende Globulin (TBG oder TBP) und zum Teil an das Albumin gebunden [11, 12, 14, 21, 22, 27].

Die Thyroxin-bindenden Proteine besitzen eine begrenzte Bindungskapazität für das Thyroxin. Sie beträgt ca. 25 μg T4/% für das TBG und ca. 100 μg T4/% für das TBPA [6, 7, 21, 22, 27].

Die Bindung von Thyroxin durch die primären Träger das TBG und TBPA ist spezifisch und stärker als die Bindung von T4 durch die Albumine [8, 27].

Ein kleiner Teil des Thyroxins zirkuliert frei im Blut. Dieser freie, nicht proteingebundene Anteil des Gesamtthyroxins, der je nach der Bestimmungsmethode $^1/_{100}$ bis $^1/_{1000}$ der gesamten Thyroxinmenge beträgt, ist für die biochemische Wirkung des Schilddrüsenhormons auf das periphere Gewebe verantwortlich [12, 17, 31, 35, 37].

Der freie Thyroxin-Anteil ist einerseits von der Menge des sezernierten Schilddrüsenhormons, anderseits von der aktuellen Bindungskapazität der Träger-Proteine abhängig [17, 20, 31, 35, 36].

Auf diese Weise erfüllen die Proteine eine passive Rolle der Trägersubstanz, gleichzeitig aber auch eine aktive Rolle eines Regulationsmechanismus, der die Thyroxin-Menge, die an die Peripherie abgegeben wird, zu beeinflussen vermag.

Gelegentlich konnte eine Bindung von Thyroxin an die Gamma-Globuline [30, 34] und an die Beta-Globuline beobachtet werden [2, 3, 18, 25, 35].

Die Bindung von Trijodthyronin an die Serum-Proteine ist wesentlich lockerer. Das T3 wird vorwiegend durch die TBG, Albumine und Beta-Globuline transportiert [13, 21, 28].

Die Bindungskapazität der Thyroxin-bindenden Proteine unterliegt großen Schwankungen.

Erkrankungen, die zu einer Hypo- oder Dysproteinämie führen, sind verständlicherweise häufig mit Änderungen der Thyroxin-bindenden Proteine verbunden [12, 16, 35, 38].

In den vergangenen 15 Jahren wurden zahlreiche Beobachtungen über kongenitale oder familiär vorkommende Abnormitäten der TBP mitgeteilt [12, 16, 21, 23, 40].

Änderungen der Thyroxin-bindenden Proteine werden während der Gravidität und unter der Behandlung mit Östrogenen und Steroiden beobachtet [5, 17, 27, 35, 41].

Eine ganze Reihe von Medikamenten führt zur Verminderung oder Zunahme der Thyroxin-bindenden Proteine durch Beeinflussung der Synthese dieser Träger-Fraktionen oder durch kompetitive Verdrängung des Thyroxins [5, 7, 17, 27, 32, 35, 41].

Erkrankungen der Schilddrüse [9] und anderer endokrinen Organe [20], bakterielle Erkrankungen der Lunge [15], postoperative Zustände [4], akuter Höheneinfluß [39], supraventrikuläre Arrhythmie [36] sowie Steigerung des partiellen CO_2-Druckes (pCO_2) führen zur Veränderung der Thyroxin-bindenden Proteine [16].

Die Zusammenhänge zwischen dem pCO_2 und den in-vitro-Funktionstesten der Schilddrüse sind aus der Literatur und der täglichen Praxis gut bekannt. Die Untersuchungen über den Einfluß von pCO_2, Wasserstoffionenkonzentration und verschiedenen technischen Untersuchungsbedingungen auf die Thyroxin-bindenden Proteine wurden meistens in einem pH-Bereich von 7,6 bis 9,0 durchgeführt [16, 28, 29].

Die folgende Arbeit ist dem Verhalten der Thyroxin-bindenden Proteine unter dem Einfluß von pH-Änderungen gewidmet.

Es wurde versucht, diese einer direkten Untersuchung kaum zugänglichen Träger-Proteine unter Einhaltung von möglichst physiologischen Bedingungen zu prüfen.

Der experimentelle Teil setzt sich zusammen aus:

1. Voruntersuchungen in vitro,
2. Untersuchungen in vivo,
3. Papierelektrophoretische Lokalisation des Thyroxins unter verschiedener Wasserstoffionenkonzentration der Pufferlösung.

1. Voruntersuchungen in vitro

Technik

Frisches Mischserum von schilddrüsengesunden Personen (Blutspendedienst des Kantonsspitals St. Gallen) haben wir mit zunehmenden Mengen von 5% HCl und 5% NaOH angesäuert und alkalisiert.

Aus jeder Probe wurde 1,0 ml für die Inkubation verwendet, der Rest diente der pH-Bestimmung.

Bei der Zugabe von HCl und NaOH haben wir auf das konstante Volumen der Proben geachtet, damit die Verdünnung des Serums in allen Proben gleich blieb. Zum Auffüllen der Serumproben auf das konstante Volumen verwendeten wir physiologische Kochsalzlösung.

Die Änderung der Thyroxin-Bindungskapazität durch Verdünnung des Serums ist somit in allen Proben konstant geblieben [8, 21].

Die pH-Bestimmungen wurden in 37 °C in einem pH-Meter („Methrom") mit Mikroglaselektrode (Komp. E 3 88) durchgeführt.

Die 7 Serumproben wurden nach Zugabe von je 2 μg ^{125}J-1-Thyroxin/% (0,02 μg T4/ml) zusammen mit dem Ionenaustauscher IRA 400 [26] während 1 Stunde in 25 °C inkubiert.

Während der Inkubationszeit haben wir die Aktivität der 7 Proben in einem Borlochkristall (Picker „Autowell") gemessen (A_1).

Nach der Inkubationszeit wurden alle Proben 6mal ausgewaschen und die Restaktivität, die an dem Ionenaustauscher geblieben ist, unter gleichen technischen Bedingungen bestimmt (A_2).

Die Restaktivität der Ionenaustauscher haben wir in % der Gesamtaktivität der einzelnen Proben ausgedrückt (A_2/A_1%).

Resultate

Die Tab. 1 und Abb. 1 geben die pH-Werte der 7 Proben sowie die Restaktivität des am Ionenaustauscher gebundenen ^{125}J-Thyroxins wieder.

Tabelle 1

Probe No.	pH	A_2/A_1 (%)
1	5,60	32,9
2	6,66	15,8
3	7,26	9,9
4	8,05	6,7
5	8,93	7,2
6	9,43	8,4
7	9,79	16,0

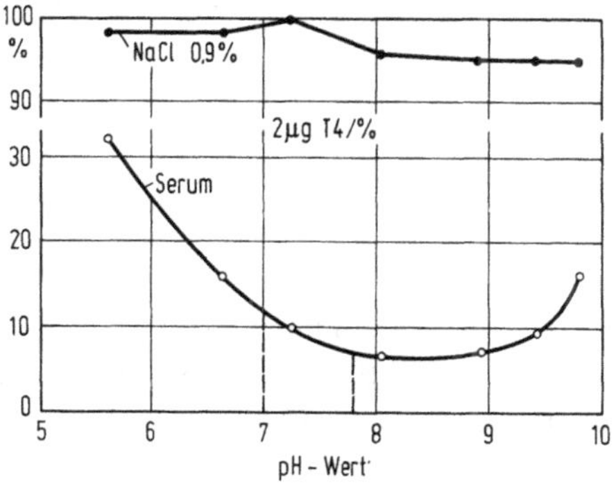

Abb. 1. ^{125}J-Thyroxin-Aktivität am Ionenaustauscher in Abhängigkeit vom pH-Wert der Serumproben

Ersetzt man das Serum durch physiologische Kochsalzlösung und wiederholt das Experiment unter gleichen Bedingungen und im gleichen pH-Bereich, so findet man am Ionenaustauscher 94,6% bis 99,7% der ^{125}J-Thyroxin-Aktivität (Abb. 1).

Um die Änderungen der Thyroxin-bindenden Proteine in einem physiologischen Bereich erfassen zu können, haben wir folgende Versuche durchgeführt.

Technik

Frisches Mischserum von schilddrüsengesunden Personen wurde mittels 1% HCl und 1% NaOH auf folgende pH-Werte eingestellt: 7,0, 7,1, 7,2 ... usw. bis pH 7,8.

Aus den 9 Serumproben verschiedener Wasserstoffionenkonzentration haben wir je 2,0 ml entnommen und nach Zugabe von 0,025 μg ^{125}J-Thyroxin/ml und 0,0025 μg ^{125}J-Thyroxin/ml das gleiche Experiment wie oben beschrieben wiederholt.

Resultate

In der Tab. 2 und Abb. 2 haben wir die Resultate der Restaktivität-Bestimmung an den Ionenaustauschern in Abhängigkeit vom pH der einzelnen Proben und der zur Inkubation verwendeten Thyroxinmenge zusammengestellt.

Tabelle 2

Probe No.	pH	A_2/A_1 (%) 2,5 μg T4/%	A_2/A_1 (%) 0,25 μg T4/%
1	7,0	16,9	9,5
2	7,1	13,9	9,1
3	7,2	14,6	9,5
4	7,3	12,6	8,9
5	7,4	13,0	8,6
6	7,5	11,9	8,9
7	7,6	11,1	8,8
8	7,7	10,5	8,3
9	7,8	11,4	8,7

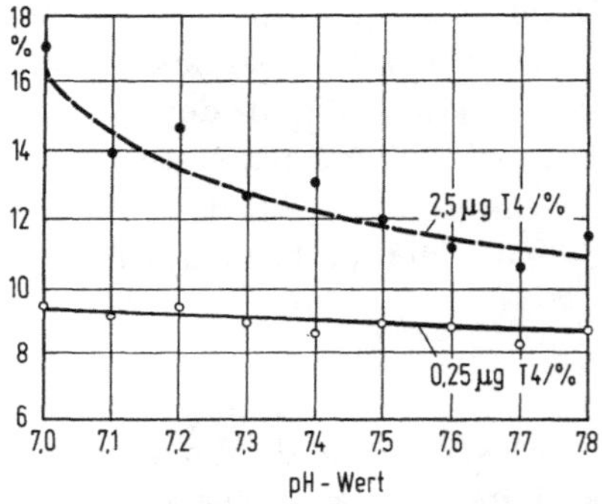

Abb. 2. ^{125}J-Thyroxin-Aktivität am Ionenaustauscher in Abhängigkeit vom pH der Serumproben im pH-Bereich von 7,0 bis 7,8

2. Untersuchungen in vivo

Technik

Ein gesunder 33jähriger Mann, dessen Schilddrüsenfunktionszustand mittels eines 2-Phasen-Radiojodstudiums überpüft wurde, atmete während 5 Minuten Sauerstoff in einer Menge von 3 l/min durch einen weichen, in ein Nasenloch eingelegten Katheter ein.

Vor der Untersuchung entnahmen wir aus der Kubitalvene die erste Blutprobe. Nach dem Beginn der O_2-Zufuhr haben wir in 1 Minute Abständen weitere 5 Blutproben entnommen.

Nach einer Ruhepause von mehreren Stunden haben wir dem gleichen Probanden 6 weitere Blutproben entnommen, nachdem er durch Atmen in einen Papiersack vermehrt CO_2 inspirierte.

Die Zeitabstände zwischen den Blutentnahmen betrugen bei dem zweiten Experiment: 1, 2, 3, 5 und 8½ Minuten.

Die Blutproben wurden genau nach 60 Minuten unter Standardbedingungen zentrifugiert und das Serum von den Erythrozyten getrennt.

Nach der Bestimmung des pH des Serums haben wir je 1,0 ml aller Serumproben mit 2,5 μg ^{125}J-Thyroxin/% und IRA 400 in 25 °C eine Stunde inkubiert.

Die Aktivitätsmessungen und das Auswaschen der Proben erfolgten wie unter Punkt 1 beschrieben.

Resultate

Die pH-Änderungen des Serums nach O_2- und CO_2-Inhalation und die Ergebnisse der Aktivitätsmessungen der einzelnen Serumproben sind in der Tab. 3 und der Abb. 3 dargestellt.

Tabelle 3

Probe No.	min O_2	min CO_2	pH	A_2/A_1 (%)
1	0		7,37	8,0
2	1		7,38	8,0
3	2		7,48	8,4
4	3		7,51	8,3
5	4		7,53	8,5
6	5		7,55	12,7(?)
7		0	7,32	8,1
8		1	7,22	8,4
9		2	7,19	8,6
10		3	7,19	9,1
11		5	7,18	verloren
12		8,5	7,17	9,3

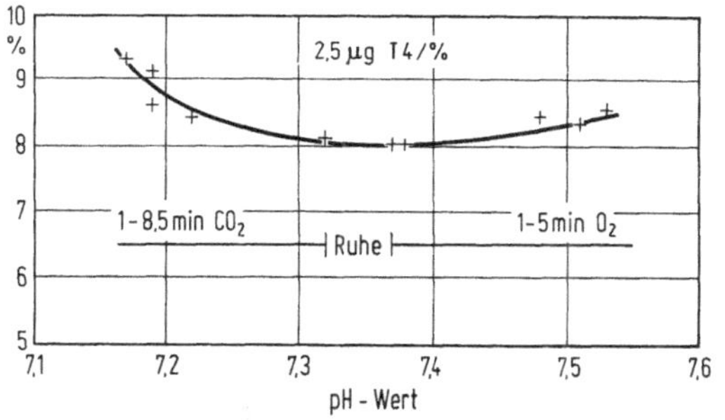

Abb. 3. Ergebnisse der ^{125}J-Thyroxin-Bindung durch den Ionenaustauscher nach O_2- und CO_2-Inhalation in vivo

3. Elektrophoretische Untersuchungen

Technik

Frisches Mischserum von schilddrüsengesunden Probanden wurde nach Zugabe von 2,0 μg ^{125}J-T4/% 2 Stunden lang bei normaler Zimmertemperatur inkubiert.

Für die elektrophoretischen Untersuchungen verwendeten wir Veronal-Natrium-Puffer, pH 8,6 und 7,0. Die kommerziell bezogene Standardlösung wurde mit Hilfe einer 1,0 n-Lösung von HCl auf einen pH-Wert von 7,0 eingestellt.

In jedem pH-Wert wurden 12 Pherogramme angefertigt. Auf die Startlinie haben wir 0,01 ml Serum aufgetragen.

Während 18 Stunden haben wir die elektrophoretische Trennung unter folgenden Bedingungen durchgeführt: 110 V, 5 mA.

Die Pherogramme wurden getrocknet, fixiert und gefärbt.

Die technisch einwandfreien Pherogramme haben wir in 5 mm breite Streifen geschnitten. Die Aktivität dieser Fragmente haben wir in einem Borlochkristall mit automatischem Probenwechsler gemessen und als Prozentanteil der Gesamtaktivität auf dem Pherogramm über den zugehörigen Protein-Fraktionen aufgetragen.

Resultate

Alle Pherogramme, welche im Veronal-Natrium-Puffer, pH 8,6 getrennt wurden, zeigten einen Aktivitätsgipfel zwischen den Alpha-1- und Alpha-2-Globulinen. Kleine Aktivitätsmengen waren über der Albumin-Fraktion und gelegentlich über den Beta-Globulinen zu finden.

Nach der Veränderung des pH-Wertes der Pufferlösung auf 7,0 zeigten die Pherogramme neben dem Aktivitätsmaximum zwischen den Alpha-1- und Alpha-2-Globulinen einen zweiten, relativ starken Aktivitäts-Peak über der Albumin-Fraktion. Auffallend war eine breite Streuung in der Intensität der beiden Gipfel (Abb. 4).

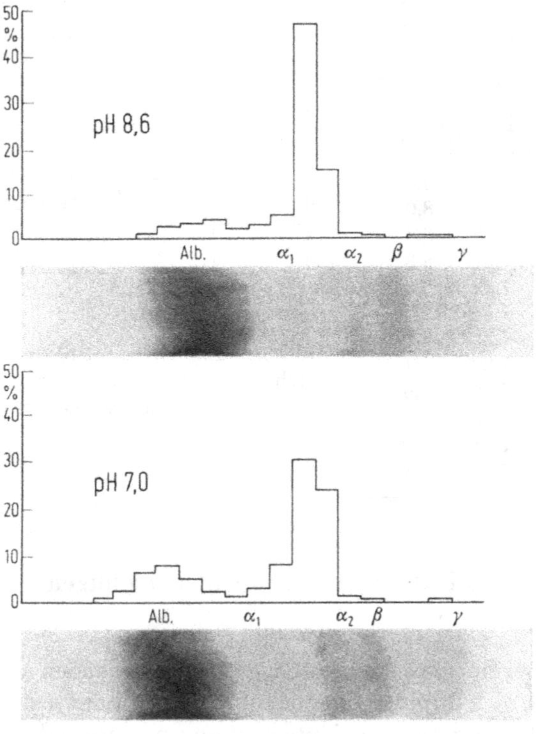

Abb. 4. ^{125}J-Thyroxin-Verteilung auf den Pherogrammen im Veronal-Natrium-Puffer, pH 8,6 und 7,0

Eine Verlagerung der Thyroxin-Aktivität aus den TBG auf die Albumine kommt zustande, wenn:

1. die Thyroxin-Menge zunimmt, so daß die aktuelle Bindungskapazität der TBG überschritten wird [3, 21, 35] oder

2. die Bindungskapazität der TBG und TBPA abnimmt, was ebenfalls zur partiellen Verlagerung der T4-Aktivität auf die Albumin-Fraktion führt.

Auf die Untersuchung der TBPA mittels Elektrophorese im Tris-Maleat-Puffer verschiedener pH-Werte konnten wir verzichten, da solche Studien bereits von MYANT und OSORIO durchgeführt wurden [29].

Die Autoren fanden eine starke Thyroxin-Bindung an die TBPA in pH-Werten der Pufferlösung von 9,2 und 8,6. Bei Veränderung der Wasserstoffionenkonzentration der Pufferlösung auf 7,6 war keine Thyroxin-Bindung an die TBPA mehr nachweisbar.

Diskussion

Das proteingebundene Thyroxin des Serums bleibt normalerweise in einem dynamischen Gleichgewicht zwischen den primären Trägerproteinen (TBG und TBPA) und dem sekundären Träger der Albumin-Fraktionen verteilt [11].

Mit Zunahme der durch die Schilddrüse sezernierten Thyroxin-Menge wird ein größerer Anteil des Hormones an die Albumin-Fraktion gebunden [3, 11].

Dieses Thyroxin kann wieder an die primären Trägerproteine zurückgebunden werden, sobald die Menge des Thyroxins abnimmt oder die Bindungskapazität der primären Trägerproteine zunimmt [11].

Die Thyroxin-Bindung durch die primären Trägerproteine ist spezifischer und stärker als die T4-Bindung an die Albumin-Fraktion [8].

Das freie, nicht proteingebundene Thyroxin des Serums bleibt im dynamischen Gleichgewicht mit dem Gesamtthyroxin des Blutes.

Durch Beeinflussung der Menge des freien Thyroxins, welches für die biologische Wirksamkeit des Schilddrüsenhormons verantwortlich ist, können die Trägerproteine eine regulatorische Funktion ausüben [17, 20, 31, 35, 37].

Die experimentellen Untersuchungen in vitro zeigen eine Änderung der Bindungskapazität der Thyroxin-bindenden Proteine in Abhängigkeit von der Wasserstoffionenkonzentration des Serums.

Die Menge des ^{125}J-Thyroxins, welche als Restaktivität am Ionenaustauscher bestimmt werden kann, hängt von folgenden Faktoren ab:

1. Qualität des Serums (Sättigung der TBG durch körpereigenes T4),
2. Menge des zur Inkubation verwendeten Thyroxins,
3. Inkubationsdauer [10, 24, 26],
4. Inkubationstemperatur [10, 24, 26].

Alle Bedingungen, bis auf das pH des Serums, haben wir bei den Untersuchungen konstant gehalten.

Man muß daher annehmen, daß sich die Thyroxin-bindenden Eigenschaften der Serum-Proteine in Abhängigkeit von der Wasserstoffionenkonzentration ändern.

Die Abb. 1 zeigt eine Kurve, die in einem pH-Bereich von etwa 7,8 abwärts (in saurer Richtung) deutlich steigt.

Im alkalischen Bereich, von pH-Werten um 8,5 zeigt sich ebenfalls ein weniger steiler Anstieg der Kurve.

Der im Bereich der alkalischen pH-Werte gelegene Schenkel unserer Kurve zeigt eine gute Übereinstimmung mit den Ergebnissen der Untersuchungen der Thyroxin-bindenden Proteine von HAMOLSKY und Mitarb. [16]. Die von den

Autoren mittels Dialyse gewonnenen Resultate zeigen eine deutliche Zunahme der Thyroxin-Durchwanderung aus dem Serum in die Pufferlösung, welche bei pH-Werten von etwa 8,3 beginnt und mit steigenden pH-Werten einen Anstieg aufweist.

Hamolsky und Mitarb. führen diese Beobachtung auf eine Verminderung der Bindungskapazität der Thyroxin-bindenden Proteine im alkalischen Milieu zurück [16].

In einem pH-Bereich, der den physiologischen Werten naheliegt, zeigt die Kurve in Abb. 1 einen leicht abfallenden Verlauf. Dieser Verlauf der „Restaktivitätskurve“ konnte auch im zweiten Versuch in vitro in einem pH-Bereich von 7,0 bis 7,8 bestätigt werden (Abb. 2). Die zur Inkubation verwendeten ^{125}J-Thyroxinmengen liegen mit 0,25 μg T4/% und 2,5 μg T4/% weit unter der physiologischen Bindungskapazität der Träger-Proteine, welche für das TBG 25 μg T4/% und für das TBPA über 100 μg T4/% beträgt [6, 7, 21, 22, 27, 35]. Interessanterweise ist die Kurve, die mit 2,5 μg T4/% gewonnen wurde, wesentlich steiler als diejenige, die aus den Meßresultaten der zehnfach geringeren ^{125}J-T4-Konzentration konstruiert wurde.

Offensichtlich wirken sich die pH-bedingten Änderungen der Thyroxin-bindenden Proteine auf sehr kleine Mengen des Schilddrüsenhormons (ca. 1% der normalen Bindungskapazität der TBG) weniger stark aus.

Der in-vivo-Test, der einen relativ kleinen pH-Bereich umfaßt und nur die Resultate von Einzelbestimmungen darstellt, zeigt ebenfalls nach Inhalation von CO_2-haltigem Gemisch einen Anstieg der Restaktivität der im tiefen pH-Bereich gelegenen Serumproben.

Das Ergebnis dieses Versuches stimmt überein mit den in vitro (Pkt. 1) gewonnenen Resultaten, muß jedoch als Einzelbestimmung mit Vorsicht betrachtet werden.

Die Verfälschung der in-vitro-Schilddrüsenfunktionstests (T3-Test, TBI, Hamolsky-Test) bei Erkrankungen, die mit einer Steigerung von pCO_2, renaler oder metabolischer Acidose und supraventrikulärer Arrhythmie verbunden sind, ist aus der Literatur [10, 12, 16] und der täglichen Praxis gut bekannt.

Die falsch positiven Resultate der Schilddrüsenfunktionstests bei den erwähnten Erkrankungen sind möglicherweise durch Änderungen der Bindungskapazität der Thyroxin-bindenden Proteine verursacht.

Anderseits zeigten nicht alle der von Hamolsky untersuchten Patienten mit erhöhtem pCO_2 und falsch positiven Resultaten der in vitro ^{131}J-T3-Erythrozyten-Aufnahme (Hamolsky-Test) eine meßbare Verschiebung der Wasserstoffionenkonzentration [16].

Die Resultate der elektrophoretischen Untersuchungen der Thyroxin-bindenden Proteine sind bekanntlich von einer ganzen Reihe technischer Untersuchungsbedingungen abhängig.

So wird die Verteilung des radioaktiv markierten Thyroxins auf dem Pherogramm von der chemischen Zusammensetzung und dem pH der Pufferlösung, z. T. Ionenstärke der Pufferlösung und der zur Inkubation verwendeten Thyroxinmenge beeinflußt [2, 3, 11, 22, 27, 28, 29, 32].

Die Qualität des Serums, insbesondere die Sättigung der Trägerproteine mit körpereigenen Hormonen sowie physiologische, pathologische oder medikamentös bedingte Änderungen der Thyroxin-bindenden Proteine können weitere Abweichungen im Verteilungsmuster des Thyroxins auf dem Pherogramm zur Folge haben.

Verwendet man zur Inkubation Thyroxinmengen, welche die physiologische Bindungskapazität der TBG nicht überschreiten, so findet man auf einem Pherogramm, das im Veronal-Puffer, pH 8,6 getrennt wurde, den Hauptanteil der Aktivität im Bereiche der Inter-Alpha-Fraktion der Globuline, den Rest der T4-Aktivität im Bereiche der Albumine und nur minimale Aktivitätsreste in den übrigen Protein-Fraktionen.

Die Ergebnisse der elektrophoretischen Untersuchungen im Veronal-Natrium-Puffer, pH 8,6 und 7,0 zeigen eine partielle Verlagerung der T4-Aktivität auf die Albumin-Fraktion bei fallenden pH-Werten der Pufferlösung.

Da die Albumin-Fraktion eine wesentlich lockerere Thyroxin-Bindung als die TBG aufweist, ist es möglich, daß der leicht dissoziierbare Thyroxin-Anteil aus der Albumin-Fraktion zur biologischen Wirksamkeit gelangen kann.

Die Diskussion über die Rolle der TBPA in der Regulation des freien Serum-Thyroxins ist noch nicht abgeschlossen [20, 21, 35, 39].

Bei schwerer Kyphoskoliose mit verminderten Atemfunktionswerten, chronischem Lungenemphysem mit pCO_2-Steigerung und anderen Erkrankungen der Luftwege fand man tatsächlich einen signifikant beschleunigten Thyroxin-Turnover mit verkürzter HWZ_{eff} des Schilddrüsenhormons [16].

Ein beschleunigter Thyroxin-Turnover wurde auch bei bakteriellen Erkrankungen der Lunge, nicht aber bei anderen Fiebererkrankungen gefunden [15]. Ein gesteigerter Verbrauch von Thyroxin in der Peripherie wurde ferner bei medikamentös erzeugtem Hypermetabolismus [1], postoperativen Zuständen [4] und akutem Höheneinfluß gefunden [39].

Unsere experimentellen Untersuchungen deuten darauf hin, daß die Thyroxin-bindenden Proteine von der Wasserstoffionenkonzentration des Serums abhängig sind.

Es ist verfrüht, zu vermuten, daß sich hinter dieser Abhängigkeit ein bisher wenig beachteter Regulationsmechanismus verbirgt, der im sauren Milieu durch Verminderung der TBG und TBPA zur Verlagerung des Thyroxins auf die Albumin-Fraktion führt, was erhöhte Thyroxinabgabe an die Peripherie zur Folge hat.

Unsere Untersuchungen sollen einen Beitrag an die umfangreichen Studien, die den TBP gewidmet sind, darstellen.

Sie zeigen auch eindrucksvoll die Relativität der gewonnenen Ergebnisse in Abhängigkeit von der verwendeten Untersuchungstechnik.

Zusammenfassung

Durch künstliche Änderungen der Wasserstoffionenkonzentration des Serums konnte eine Veränderung der Thyroxin-bindenden Proteine nachgewiesen werden.

Diese zeigte sich in einer Verminderung der Bindungskapazität der TBP in einem pH-Bereich unter 7,5 und über 8,5.

Durch Verschiebungen der pH-Werte des Serums durch intensive O_2- und CO_2-Inhalation konnten auch in vivo diskrete Abweichungen der Thyroxin-bindenden Proteine beobachtet werden.

Die elektrophoretischen Untersuchungen wurden im Veronal-Natrium-Puffer von pH 7,0 und 8,6 durchgeführt.

Das ^{125}J-Thyroxin, welches in einer Menge von 2,0 μg/% 2 Stunden vor der Elektrophorese dem Serum zugefügt wurde, fand sich in größeren Mengen auf der Albumin-Fraktion, wenn man das pH der Pufferlösung von 8,6 auf 7,0 veränderte.

Die Möglichkeit einer physiologischen Bedeutung der experimentellen Ergebnisse wird erwähnt.

Für die Durchführung der pH-Bestimmungen danke ich herzlich Dr. Gurtner, Med. Klinik des Kantonsspitals St. Gallen.

Das ^{125}J-1-Thyroxin mit einer spez. Aktivität von 20–50 mCi/mg in 50% Propylalkohol wurde vom Radiochemical Centre, Amersham bezogen.

Literatur

1. Austen, F. K., Rubini, M. E., Meroney, W. H., Wolff, J.: Salicylates and the Thyroid Function. I. Depression of Thyroid Function. J. Clin. Invest. **37,** 1131 (1958).
2. Bednar, J., Nemec, J., Soutorova, M., Röhling, S.: Radiogen verursachte Veränderungen in der Verteilung der an Serumproteine gebundenen Aktivität bei mit ^{131}J behandelten Patienten. Nuclear Medizin **7,** 73 (1968).
3. Bekier, A.: Thyroxinbindende Beta-Globuline während einer subacuten Thyreoiditis de Quervain. Nuclear Medizin **8,** 75 (1969).
4. Bernstein, G., Hasen, J., Oppenheimer, J. H.: Turnover of ^{131}I-Thyroxine in Patients Subjected to Surgical Trauma. J. Clin. Endocr. **27,** 741 (1967).
5. Blomstedt, B., Einhorn, J.: Effect of Cortisone on the Peripheral Degradation of ^{131}I-Thyroxine. J. Clin. Endocr. **25,** 181 (1965).
6. Bravermann, L. E., Dawber, N. A., Ingbar, S. H.: Observations Concerning the Binding of Thyroid Hormones of Normal Subjects of Varying Ages. J. Clin. Invest. **45,** 1273 (1966).
7. — Forster, A. E., Ingbar, S. H.: Sex-Related Differences in the Binding Serum of Thyroid Hormones. J. Clin. Endocr. **27,** 227 (1967).
8. Danowski, T. S., D'Ambrosia, R. D., Dorcak, J., Nakano, M.: Effect of Thyroxine and other Iodinated Compounds upon Binding of Triiodthyronine to Human Serum Proteins. Metabol. **11,** 443 (1962).
9. DeGroot, L. J.: Current Views on Formation of Thyroid Hormones. New Engl. J. Med. **278,** 243, 297, 355 (1965).
10. Freedberg, A. S., Hamolsky, M. W.: The Usefulness of the In Vitro Erythrocyte Uptake of ^{131}I Triiodthyronine in the Diagnosis and Therapy of Thyroid Disorders. In: Radioaktive Isotope in Klinik und Forschung, Band 5. München-Berlin: Urban & Schwarzenberg 1963, S. 161.
11. Freinkel, N., Dowling, J. T., Ingbar, S. H.: The Interaction of Thyroxine with Plasma Proteins: Localisation of Thyroxine-binding Protein in Cohn Fractions of Plasma. J. Clin. Invest. **34,** 1968 (1955).
12. Freischmidt, P.: Die Schilddrüsenerkrankungen. Stuttgart: Georg Thieme 1968, S. 7.
13. Gebraulet, K., Maurer, W.: Papierelektrophoretische und papierchromatographische Untersuchungen der organischen Jodverbindungen des Serums. In: Radioaktive Isotope in Klinik und Forschung, Band 33. München: Urban & Schwarzenberg 1955, S. 116.
14. Gordon, A. H., Gross, J., O'Connor, J., Pitt-Riviers: Nature of Circulating Thyroid Hormon Plasma Protein Complex. Nature **169,** 19 (1952).
15. Gregerman, R. I., Solomon, N.: Acceleration of Thyroxine and Triiodthyronine Turnover During Bacterial Pulmonary Infections and Fever: Implications for the Functional State of the Thyroid During Stress and in Senescense. J. Clin. Endocr. **27,** 93 (1967).
16. Hamolsky, M. W., Stein, M., Fischer, D. B., Freedberg, A. S.: Further Studies of Factors Affecting the Plasma Protein-Thyroid Complex. Endocrinology **68,** 662 (1961).
17. Hollander, Ch. S., Scott, R. L., Burges, J. A., Rabinowitz, D., Merimee, T. J., Oppenheimer, J. H.: Free Fatty Acids: A Possible Regulator of Free Hormone Levels in Man. J. Clin. Endocr. **27,** 1219 (1967).

18. HORST, W.: Neue Ergebnisse der Anwendung von J^{131} in Diagnostik und Therapie von Schilddrüsenerkrankungen. In: Vrhdl. der Dtsch. Ges. für Verdauungs- und Stoffwechselkrankheiten, W. H. BANSI (Hrsg.). Stuttgart: Georg Thieme 1954, S. 152.
19. — HEUWIESER, H.: Radiopapierchromatographische Untersuchungen zur Frage der endokrinen Leistung des Schilddrüsenkarzinoms als Basis der Radiojodtherapie. Strahlentherapie **102,** 379 (1957).
20. INADA, M., STERLING, K.: Thyroxine Turnover and Transport in Active Acromegaly. J. Clin. Endocr. **27,** 1019 (1967).
21. INGBAR, S. H., FREINKEL, N.: Thyroid Hormones. In: Hormones in Human Plasma, H. N. ANTONIADES (Ed.). Boston: Little. Brown and Co., S. 513.
22. —: Pre-Albumin, a New Thyroxin-Binding Protein of Human Plasma: Its Isolation and Physiologic Activity in Normal and Abnormal States. J. Clin. Invest. **37,** 903 (1958).
23. —: Clinical and Physiological Observations in a Patient with Idiopathic Decrease in the Thyroxin-Binding Globulin of Plasma. J. Clin. Invest. **40,** 2053 (1961).
24. KEIDERLING, W., EMRICH, D., PFANNENSTIEL, P., KLEINE, N., HOFFMANN, G.: Klinische Brauchbarkeit der in vitro Erythrozyten-Aufnahme von ^{131}J-markiertem Trijodthyronin zur Beurteilung der Schilddrüsenfunktion. In: Radioaktive Isotope in Klinik und Forschung, Band 5. München-Berlin: Urban & Schwarzenberg 1963, S. 178.
25. MAGAS, S.: Thyroid Sensitivity to ^{131}J and the Distribution of Radioiodine among the Plasma Protein Fractions in Hyperthyroid Subjects treated with ^{131}J. Polish Endocrinology **16,** 5 (1965).
26. MITCHELL, M. L., O'ROURKE, M. E.: Resin Uptake of Radiothyroxine in Sera from Non-Pregnant and Pregnant Women. J. Clin. Endocr. **18,** 1437 (1957).
27. MURPHY, B. E., PATTEE, CH. J.: Determination of Thyroxine Utilizing the Property of Protein Binding. J. Clin. Endocr. **24,** 187 (1964).
28. MYANT, N. B., OSORIO, C.: Competition for Triiodthyronine between Serum Proteins and Red Cells. J. Physiol. **157,** 529 (1961).
29. — —: Paper Electrophoresis of Thyroxine in Tris-Maleate Buffer. J. Physiol. **152,** 601 (1960).
30. PREMACHANDRA, B. N., BLUMENTHAL, H. T.: Abnormal Binding of Thyroid Hormone in Sera from Patients with Hashimoto's Disease. J. Clin. Endocr. **27,** 931 (1967).
31. REICHLIN, S., UTIGER, R. D.: Regulation of the Pituitary-Thyroid Axis in Man: Relationship of TSH Concentration to Concentration of Free and Total Thyroxine in Plasma. J. Clin. Endocr. **27,** 251 (1967).
32. ROBBINS, J., RALL, J. E.: Effects of Triiodthyronine and other Thyroxine Analogues on Thyroxin-binding in Human Serum. J. Clin. Invest. **34,** 1331 (1955).
33. — WOLFF, J., RALL, J. E.: Iodoproteins in Normal and Abnormal Human Thyroid Tissue and in Normal Sheep Thyroid. Endocrinology **64,** 37 (1959).
34. — RALL, J. E., RAWSON, R. W.: An Unusual Instance of Thyroxin-binding by Human Serum Gamma Globulin. J. Clin. Endocr. **16,** 573 (1956).
35. SCAZZIGA, B. R., LEMARCHAND-BÉRAUD, TH.: Die Pathophysiologie der Schilddrüse. Acta Clinica **5** (1966).
36. SCHATZ, D. L.: Serum Free Thyroxine and Thyroxine-binding Protein Studies in Patients with Supraventricular Tachycardias. J. Clin. Endocr. **27,** 165 (1967).
37. SCHUSSLER, G. C., PLAGER, J. E.: Effect of Preliminary Purification of ^{131}J-Thyroxine on the Determination of Free Thyroxine in Serum. J. Clin. Endocr. **27,** 242 (1967).
38. SODEE, D. B., BERMAN, A. R.: The Value of the I^{131} Labeled Triiodthyronine Red Cell Uptake in the Study of Thyroid Physiology. In: Radioaktive Isotope in Klinik und Forschung, Band 5. München-Berlin: Urban & Schwarzenberg 1963, S. 187.
39. SURKS, M. I., BECKWITT, H. J., CHIDSEY, CH. A.: Changes in Plasma Thyroxine Concentration and Metabolism, Catecholamine Excretion and Basal Oxygen Consumption in Man During Acute Exposure to High Altitude. J. Clin. Endocr. **27,** 789 (1967).
40. WERNER, S. C., BIOCK, R. J., MANDL, R. H.: Circulating Iodoproteins in a Non Goitrous Adult with Primary Amenorrhea, Bony Deformities and Normal Levels of Serum Precipitable Iodin and Thyroid ^{131}I Uptake. J. Clin. Endocr. **17,** 1141 (1957).
41. — PLATMAN, S. R.: Remission of Hyperthyroidism (Graves Disease) and Altered Pattern of Serum-Thyroxine Binding Induced by Prednisone. Lancet 7416, 751 (1965 II).

Tierexperimentelle und klinische Untersuchungen zum Nachweis des sog. „alveolo-vasculären Reflexes" (Euler-Liljestrandscher Reflex) *

Animal Experiments and Clinical Investigations to Demonstrate the Alveolo-Vascular (Euler-Liljestrand) Reflex

H. SCHLEHE und K. J. WOLF

Nuklearmedizinische Sektion und Pulmonologische Sektion des Zentrums für Innere Medizin und Kinderheilkunde der Universität Ulm
und Nuklearmedizinisches Institut des Deutschen Krebsforschungszentrums Heidelberg

Summary

A method of perfusion scintigraphy using ^{99m}Tc and ^{113}In was worked out on dogs with experimentally produced airways obstruction in an attempt to obtain a quantitative evaluation of tests for the Euler-Liljestrand reflex. The method was successfully reproduced in patients, although results must be regarded as preliminary only.

Der von EULER und LILJESTRAND 1946 an Tieren entdeckte funktionale Zusammenhang zwischen alveolärem O_2- und CO_2-Partialdruck und pulmonaler Durchblutungsgröße wurde 1951 von MOTLEY et al. auch beim Menschen nachgewiesen. Das Erfolgsorgan des Reflexes liegt im arteriellen pulmonalen Gefäßbereich, der bei einer Ventilationsstörung eingeengt wird und eine Durchblutungsdrosselung bei gleichzeitiger Druckerhöhung zur Folge hat. So führen partielle Verlegungen der Atemwege zu Perfusionsstörungen des Lungenparenchyms mit daraus resultierender Herz-Kreislauf-Belastung. Als Prototyp des polyätiologischen Syndroms kann die carcinomatöse Verlegung eines größeren Bronchus angesehen werden.

Problemstellung

Die Treffsicherheit der Perfusionsszintigraphie zum Nachweis eines zentralen Bronchialcarcinoms ist erstaunlich groß, sie wird vielfach höher als die der Lungenröntgenuntersuchung eingeschätzt. Auch sehr kleine Tumoren markieren sich bereits durch eine ausgeprägte Minderfixation der radioaktiv markierten Partikel als Ausdruck einer verminderten Perfusion des betroffenen Lungenteilbereiches. In diesen Fällen kann eine Infiltration der Gefäßwandungen mit Einengung der Gefäßlumina und daraus resultierendem vermindertem Blutdurchfluß ausgeschlossen werden; vielmehr muß eine Wirksamkeit anderer Faktoren unterstellt werden. Damit rückt aber die von EULER und LILJESTRAND beschriebene Ventilations-Perfusionsbeziehung in den Mittelpunkt nuklearmedizinischen Interesses. In der vorliegenden Arbeit sollen die Beziehungen von Ventilation und

* Mit Unterstützung der Deutschen Forschungsgemeinschaft.

Perfusion tierexperimentell und an Patienten untersucht werden, und zwar unter weitgehender Imitation der im beschriebenen Falle eines Bronchialcarcinoms tatsächlich vorliegenden Verhältnisse. Als Modell für einen intrabronchialen raumfordernden Prozeß wählten wir tierexperimentell nach dem Vorschlag von FLOHR und WÜRDINGER (1967) die rückbeatmete Hundelunge. Als Modell in der Humanmedizin bot sich der bronchographierte Patient an. Der Eingriff bewirkt in beiden Fällen eine temporäre Senkung des O_2-Partialdruckes sowie eine Erhöhung des CO_2-Partialdruckes einer Lunge. Beide Modelle gestatten, die reaktiven Veränderungen der Lungenperfusion auf die verminderte Ventilation hin zu beobachten. Die von uns erarbeitete Methodik der Versuchsdurchführung, insbesondere der Datengewinnung und -verarbeitung, erlaubt darüber hinaus, quantitative Aussagen über das Ausmaß der Perfusionsreduktion zu machen. Ferner können Aussagen über den ungefähren Zeitpunkt des Eintritts und über die Dauer der Reaktion gemacht werden. Von praktisch klinischer Bedeutung scheint auch die Frage zu sein, wie weit ein typisch pulmonologischer Eingriff, wie die Bronchographie ihn darstellt, eine zusätzliche Belastung für den lungenkranken Patienten bedeutet. Schließlich sollen die Untersuchungen einen Beitrag zur Diskussion über die zugrunde liegenden mechanischen Vorgänge im Gefäßbereich erbringen. Theoretisch kann ja einer Minderfixation von Makroalbuminen in Lungenteilbereichen neben einer Gefäßengstellung mit vermindertem Antransport von markierten Partikeln auch eine Weitstellung der Gefäßbahnen über den Durchmesser der Partikelgröße hinaus zugrunde liegen. In einem solchen Falle können die Partikel die Lungenperipherie frei passieren (OESER, ERNST und KRÜGER 1967, FEINE, HAYDUK und HOFFMANN 1967).

Material und Methode

Zur Beurteilung des Ausmaßes von Perfusionsstörungen in Lungenteilbereichen müssen quantitative Auswerteverfahren angewandt werden. Eintritt und Dauer der Blutverteilungsstörung lassen sich lediglich durch wiederholte perfusionsszintigraphische Untersuchungen feststellen. Damit sind aber materielle und methodische Voraussetzungen der Untersuchungen definiert: Als Tracersubstanzen lassen sich lediglich kurzlebige Nuklide verwenden. Bei den Tierexperimenten applizierten wir ^{99m}Tc-Makro-Albumin-Aggregate (hergestellt nach der Methode von KAZEM und MAIER-BORST 1966), die Patientenuntersuchungen wurden mit ^{113m}In-Chelat durchgeführt (hergestellt nach der Methode von KRAMER et al. 1968).

Durchführung der Tierversuche: 6 ca. 20 kg schwere Hunde wurden in Trapanal-Kurznarkose auf einem Operationstisch unter dem Detektor der Szintillationskamera gelagert. Kurz vor der intravenösen Injektion von 1 mCi ^{99m}Tc-Makro-Albumin-Aggregaten wurde ein Analogmagnetbandgerät (Data Tape VR 3300) gestartet, das die von der Szintillationskamera kommenden Informationen (x-, y-, und z-Impuls) auf drei Spuren speichert (Abb. 1) (ADAM, LORENZ und SCHEER 1967). Spätestens fünf Minuten nach Injektion wurde die Untersuchung abgeschlossen. In einer zweiten Serie wurden 5 der insgesamt 6 Hunde unter den gleichen Aufnahmebedingungen erneut untersucht, und zwar nach Durchführung folgender Prozedur: Ein Carlens-Bronchialkatheter wurde bis zur Bifurkation der

Trachea vorgeschoben. Damit konnte eine getrennte Beatmung beider Lungen erzielt werden. Die Auslaßöffnung des mit der linken Lunge kommunizierenden Katheters wurde mit einem Gummiballon verschlossen, während die rechte Lunge frei mit der Außenluft kommunizierte. Im Verlauf von 20 Minuten kam es zu einer Senkung des Sauerstoffpartialdruckes und einer Erhöhung des CO_2-Partialdruckes in der linken Lunge. Anschließend wurde das markierte Makro-Albumin injiziert und die gesamte Information erneut auf dem Magnetband gespeichert.

Durchführung der Patientenuntersuchungen: Am Tage vor der Bronchographie erfolgte die erste perfusionsszintigraphische Untersuchung unter den gleichen Bedingungen wie oben beschrieben. Jedoch mußte, um beide Lungen zu erfassen, ein divergierender Kollimator benutzt werden. Die zweite szintigraphische Untersuchung erfolgte im unmittelbaren Anschluß an die Bronchographie sowie 24 Stunden und 4 Tage nach dem Eingriff.

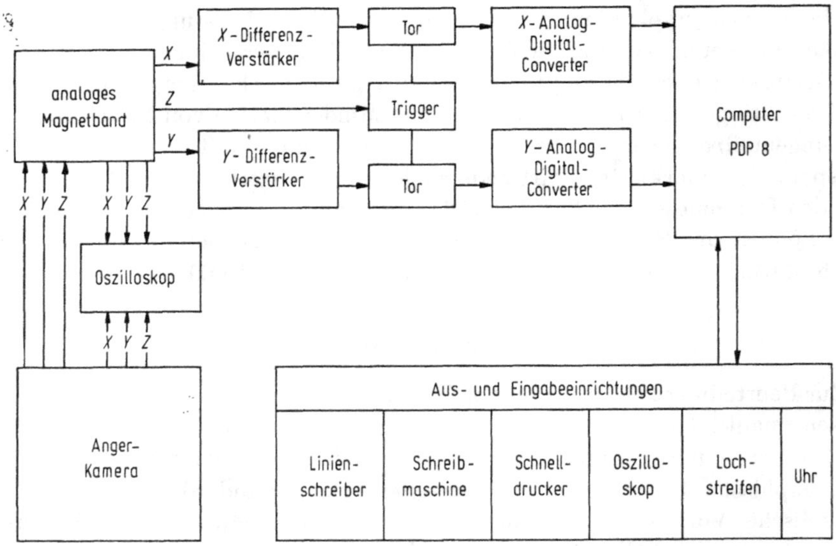

Abb. 1. Schematische Darstellung der Datenverarbeitung

Datenverarbeitung zur Gewinnung quantitativer Aussagen: Die auf dem Magnetband in analoger Form vorliegenden Daten werden über Analog-Digital-Wandler in eine 32×32-Matrix des Computers PDP 8 eingespielt. Per Programm und mit Hilfe eines homogenen Phantoms werden Inhomogenitäten der Sensitivität des Detektors korrigiert. Die Datenausgabe erfolgt am Computer-Display (Konturszintigramme, isometrische Darstellungen), über die Schreibmaschine (Teletypewriter) als Zahlenszintigramm oder über einen Plotter, soweit Zeit-Aktivitätskurven über interessierenden Lungenteilbereichen gewonnen werden sollen. Die Zeit-Aktivitätskurven können mit vorgewählten Zeitschritten ausgedruckt werden. Die über Lungenteilbereichen gemessenen Impulssummen können jedoch nicht direkt für eine Beurteilung des Ausmaßes der Belüftung verwertet werden; es handelt sich dabei um eine Schwierigkeit, die bei klinischen Beurteilungen perfusionsszintigraphischer Bilder vielfach nicht gewürdigt wird. Mit der Methodik wird ja nicht die Perfusion pro Volumeneinheit Lunge erfaßt,

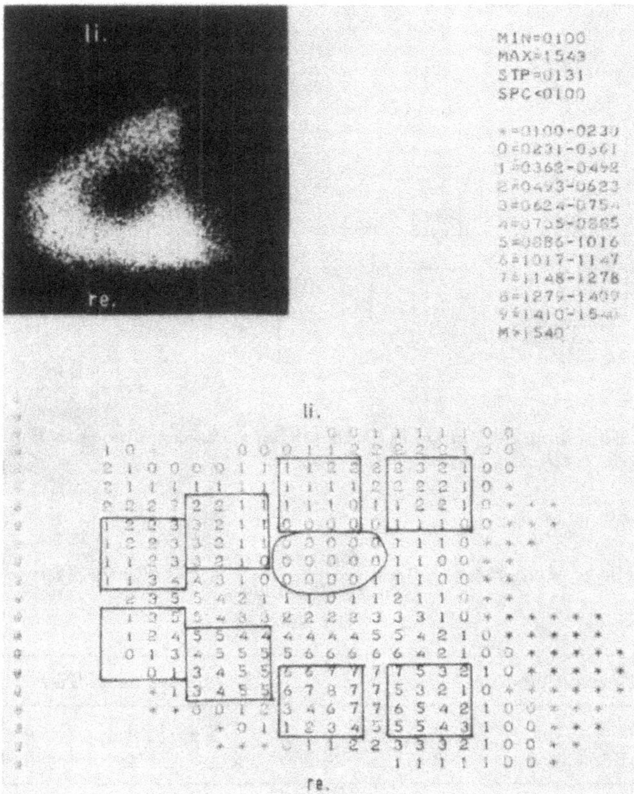

Abb. 2. Normale Hundelunge, Anger-Kamera-Bild und Computer-Zahlenszintigramm

sondern in etwa das Produkt aus Lungenvolumen und Perfusion. Der Bezugswert „Lungenvolumen“ läßt sich nicht bestimmen. Voraussetzung für eine quantitative oder zumindest semiquantitative Behandlung der perfusionsszintigraphisch gewonnenen Daten muß darum die Festlegung eines Normalbereiches des Verhältnisses der Aktivitätsverteilung in symmetrischen Lungenpartien sein. Diese Normalverhältnisse sind in Tab. 1 dargestellt. Es zeigt sich, daß die linke Lunge nur 82—94% der Aktivität der rechten Lunge speichert. Diese Differenz ist besonders deutlich in den basalen Lungenpartien, bedingt durch das einen großen Teil des basalen linken Thoraxraumes einnehmende Herz (vgl. Feld 8/7, Tab. 1a). Eine quantitative Aussage über die Umverteilung der Perfusion im pathologischen Falle muß sich auf die an einem Normalkollektiv gewonnenen Verhältniszahlen beziehen. Da neben dem Mittelwert auch die Abweichung vom Mittelwert bestimmt wurde, läßt sich die Signifikanz einer Abweichung vom Normwert überprüfen (Tab. 1). Darüber hinaus erlaubt aber auch die Änderung der Verhältniszahl gegenüber dem Normalkollektiv eine Abschätzung der Minderdurchblutung des betroffenen Lungenteilbereiches. Eine Reduktion der Durchblutung der linken Lunge z. B. um 50% bei unverändertem Minutenvolumen hat in der linken Lunge eine Aktivitätsminderung um 50% und in der rechten Lunge eine Aktivitätsver-

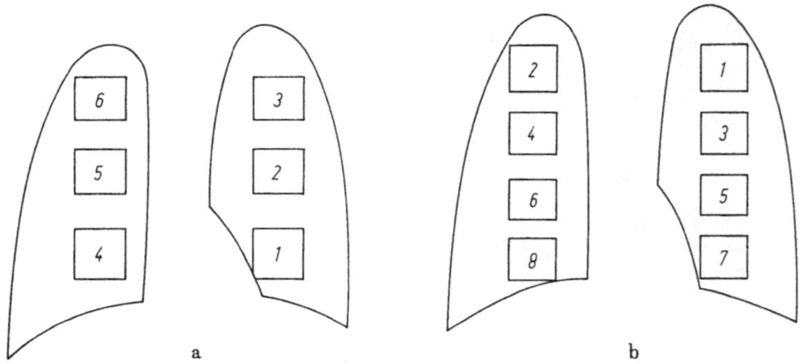

Abb. 3. Einteilung der Lunge in Teilbereiche. a Patientenlunge, b Hundelunge

Tabelle 1. *Aktivitätsverhältnis* $\frac{\textit{linke Lunge}}{\textit{rechte Lunge}} \times 100\ (\%)$ *im Tierversuch. Feldeinteilung siehe Abb. 3.*

Hund Nr.	a) Normaltiere				b) rückbeatmete Tiere			
	Feld 1 / Feld 2	3/4	5/6	7/8	1/2	3/4	5/6	7/8
6	100	99	70	93	67	39	45	55
10	100	65	46	72	52	36	20	27
4	95	111	96	90	48	50	65	57
3	82	71	60	93	54	63	35	72
2	96	83	57	63	70	64	34	57
5	90	83	53	83				
Mittel	94	85	64	82	58	51	40	54
Standard-abweichg.	± 6,9	± 17,1	± 17,7	± 12,5	± 9,6	± 12,9	± 16,5	± 16,4

mehrung um etwa 40% zur Folge, wenn man das größere Volumen der rechten Lunge berücksichtigt. Anhand der Tab. 1b seien die Verhältnisse noch einmal erläutert: Die linke Lunge zeigt bei Hunden eine Aktivitätsakkumulation von 81% der rechten Lunge (Mittelwert von 6 Hunden). Nach Rückbeatmung sinkt der Wert auf 50%. Das bedeutet aber, daß nach Rückbeatmung die Durchblutung der linken Lunge um etwa 12–14% gedrosselt wird (vgl. auch Abb. 6 u. 7).

Ergebnisse

Tierexperimentelle Untersuchungen: Der Vorgang der Einschwemmung der Makro-Aggregate in die Lunge läßt sich, wie Abb. 4a zeigt, sehr exakt mit dem oben beschriebenen Verfahren erfassen. Beim Normalhund zeigen die über verschiedenen Lungenpartien gewonnenen Zeit-Aktivitätskurven einen allmählichen Anstieg bis zu einem Plateau. Eine Ausnahme macht dabei lediglich das obere Segment der rechten Lunge, das einen vorübergehenden Anstieg in der ersten Phase, dem Durchfluß der Aktivität durch die rechte Vena subclavia entsprechend erkennen läßt (Abb. 4a, Kurve 2). Die über dem Teilbereich „Herz" gewonnene

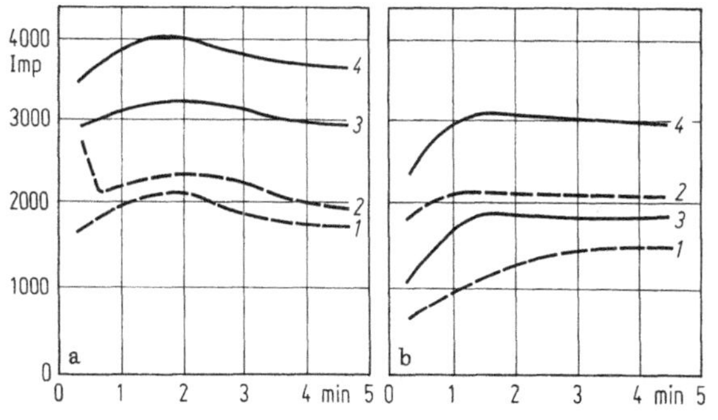

Abb. 4. Hund Nr. 2. Zeit-Aktivitätskurven. a vor, b nach Rückbeatmung links. Die Kurvenziffern entsprechen den Teilbereichen in Abb. 3

Zeit-Aktivitätskurve läßt lediglich einen Peak erkennen, der dem Durchfluß der Aktivität durch die rechte Herzhälfte entspricht (Abb. 5). Ein zweiter Peak als Ausdruck einer freien Lungenpassage der Aktivität ist nicht nachweisbar. Die rückbeatmeten Hunde lassen eine den Normalhunden entsprechende Aktivitätseinschwemmung erkennen. Allerdings liegt das Plateau im Bereich der rückbeatmeten Lunge insgesamt niedriger, der Anstieg erfolgt flacher und gering

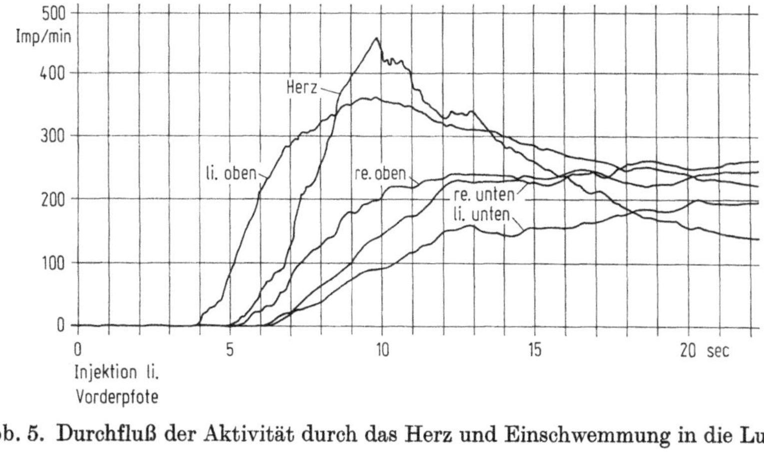

Abb. 5. Durchfluß der Aktivität durch das Herz und Einschwemmung in die Lunge

verzögert (Abb. 4b). Ein Abfall des Plateaus im weiteren Verlauf ist dabei ebensowenig zu beobachten wie bei einer normalbelüfteten Lunge. Der Nachweis einer Durchflußkurve als Äquivalent einer freien Lungenpassage gelingt somit nicht.

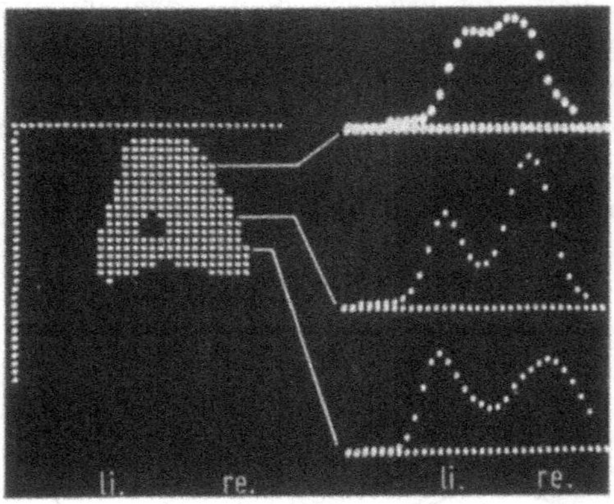

Abb. 6. Normale Hundelunge. Konturszintigramm und Profilschnitte

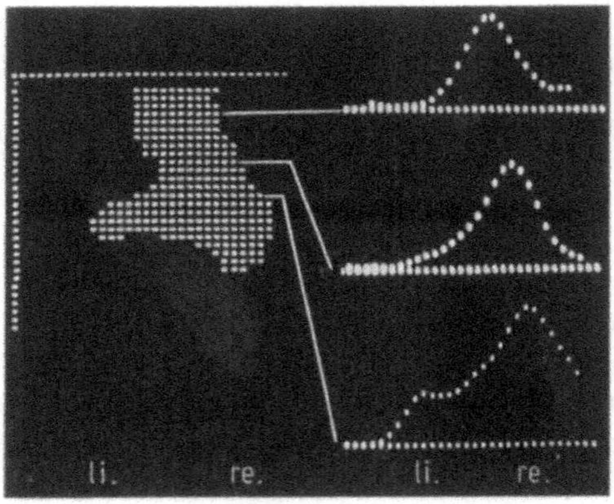

Abb. 7. Rückbeatmete Hundelunge links. Darstellung wie Abb. 6

Das Ausmaß der Minderdurchblutung der rückbeatmeten Lunge läßt sich anhand der im vorangegangenen Kapitel beschriebenen Methodik der quantitativen Auswertung abschätzen: Nach Rückbeatmung akkumuliert die betroffene Lunge nur noch 50% der rechten Lunge gegenüber 81% im Normalfalle. Die Durchblutung der linken Lunge muß also um 12—14% gedrosselt worden sein.

Klinische Untersuchungen: Die Analyse des Einschwemmungsvorganges ergibt bei den untersuchten Patienten vor der Bronchographie im Prinzip die gleichen Verhältnisse wie beim Tierkollektiv, lediglich mit einer Einschränkung: Das von uns verwandte ^{113m}In-Chelat enthielt auch Partikel, die kleiner als der mittlere Durchmesser der Lungenkapillaren waren. Infolgedessen kam es bei einem Teil der Patienten zu einer Darstellung der Leber und der Milz sowie zu einer Durchflußkurve über den Lungen. Die direkt nach Bronchographie durchgeführte

Szintigraphie zeigt Verhältnisse wie bei der rückbeatmeten Hundelunge: tieferliegendes Plateau und verzögerte Akkumulation in der minderdurchbluteten Lunge (Abb. 8 u. 9). Das Verhältnis der Radioaktivitätsverteilung entspricht bei

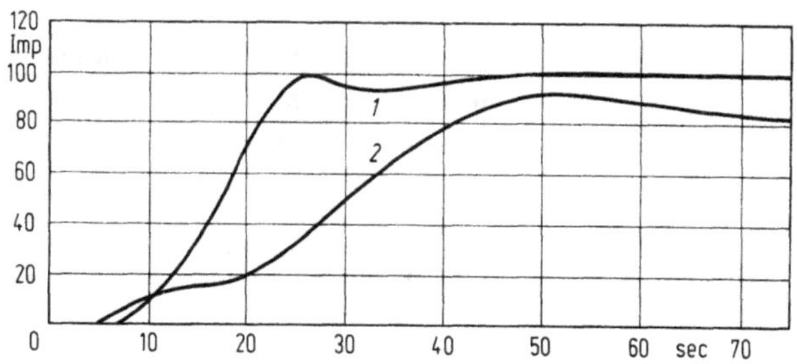

Abb. 8. Patient F. H. Aktivitätseinstrom in rechtes Unterfeld. Kurve 1: vor, Kurve 2: nach Bronchographie rechts

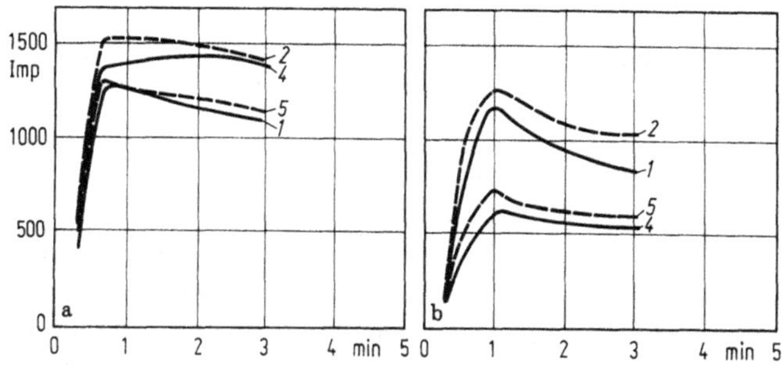

Abb. 9. Patient F. H. Zeit-Aktivitätskurven in symmetrischen Lungenteilbereichen (vgl. Abb. 3). a vor, b nach Bronchographie rechts

unseren Patienten vor der Bronchographie (Tab. 2) nicht den Normalwerten, da es sich in allen Fällen um pathologische Lungenbefunde handelte (Diagnosen siehe Tab. 3). Wir können uns also nicht auf ein Normalkollektiv beziehen: Das ist bei unseren Untersuchungen an Patienten auch gar nicht notwendig, da wir für jeden Fall Verlaufsuntersuchungen durchführten und somit den vor Broncho-

Tabelle 2. *Aktivitätsverhältnis* $\frac{\textit{rechte Lunge}}{\textit{linke Lunge}} \times 100\ (\%)$ *bei Patientenuntersuchung. Feldeinteilung siehe Abb. 3*

Pat.	vor Bronchographie			direkt nach Bronchographie			24 h nach Bronchographie			4 Tage nach Bronchographie		
	Feld 4 / Feld 1	5/2	6/3	4/1	5/2	6/3	4/1	5/2	6/3	4/1	5/2	6/3
G.O.	71	72	80	58	68	73	63	77	80	64	72	79
F.H.	123	84	80	64	41	66	92	85	97	104	82	169
R.J.	122	92	100	153	116	92	146	104	107	127	96	99
G.S.	107	107	111	191	193	199	144	116	117	117	105	111

graphie erhobenen Befund als Ausgangswert zugrunde legen können. In allen untersuchten Fällen zeigt sich nach der Bronchographie eine verminderte Perfusion der untersuchten Lunge. Die Perfusionsreduktion reicht von etwa 6%

Tabelle 3. *Rückbildung der Perfusionsstörung nach Bronchographie.*

$$\textit{Aktivitätsverhältnis}\ \frac{\textit{bronchographierte Lunge}}{\textit{normale Lunge}} \times 100\ (\%)$$

Pat.	bronchogr. Lunge	Diagnose	vor Bronchogr.	direkt nach Bronchogr.	24 h nach Bronchogr.	4 Tage nach Bronchogr.
G.O.	rechts	Emphysem re. U.F.	74	66	73	72
F.H.	rechts	Infiltr. re. O.F.	102	57	91	118
R.J.	links	Hämoptoe	97	86	86	95
G.S.	links	Bronchiektasen ?	91	50	70	90

(Patient G.O.) bis zu 19% (Patient G.S.). Ein ähnlich buntes Bild ergeben die Kurven, die die Dauer der Perfusionsstörung beschreiben. In allen Fällen ist die Erholung nach 4 Tagen abgeschlossen, in einem Falle (F.H.) wird der Anfangswert der Aktivitätsakkumulation der erkrankten Lungenseite überschritten. Dabei handelte es sich um eine Einengung der Bronchiallichtung entzündlicher Natur, die sich im Verlauf der 4 Tage zurückbildete (siehe Abb. 10, vgl. auch Abb. 11a u. b).

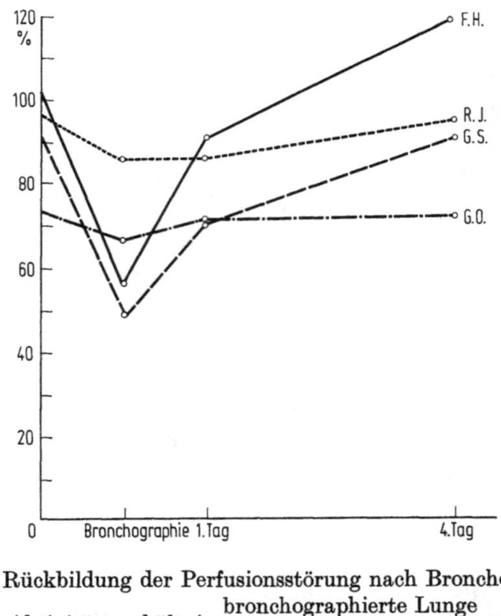

Abb. 10. Rückbildung der Perfusionsstörung nach Bronchographie.
Ordinate: Aktivitätsverhältnis $\frac{\text{bronchographierte Lunge}}{\text{normale Lunge}} \times 100(\%)$

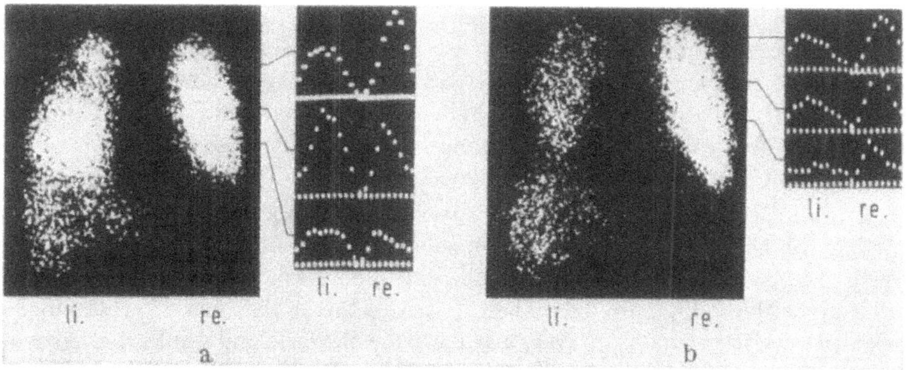

Abb. 11. Patient F. H. Szintillationskamerabilder und horizontale Konturschnitte der Ober-, Mittel- und Unterfelder. a vor, b nach Bronchographie rechts

Diskussion

Die vorgelegten Ergebnisse stimmen mit denen von EULER und LILJESTRAND überein. Sie unterstützen nicht nur die These der Existenz eines sog. alveolo-vasculären Reflexes, sondern darüber hinaus auch die von den Autoren vorgetragene Deutung des Vorganges zur Optimierung des Gasaustausches in den Lungen: Lediglich belüftete Lungenpartien werden auch perfundiert, die Drosselung nicht ventilierter Lungenbereiche verhindert eine mangelhafte Arterialisierung des venösen Blutes. Dieser Effekt kann aber lediglich durch eine Engstellung des arteriellen Schenkels der Lungengefäße erzielt werden. Daß bei der Minderbelüftung in der Tat eine Drosselung der Blutzufuhr zu den minderbelüfteten Lungenbereichen erfolgt, ergibt sich aus unserer Analyse des Einschwemmungsvorganges der radioaktiven Substanzen in die Lunge. Die Tatsache, daß bereits 20 Minuten nach Beginn der Rückbeatmung beim Hund eindeutig eine Minderdurchblutung nachweisbar ist, läßt sich darüber hinaus als Argument für den Reflexcharakter des Regulationsvorganges verwerten.

Über diesen Beitrag zur Beschreibung und Deutung des alveolo-vasculären Reflexes hinaus sollen jedoch insbesondere klinisch relevante Gesichtspunkte unserer Untersuchungen hervorgehoben werden: Erstaunlich ist das Ausmaß der verminderten Partikelfixation im Tierexperiment und bei Patienten bei relativ geringen Eingriffen in die Ventilationsvorgänge. Das stimmt überein mit der eingangs erwähnten Empfindlichkeit perfusionsszintigraphischer Verfahren beim Nachweis von zentralen Bronchialcarcinomen. Dabei muß berücksichtigt werden, daß auf der erkrankten Seite nicht nur eine verminderte Aktivitätsakkumulation entsprechend der verminderten Durchblutung resultiert, sondern auch eine vermehrte Akkumulation in der gesunden Lunge, der vermehrten Durchblutung dieser Seite entsprechend. Damit verstärkt sich die Aktivitätsdifferenz zwischen beiden Lungen, so daß auch geringe Drosselungen der Durchblutung einer Lunge deutlich sichtbar werden. Wir errechneten im Mittel sowohl für die rückbeatmete Hundelunge als auch für die bronchographierte Patientenlunge eine Reduktion der Perfusion um 10—15%. Diese Feststellung erlaubt selbstverständlich keine exakte Korrelierung zu Ventilationsgrößen, da uns das Ausmaß der O_2- und CO_2-Partialdruckänderungen nicht bekannt ist. Die Feststellung, daß die Perfusionsreduktion so gering ist und die an den beigefügten Bildern belegbare Tatsache, daß die Aktivitätsdifferenz in allen Fällen evident erscheint, erhärtet die Bedeutung der oben angeführten Überlegungen für die Erklärung der Treffsicherheit der Perfusionsszintigraphie in der Diagnostik zentraler Bronchialcarcinome.

Von klinischer Bedeutung erscheint weiterhin die Dauer der Durchblutungsstörung, die nach unseren Untersuchungen etwa 4 Tage beträgt (Abb. 10). Diese Werte stimmen sehr gut mit denen überein, die CHRISTOFORIDIS et al. (1962) mit anderer Methodik gefunden haben. Dabei gestattet aber unser Verfahren eine wesentlich weitergehende Analyse, und zwar die Betrachtung des Verhaltens von Lungenteilbereichen und nicht lediglich die Beobachtung der Globalfunktion der rechten und linken Lunge. So ist im Falle G.O. die Minderdurchblutung nach Bronchographie besonders im Bereich des rechten Unterlappens ausgeprägt nachweisbar mit verspäteter Erholung dieses Bereiches, während mittlere und obere Lungenpartien eine deutlich geringere Perfusionsstörung mit rascher Erholung erkennen lassen (vgl. Tab. 2). Es würde an dieser Stelle zu weit führen, aus Einzelbeobachtungen bei wenigen Patienten allgemeine Rückschlüsse zu ziehen. Die Ausführungen sollten lediglich ein Schlaglicht auf die Vorteile nuklearmedizinischer Methodik werfen, die sich nicht nur darum bemüht, Meßresultate durch Außenmessungen, also ohne ergebnisverfälschende Eingriffe zu gewinnen, sondern auch noch weitere Details zur Diagnostik hinzuzufügen, die auf keinem anderen diagnostischen Wege zu erzielen sind.

Zusammenfassung

Es wird über perfusionsszintigraphische Untersuchungen mit quantitativer Auswertung der Ergebnisse zum Nachweis des Euler-Liljestrandschen Reflexes berichtet. An Hunden wird eine Methodik entwickelt, die es erlaubt, eine Analyse des Einschwemmungsvorganges radioaktiv markierter Partikel in beide Lungenhälften durchzuführen. Es zeigt sich, daß bei tierexperimenteller Imitation eines obliterierenden Prozesses im Bronchialbaum das Ausmaß der Durchblutungsreduktion und die Dauer der gestörten Perfusionsverteilung erfaßt werden können. Die gewonnene Methodik wird auf Patienten übertragen, die aus diagnostischen Gründen einer Bronchographie unterzogen wurden. An diesem Krankengut können klinisch interessante Ergebnisse bezüglich der Existenz, der Wirksamkeit und der Dauer der durch den Euler-Liljestrandschen Reflex bedingten Perfusionsstörung gewonnen werden.

Literatur

1. ADAM, W. E., LORENZ, W. J., SCHEER, K. E.: Quantitative Untersuchungen mit der Szintillationskamera. IV. Jahrestagung der Ges. f. Nuclearmedizin in Heidelberg, 6.—8. 10. 1966: Radioisotope in der Lokalisationsdiagnostik. Stuttgart: Schattauer 1967, S. 55—62.
2. AMANN, W., ADAM, W. E., LORENZ, W. J., SCHEER, K. E.: Der Einsatz eines Computers in der Szintigraphie. Vortrag 12. Kongreß d. Ges. f. Radiologie, Berlin 1967.

3. Christoforidis, A. J., Nelson, S. W., Tomashefski, J. F.: Effects of bronchography on pulmonary function. Amer. Rev. Resp. Dis. **85**, 127 (1962).
4. Euler, U. S. v., Liljestrand, G.: The regulation of respiration during muscular work. Acta physiol. scand. **12**, 268 (1946).
5. Feine, U., Hayduk, K., Hoffmann, K.: Lungenfunktion und Lungenszintigramm. Radioisotope in der Lokalisationsdiagnostik. Stuttgart: Schattauer 1967, S. 231–238.
6. Flohr, H., Würdinger, H.: Szintigraphische Untersuchungen der pulmonalen Durchblutung bei experimenteller Ausschaltung einer Lunge vom Gasaustausch. Klin. Wschr. **45**, 322 (1967).
7. Gemill, W. J., Scott, D. M., Kramer, H. H.: Alternative ray to produce ^{113m}In-Macro-Iron Hydroxide. J. Nucl. Med. **9**, 170 (1968).
8. Kazem, I., Maier-Borst, W.: Zur Herstellung von ^{99m}Tc-markiertem Humanserumalbumin. Nuclear-Medizin **5**, 285–288 (1966).
9. Motley und Mitarb., zit. n. Rossier, P. H., Böhlmann, P. H., Wiesinger, K.: Physiologie und Pathophysiologie der Atmung. Berlin-Göttingen-Heidelberg: Springer 1956, S. 50.
10. Oeser, H., Ernst, H., Krüger, J.: Das normale und das von der Norm abweichende Lungenszintigramm. Röfo **106**, 549–554 (1967).

Die heutigen Möglichkeiten der positiven Tumorszintigraphie

Present Prospects for Positive Tumor Localization by Scintiscan

H.-J. Altenbrunn

Isotopen-Abteilung, Robert-Rössle-Klinik der Deutschen Akademie
der Wissenschaften zu Berlin

Summary

In order to determine the feasibility of positive tumor localization with existing equipment, laboratory tests were performed with hollow cubes of 10, 15, 20 and 40 mm side, some filled with radioactive material. Results were encouraging. Animal experiments were carried out on Wistar rats with Yoshida sarcoma to test the tumor-enriching potential of eleven substances labelled with various radioisotopes. The most successful of these, ^{131}I human fibrinogen and ^{131}I antihuman fibrinogen, gave an indication of malignancy in 21 out of 26 patients in whom their diagnostic potential was tested. Preliminary trials with ^{131}I-labelled tumor antibodies from which 90—95% of cross-reacting antibodies had been removed showed that an enrichment of up to 70 times could be obtained in UV-induced tumors in mice and up to 30 times in lymph-node metastases in a human patient. Such a procedure offers in addition the exciting prospect of treating distant metastases by the injection of γ-labelled antibodies.

In vielen Staaten sind Geschwulstkrankheiten die zweithäufigste Todesursache. Die Diagnostik maligner Tumoren hat große praktische Bedeutung, da das Schicksal des Patienten weitgehend vom klinischen Stadium der Tumorerkrankung zum Zeitpunkt der Diagnose abhängt (World Health Organization, 1966 [27]).

Ein direkter Tumornachweis mit gammamarkierten Verbindungen, entsprechend der szintigraphischen Darstellung von Leber- oder Nierenparenchym, ist leider nicht möglich [26], da es bisher noch nicht gelungen ist, tumorspezifische Reaktionen nachzuweisen und durch Gammamarkierung der Precursor derartige Reaktionen zur spezifischen Tumordiagnostik auszunutzen.

Die nuklearmedizinische Tumordiagnostik wird deshalb heute hauptsächlich als negative Tumordiagnostik ausgeführt. Maligne Tumoren haben ihre spezifische Organfunktion völlig oder weitgehend verloren und sind daher nicht in der Lage, gammamarkierte Verbindungen, die im Organparenchym gespeichert werden, aufzunehmen. Der Bereich des malignen Tumors wird deshalb im Szintigramm als Ausfallbezirk nachgewiesen.

Für den negativen Tumornachweis muß die Impulsrate über dem Tumor um 10—15% niedriger als über dem umgebenden Gewebe sein [5, 9, 24]. Durch den Einsatz der Methoden der elektronischen Datenverarbeitung werden die Möglichkeiten des negativen Tumornachweises verbessert. Dennoch ist auch hier noch eine etwa 10% geringere Impulsrate über dem Tumor notwendig [14, 17]. Der negative Tumornachweis ist um so schwieriger, je größer das zu untersuchende Organ ist, je kleiner der maligne Tumor ist und je tiefer der Tumor in dem gesunden Gewebe liegt.

Der positive szintigraphische Tumornachweis ist aus meßtechnischen Gründen viel leichter zu führen. Mit modernen Isotopenlokalisationsgeräten ist es möglich, ein μCi eines Gammastrahlers mittlerer Energie im Körper nachzuweisen und zu lokalisieren, wenn die Umgebung praktisch frei von Radioaktivität ist. Die positive Tumordiagnostik hat ihre Leistungsfähigkeit durch den Nachweis von jodspeichernden metastasierenden Schilddrüsentumoren, die Hirntumordiagnostik und die Diagnostik von Knochentumoren bewiesen.

Die Aufgabe der Arbeit ist es, die physikalischen, tierexperimentellen, immunologischen und medizinisch-diagnostischen Aspekte und Grenzen einer nuklearmedizinischen, positiven Tumordiagnostik an Hand eigener Untersuchungsergebnisse zu zeigen. Methodische Einzelheiten sind an anderer Stelle [1] eingehend beschrieben. Über einige tierexperimentelle und klinische Befunde wurde bereits berichtet [2, 3].

Physikalische Untersuchungen

Zum Vergleich der Erkennbarkeit von aktiven und inaktiven Bezirken verwendeten wir mit ^{131}J gefüllte kubische Hohlkörper aus Piacryl mit 10, 15, 20 und 40 mm Kantenlänge. Die Hohlkörper füllten wir mit einer wäßrigen Lösung, die 0,5 μCi ^{131}J/ml enthielt und führten Phantommessungen in inaktiver und radioaktiver Umgebung mit unterschiedlichem radioaktiven Untergrund durch. Die inaktiven Bezirke bestanden aus massiven PVC-Würfeln mit den gleichen Dimensionen. Mit dem Scanner PHODOT (Nuclear Chicago) wurde mit dem 3″ NaJ-Kristall und dem 37-Loch-Kollimator die Impulsrate über den heißen bzw. kalten Bezirken bei unterschiedlicher Tiefe des Würfels und bei verschieden hoher Flüssigkeitsschicht bestimmt. Die Impulsraten über dem Würfel und der Umgebung wurden zueinander in Beziehung gesetzt. Die Ergebnisse sind in Abb. 1 dargestellt.

40 mm große Würfel können in jeder Tiefe sowohl mit der positiven als auch mit der negativen Darstellung nachgewiesen werden. Bei der positiven Szintigraphie wird bei dem an der Oberfläche gelegenen 40-mm-Würfel eine 30fach höhere Radioaktivität als über der Umgebung gemessen. Demgegenüber ist bei der negativen Szintigraphie die Impulsrate über dem gleichen Würfel bei einer Schichtdicke von 120 mm nur um 45% geringer als über der Umgebung. Die Leistungsfähigkeit der positiven Szintigraphie zeigt sich besonders bei der Darstellung kleiner Bezirke. Mit dieser Methode können Testwürfel von 10 mm Kantenlänge auch in 7 cm Tiefe nachgewiesen werden, wenn die Umgebung frei von Radioaktivität ist. Die positive Szintigraphie gelingt auch, wenn die Umgebung des Würfels eine Untergrundaktivität aufweist. Diese Feststellung ist wichtig, da sich bei der positiven Tumordarstellung ein unterschiedlicher radioaktiver Untergrund nicht vermeiden läßt. Ist das Verhältnis von Würfelaktivität zur Umgebungsaktivität wie 5 : 1, so können 20 mm große Würfel auch in 8 cm

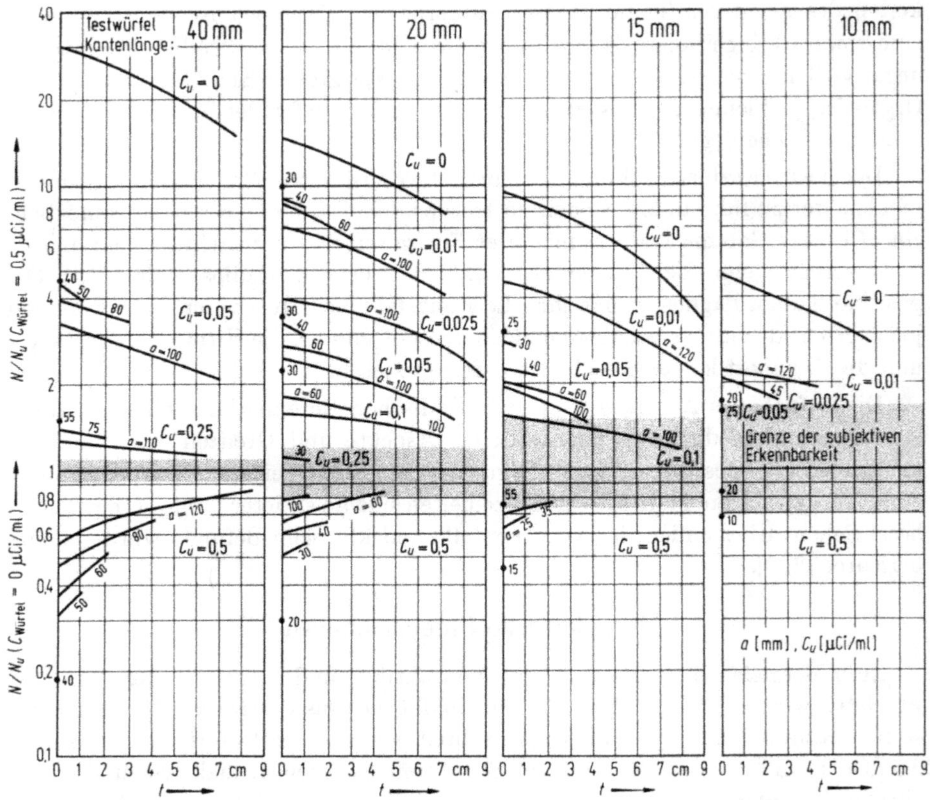

Abb. 1. Darstellung des Verhältnisses N/N_u als Funktion der Tiefe t des Testwürfels in der aktiven Schicht der Dicke a mit der Konzentration C_u. Die Abbildung zeigt oben die Ergebnisse für heiße und unten für kalte Bereiche. In dem grau markierten Gebiet war ein Nachweis im Scan nicht möglich

Tiefe gemessen werden. 15 mm große Würfel können noch bei einem Aktivitätsverhältnis von 10 : 1 in 5 cm Tiefe nachgewiesen werden.

Demgegenüber ist bei der negativen Szintigraphie ein Testwürfel von 20 mm Kantenlänge in 10 cm Schichtdicke nur noch an der Oberfläche nachweisbar. Ein 10 mm großer Würfel ist selbst in 20 mm Schichtdicke nicht mehr durch die negative Szintigraphie darzustellen.

Tierexperimentelle Untersuchungen

Mit markierten Verbindungen, von denen eine Anreicherung im Tumor bekannt oder zu erwarten war, prüften wir die Anreicherung im Tumor und die Verteilung in den Organen. Wir verwendeten dazu 150—200 g schwere Wistarratten mit soliden Impftumoren von Yoshida-Sarkomen (Y.Sa.) am Hinterschenkel. Je 50 Ratten erhielten etwa 50 μCi der zu untersuchenden Verbindungen i.v. Nach 10 min, 30 min, 1, 2, 4, 24, 48, 96 und 168 Stunden töteten wir je 5 Tiere und bestimmten die Radioaktivität von Blut, Lunge, Niere, Leber, Herz, Muskel und Tumor von getrockneten Organproben. Die Mittelwerte der Organaktivität/g Organtrockengewicht in % der applizierten Dosis wurden durch einen ZAR 1-Rechner bestimmt. Im einzelnen prüften wir die Tumoranreicherung und die Organkinetik von folgenden Verbindungen:

1. ^{131}J-Human-Fibrinogen-Antikörper (^{131}J-HFAK) vom Kaninchen. Die HFAK wurden nach einer von uns modifizierten Methode nach Spar, Bale, Goodland und Izzo, 1964 [25], isoliert und nach der Chloramin T-Methode nach Banerjee und Gibson, 1962 [4], markiert.

2. ^{131}J-Human-Fibrinogen der Firma Sorin, Saluggia, Italien.

3. ^{131}J-Ratten-Fibrinogen, das nach Blombäck und Blombäck, 1956 [7], isoliert und in Berlin-Buch markiert wurde.

4. ^{32}P als Na_2HPO_4 trägerfrei vom Institut für Kernforschung, Dresden-Rossendorf.

5. ^{203}Hg-Neohydrin vom Institut für angewandte Isotopenforschung.

6. ^{203}Hg-Azetat aus Saclay, Frankreich.

7. ^{99m}Tc als $NaTcO_4$ aus ^{99}Mo-Generatoren aus Dresden-Rossendorf.

8. ^{131}J-Dijodfluoreszin von Amersham, England.

9. ^{131}J-Ölsäure vom Institut für angewandte Isotopenforschung, Berlin-Buch.

10. ^{131}J-Lezithin. Ein speziell präpariertes Ei-Lezithin der Firma Calbiochem, Zürich, das im Molekül Ölsäure enthielt, wurde im Institut für angewandte Isotopenforschung nach der JCl-Methode ohne Schwierigkeiten markiert.

11. ^{131}J-Monojodazetat wurde im Institut für angewandte Isotopenforschung durch Austausch des Chlors in ^{131}J aus Monochloressigsäure hergestellt.

Abb. 2 zeigt die Anreicherung von verschiedenen markierten Substanzen im Yoshida-Sarkom von Ratten. Die meisten geprüften Substanzen — ^{203}Hg-Azetat, ^{203}Hg-Neohydrin, ^{99m}Tc-Pertechnetat, ^{131}J-Dijodfluoreszin, ^{131}J-Monojodazetat, ^{131}J-Lezithin und ^{131}J-Stearat — wiesen maximal eine Anreicherung von 3%/g Tumorgewicht auf. Die höchste Radioaktivität wurde dabei nach verschiedenen Zeiten gemessen. Während Monojodazetat und ^{99m}Tc-Pertechnetat sehr bald ihre maximale Konzentration im Tumor erreichten, wurde sie bei ^{203}Hg-Azetat erst nach 24 Stunden beobachtet. Eine erheblich höhere Anreicherung im Tumor wurde mit ^{32}P erzielt. Sie war etwa doppelt so hoch wie bei den eben aufgeführten Verbindungen. Die stärkste Aktivitätsanreicherung erhielten wir nach Applikation von ^{131}J-HF und ^{131}J-HFAK. Beide Verbindungen wiesen eine maximale Anreicherung von etwa 12% der applizierten Dosis/g Tumortrockengewicht auf. ^{131}J-HF und ^{131}J-HFAK wiesen ein unterschiedliches Verhalten auf. ^{131}J-HFAK war nach 4 Stunden maximal im Tumor angereichert. Danach kam es zu einer sekundären Freisetzung der Radioaktivität im Tumor, so daß die Radioaktivität des Tumors mit der Zeit abfiel. ^{131}J-HF erreichte die maximale Konzentration erst nach 24 Stunden. Die Radioaktivität fiel in der Folgezeit jedoch nicht ab.

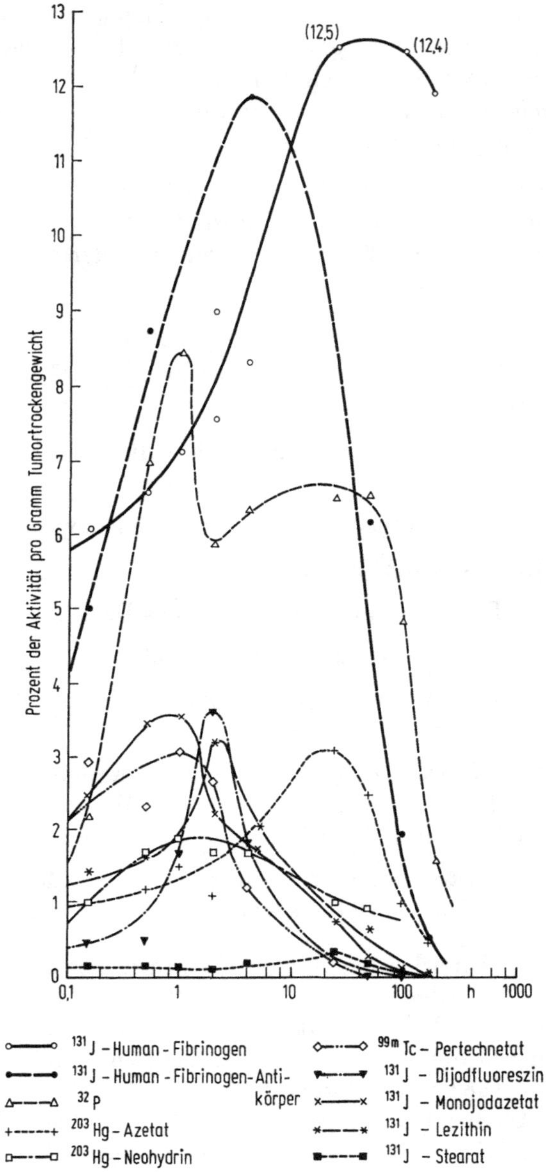

Abb. 2. Anreicherung von verschiedenen markierten Substanzen im soliden Yoshida-Sarkom von Ratten

In der Abb. 3 ist die Kinetik von ^{131}J-HFAK in Ratten mit Yoshida-Sa. dargestellt.

Die hier aufgeführten Organe Blut, Niere, Milz und Muskel zeigten eine unterschiedliche Gammastrahlung, die mit der Zeit in allen Organen gleichsinnig abnahm. Nur die Radioaktivität im Tumor verhielt sich anders, 10 min nach der Injektion betrug sie etwa 6%/g Tumortrockengewicht und stieg auf Werte um

10%/g Tumortrockengewicht an. Nach 24 Stunden war die Tumoraktivität höher als die Blutaktivität. Danach kam es jedoch im Tumor zu einem stärkeren Radioaktivitätsabfall als in den übrigen Organen, wodurch die Nachweisbedingungen verschlechtert wurden. Als Maß für die Nachweismöglichkeit von Tumoren an den Extremitäten wurde ein Anreicherungsfaktor aus Tumoraktivität durch

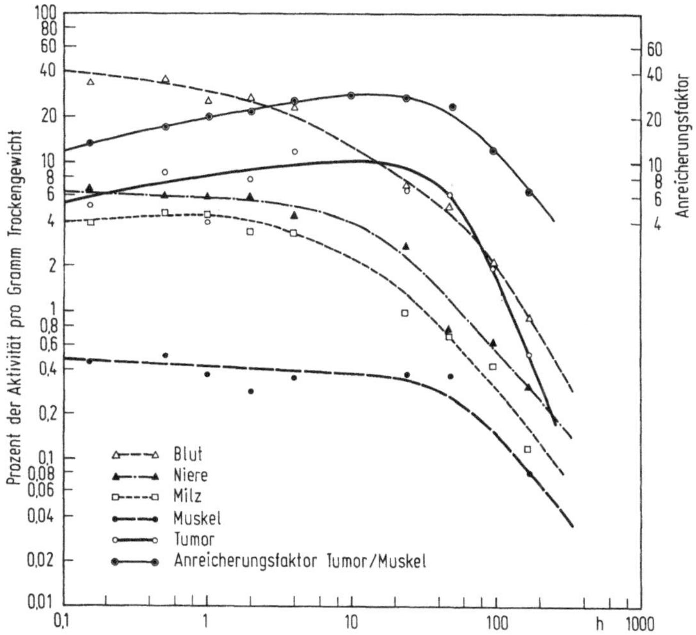

Abb. 3. Verteilung von ^{131}J-Human-Fibrinogen-Antikörpern (^{131}J-HFAK) vom Kaninchen in Ratten mit Yoshida-Sarkom

Muskelaktivität gebildet. Dieser Anreicherungsfaktor betrug nach 10 min etwa 11 und erreicht nach 4 Stunden maximale Werte von etwa 34. Durch die starke Aktivitätsfreisetzung im Tumor lag er am Ende der Untersuchungszeit bei 5. Abb. 4 zeigt die entsprechenden Ergebnisse nach i. v.-Injektion von ^{131}J-HF.

Die hier dargestellten Organaktivitäten von Blut, Niere, Milz und Muskel zeigten die gleiche Organaktivität und die gleiche Kinetik wie nach Injektion von ^{131}J-HFAK. Auch die Tumoraktivität war mit 6%/g Tumortrockengewicht nach 10 min gleich hoch wie nach ^{131}J-HFAK-Injektionen. Anreicherungsgeschwindigkeit und -ausmaß waren gleich groß wie nach ^{131}J-HFAK. Unterschiedlich war lediglich die Fixierung der Radioaktivität im Tumor. Die Radioaktivität des Tumors lag nach 1 bis 7 Tagen wesentlich höher als die aller anderen Organe. Am Ende der Untersuchung betrug dieser Unterschied zwei Größenordnungen. Diese unterschiedliche Fixierung der Radioaktivität im Tumor fand auch im Anreicherungsfaktor Tumoraktivität durch Muskelaktivität ihren Ausdruck. Während der Anreicherungsfaktor nach 10 min mit einem Wert um 10 gleich hoch wie bei ^{131}J-HFAK lag, erreichte er am Ende der Versuchsperiode Werte über 200.

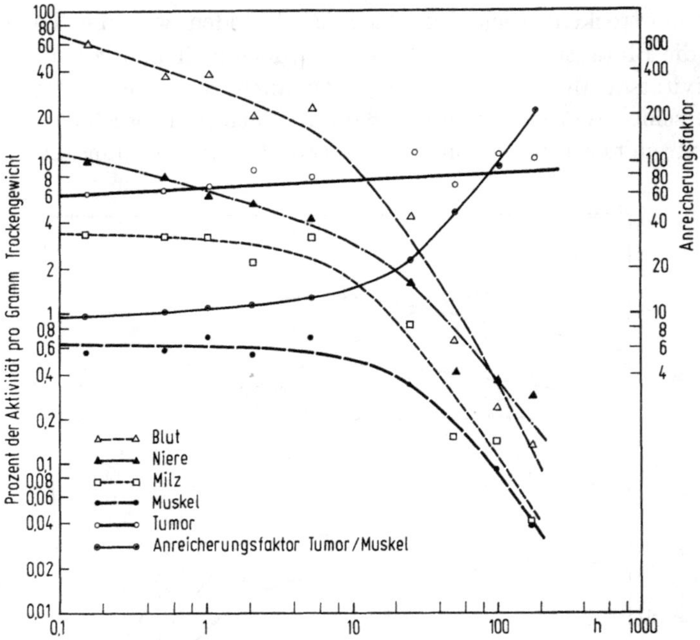

Abb. 4. Verteilung von ^{131}J-Human-Fibrinogen (^{131}J-HF) in Ratten mit solidem Yoshida-Sarkom

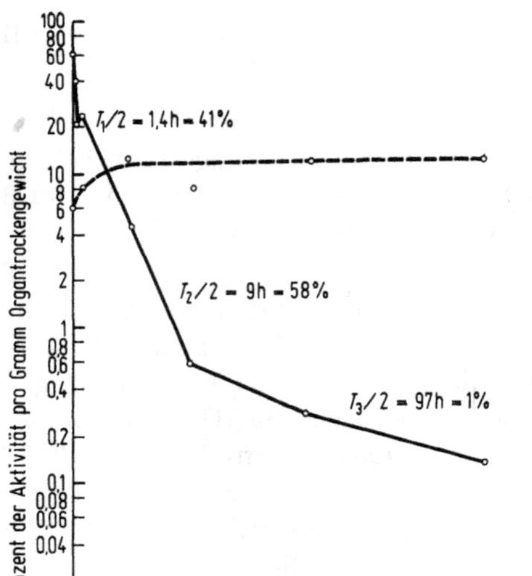

Abb. 5. Blut und Tumoraktivität nach i. v.-Injektion von ^{131}J-Human-Fibrinogen bei Ratten mit solidem Yoshida-Sarkom

Abb. 5 zeigt das Verhalten der Aktivität im Blut und im Tumor nach Applikation von ^{131}J-Humanfibrinogen bei Ratten mit Yoshida-Sarkom. Aus der halblogarithmischen Darstellung ist ersichtlich, daß die Blutaktivität mit drei unterschiedlichen Geschwindigkeitskonstanten absank. Es ist zu vermuten, daß die Radioaktivität im Tumor mit einer mittleren Geschwindigkeitskonstante angereichert wird. Durch Auftragung der Tumoraktivität gegen die Zeit im halblogarithmischen Maßstab konnten wir die Anreicherungsgeschwindigkeit im Tumor mit 9,6 h bestimmen, sie stimmt damit mit der mittleren $T/2$ von 9 h, die 58% der Radioaktivität aufwies, sehr gut überein.

Tab. 1 enthält eine Zusammenstellung der biologischen Halbwertszeiten von ^{131}J-Fibrinogen und ^{131}J-Fibrinogen-Antikörpern im strömenden Blut.

Tabelle 1. *Biologische Halbwertszeiten von ^{131}J-Fibrinogen und ^{131}J-Fibrinogen-Antikörpern im strömenden Blut*

Markierte Verbindung	Organismus	Tumor	Biologische Halbwertszeiten					
			kurz		mittel		lang	
			Std.	%	Std.	%	Std.	%
Human-FAK	Ratte	Y. Sa.	—	—	10+	64	42	36
Human-FAK	Ratte	W. Ca.	3	45	25+	30	50	25
Human-F	Ratte	Y. Sa.	1,4	41	9+	58	97	1
Human-F	Ratte	W. Ca.	1,6	54	13+	46	—	—
Human-F	Mensch	Spontan-Tumor	—	—	10	38	55	62
Ratten-F	Ratte	W. Ca.	2,0	90	—	—	54	10

+ Fraktion, die eine Anreicherung im Tumor zeigt.
Y.Sa. = Yoshida-Sa., W.Ca. = Walker-Ca.

Die mit + gekennzeichneten Fraktionen wiesen eine Anreicherung im Tumor auf. Die Tabelle zeigt, daß nur Fraktionen mit einer mittleren biologischen Halbwertszeit von 9 bis 25 Stunden eine Anreicherung im Tumor aufweisen.

In Abb. 6 sind die Anreicherungsfaktoren Tumor/Muskelgewebe nach Applikation verschiedener markierter Substanzen bei Ratten mit Yoshida-Sarkom unterschiedliche Zeiten p.i. dargestellt.

Die Verbindungen, die bereits eine schlechte Anreicherung im Tumor aufwiesen, haben auch einen geringen Anreicherungsfaktor. Für ^{99m}Tc-Pertechnetat, ^{131}J-Dijodfluoreszin, ^{131}J-Monojodazetat, ^{131}J-Lezithin, ^{131}J-Stearat erreichen die Anreicherungsfaktoren maximal den Wert 3. Da diese Verbindungen gleichzeitig eine erhebliche Speicherung in parenchymatösen Organen aufweisen, sind sie zur Szintigraphie nicht geeignet. Bei ^{32}P und ^{203}Hg-Neohydrin ist der Anreicherungsfaktor doppelt so hoch wie bei den eben genannten Verbindungen. Für ^{203}Hg-Azetat ermittelten wir beim Yoshida-Sarkom einen Anreicherungsfaktor von 15. Nach unseren Versuchen können wir feststellen, daß eine Darstellung maligner Tumoren mit Hg-Neohydrin möglich ist, wie sie bereits 1964 von SODEE beschrieben wurde [24]. Das Quecksilber-Ion weist eine höhere Affinität zum Tumor auf. Leider ist die Anreicherung in der Niere 40mal höher als im Tumor, so daß die daraus resultierende Strahlenbelastung eine Tumordiagnostik mit Hg-Azetat beim Menschen verbietet. Wesentlich höhere Anreicherungsfaktoren erzielten wir mit ^{131}J-Fibrinogen-Antikörpern (maximal über 25) und ^{131}J-Fibrinogen (Werte über 200).

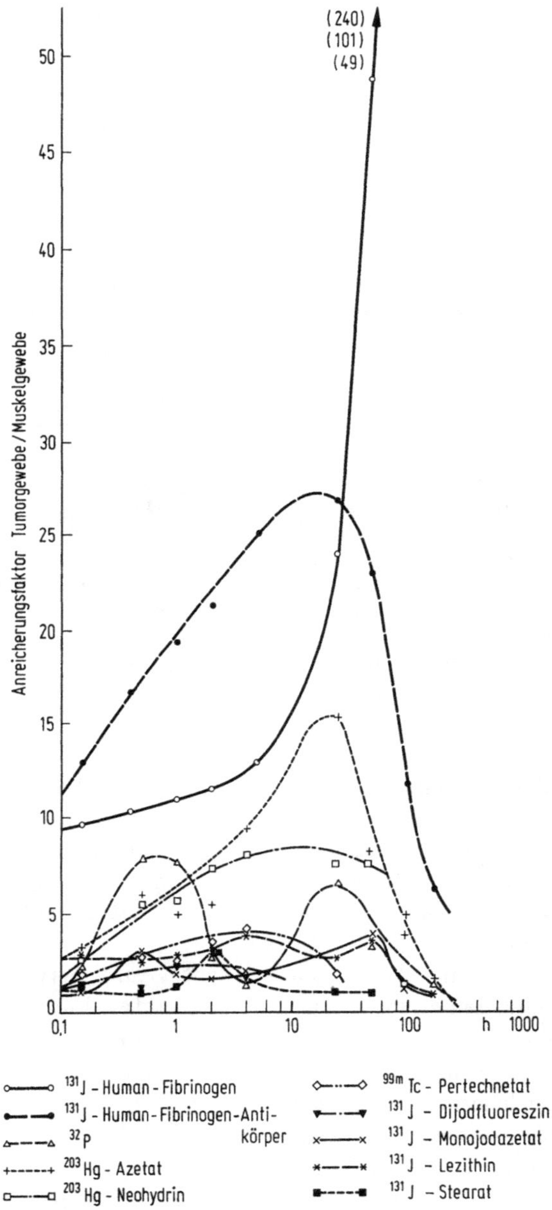

Abb. 6. Anreicherungsfaktoren Tumor/Muskelgewebe nach Applikation verschiedener markierter Substanzen bei Ratten mit Yoshida-Sarkom

Medizinisch-diagnostische Untersuchungen

Die am besten im Tumor angereicherten Verbindungen ^{131}J-HF und ^{131}J-HFAK wurden auf ihre Brauchbarkeit in der Diagnostik maligner Tumoren beim Menschen überprüft. Bei vier Patienten versuchten wir nacheinander, Tumoren mit ^{131}J-HF und ^{131}J-HFAK darzustellen (Tab. 2).

Tabelle 2. *Szintigraphischer Nachweis pathologischer Prozesse mit ^{131}J-Human-Fibrinogen und ^{131}J-Human-Fibrinogen-Antikörper*

Krankenbl. Nr.	Diagnose	HF		HFAK	
		Dosis μCi	Darst.	Dosis μCi	Darst.
765/67	Fibrosarkom re. Unterschenkel	75	—	1040	++
592/67	patholog. Fraktur re. Oberarm bei Bronchial-Ca.	790	—	1350	++
359/67	eosinophiles Granulom li. Schulter	400	+	880	++
569/67	osteolytische Meta u. Spontanfraktur li. Oberschenkel nach Bronchial-Ca.	495	—	800	++

++ gut darstellbar, + schwach darstellbar, — nicht darstellbar.

Nur in einem Fall, bei einem Patienten mit einem eosinophilen Granulom der linken Schulter, erzielten wir mit HF eine schwache Tumordarstellung. Die drei malignen Tumoren konnten mit HF nicht nachgewiesen werden. Mit ^{131}J-HFAK erreichten wir in allen Fällen eine gute szintigraphische Darstellung.

Abb. 7 zeigt oben das Scan beider Unterschenkel einer 16jährigen Patientin mit einem Fibrosarkom am rechten Unterschenkel 5 Tage nach Applikation von ^{131}J-HF. Während in diesem Falle innerhalb des Tumors nur die Untergrundstrahlung nachweisbar war, konnten wir bei der gleichen Patientin 5 Tage nach der Applikation von ^{131}J-HFAK eine deutliche Anreicherung im Tumor beobachten.

Abb. 8 enthält die Befunde einer 66jährigen Patientin mit einem Chordom. Links oben ist auf den Röntgenbildern eine Aufhellung der unteren Kreuz- und oberen Steißbeinregion durch das Chordom zu erkennen. Links unten ist das seitliche Scan 5 d nach der i.v.-Injektion von 1,2 mCi ^{131}J-HFAK abgebildet. Außer der Körperbegrenzung ist der hintere palpable Rand des Tumors eingezeichnet. Die nicht homogene Speicherungsfigur entspricht der Größe des

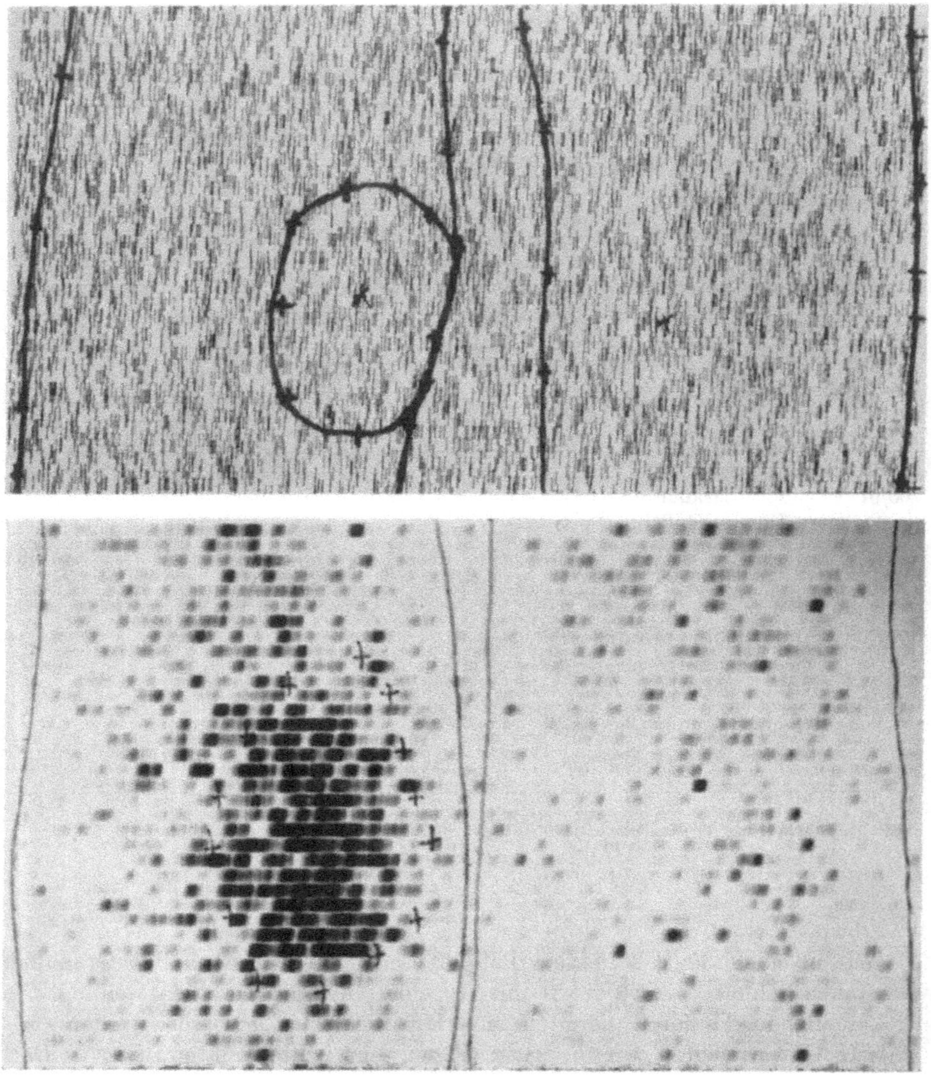

Abb. 7. Scan beider Unterschenkel bei einer 16jährigen Patientin mit einem Fibrosarkom am rechten Unterschenkel; oben: 5 Tage nach Applikation von 75 μCi ^{131}J-HF; unten: 5 Tage nach Applikation von 1040 μCi ^{131}J-HFAK. Die Fibrinolyse wurde in beiden Fällen durch Epsilon-Aminocapronsäure gehemmt

Tumors im Röntgenbild und dem Op.-Präparat. Auf der Schnittfläche des Tumors sind auf der Makroautoradiographie (rechts oben) stecknadel- bis glasstecknadelkopfgroße Bezirke zu erkennen, die entsprechend ihrer Schwärzung eine wesentlich höhere Radioaktivität als die Masse des Tumors aufweisen. Die Mikroautoradiographie (rechts unten) zeigt bei 400facher Vergrößerung, daß die Radioaktivität an den Grenzen der im Chordom charakteristischen Gewebsspalten abgelagert ist.

Immunologische Untersuchungen

Trotz der in der nuklearmedizinischen Diagnostik üblichen Tracertechnik ist ein positiver spezifischer Tumornachweis nicht möglich, da keine biochemischen Reaktionen bekannt sind, die ausschließlich im malignen Tumor ablaufen. Der einzige zur Zeit denkbare Weg für eine spezifische Tumordiagnostik mit nuklearmedizinischen Methoden wäre, Antikörper gegen Tumorgewebe zu erzeugen, sie mit Gammastrahlern zu markieren und den Tumor, den Ort der Antigen-Antikörperreaktion, mit Isotopenlokalisationsgeräten darzustellen. Die negativen Ergebnisse der 20jährigen Bemühungen der amerikanischen Arbeitsgruppen um Pressman [23] und Day [10, 11, 12] zeigen anschaulich, daß die Schwierigkeiten hierbei fast unüberwindlich sind. Bisher wurde auf zwei Wegen versucht, zu markierten Antikörpern gegen Tumorgewebe zu gelangen:

1. Reinigung des Antiserums gegen Tumorgewebe von kreuzreagierenden Antikörpern mit ^{131}J. Die Reinigung erfolgte durch Absorption der Antikörper an Tumorantigen sowie Sprengung des Antigen-Antikörperkomplexes im alkalischen Milieu bei pH-Werten von etwa 11. Da dabei Antigeneiweiß in Lösung geht, wird bei der abschließenden Markierung neben den Antikörpern gegen Tumorgewebe das gelöste Antigeneiweiß mitmarkiert.

2. Markierung der Gammaglobuline des Antiserums gegen Tumorgewebe und nachfolgende Isolierung der Antikörper. Bei diesem Verfahren ist eine nachträgliche Verunreinigung des Antiserums mit löslichem Antigeneiweiß unwichtig, da das gelöste Antigeneiweiß nicht markiert wird. Bei diesem von Day [10] eingeschlagenen Weg ist jedoch nur 0,002% der eingesetzten Aktivität im isolierten Antikörper vorhanden, so daß ein szintigraphischer Tumornachweis nicht mehr gelingt.

Da die Antigenität von Tiertumoren bekannt ist [21, 22] und bei menschlichen Tumoren vermutet wird [20] und außerdem die Antigenität der Transplantationstumoren sehr gering ist [21], erschien uns lediglich das nachfolgend beschriebene Verfahren zur Gewinnung von Antikörpern gegen Tumorgewebe, die frei von kreuzreagierenden Antikörpern und frei von gelösten Serumproteinen sind, geeignet. Nach einem Verfahren von Campbell, Luescher und Lerman, 1951 [8], modifiziert nach Drislich und Gurwitsch, 1964 [13], absorbierten wir aus dem Antiserum gegen Tumorgewebe zunächst einmal die in großer Menge vorhandenen kreuzreagierenden Antikörper. Diese Absorption erfolgte durch diazotierte Zellulose, an die Serum- bzw. Organeiweiß gekoppelt war. Mit diesen diazotierten Zellulose-Eiweißpartikeln gelang es, die kreuzreagierenden Antikörper aus dem Serum zu entfernen. Erst jetzt absorbierten wir mit diazotierten Zellulose-Tumoreiweißpartikeln die gegen das Tumoreiweiß gerichteten Antikörper aus dem Serum und spalteten sie anschließend in Zitratpuffer wieder ab. Mit diesem Verfahren gelang es uns, etwa 5% der Gesamtantikörpermenge im Serum als Antikörper, die nur gegen Tumorgewebe gerichtet sind, zu isolieren. Die Antikörper wurden anschließend mit ^{131}J nach der JCl-Methode [4] markiert.

Abb. 9 enthält Scans von Mäusen, denen gereinigte ^{131}J-markierte Antikörper gegen Tumorgewebe injiziert wurden, die von einem durch UV-Strahlung erzeugten Tumor beim Kaninchen gewonnen worden waren.

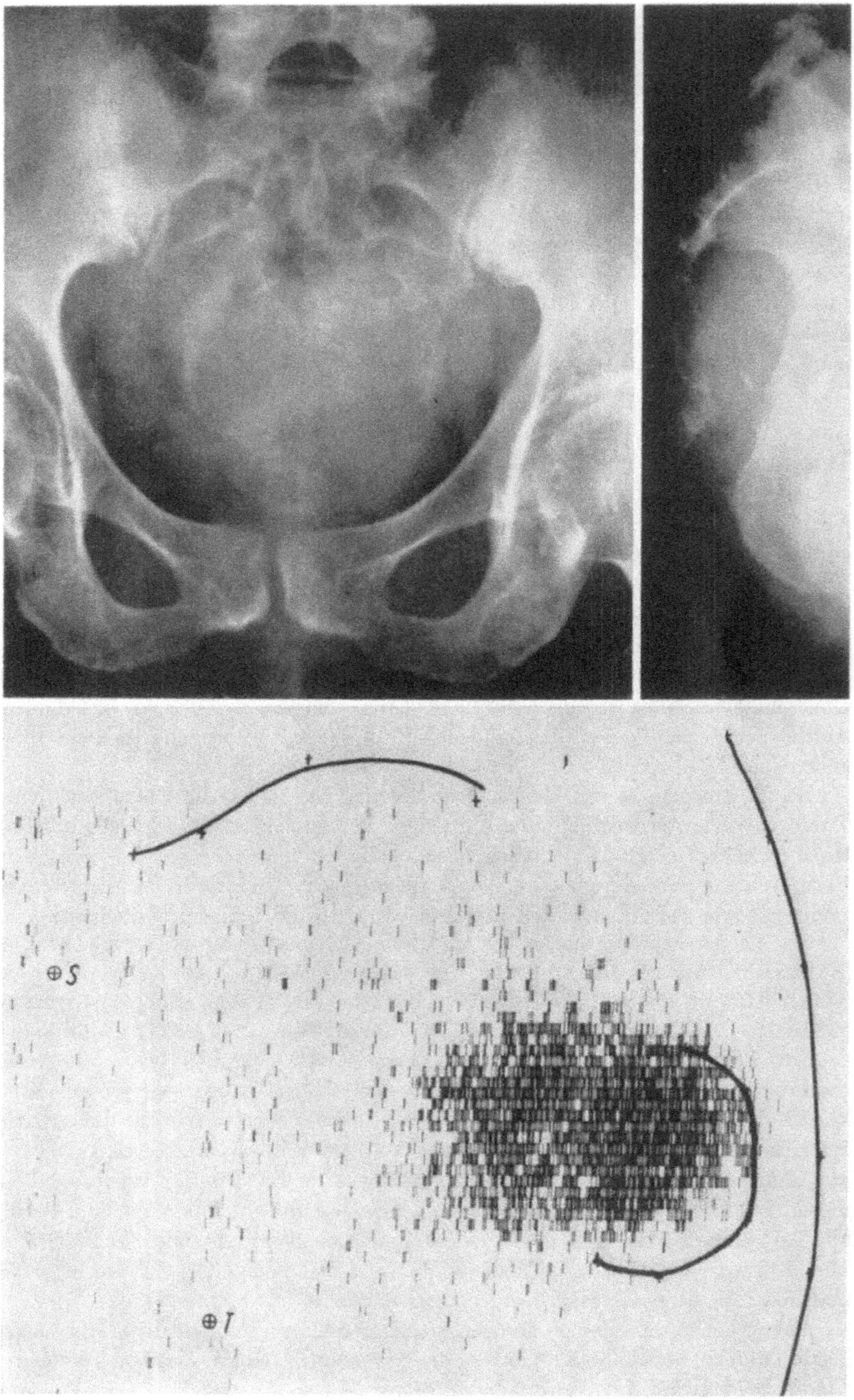

Abb. 8

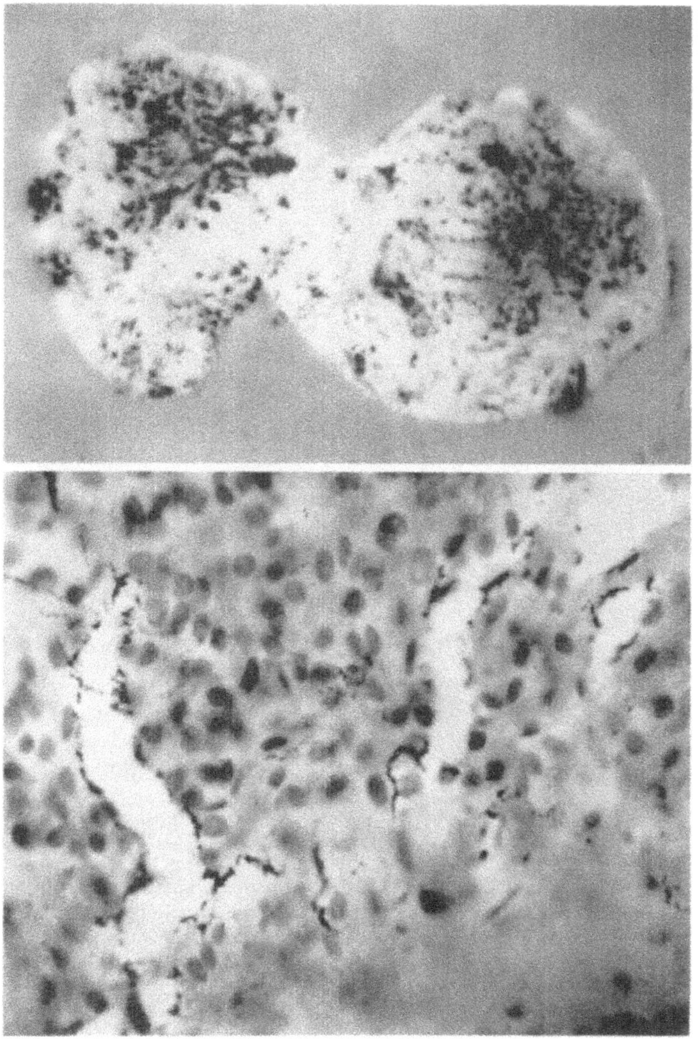

Abb. 8

Abb. 8. Befunde einer 66jährigen Patientin mit einem Chordom

← *Linke Seite*

oben: a. p. und seitliches Röntgenbild der Kreuz- und Steißbeinregion; unten: seitliches Scan 5 d nach i. v.-Injektion von 1,2 mCi ^{131}J-HFAK (S = Sp. il. ventralis, T = Trochanter major)

↑ *Rechte Seite*

oben: Makroautoradiographie des aufgeschnittenen Op.-Präparates; unten: Histoautoradiographie, 400fach vergrößert

Nach 30 min war die Radioaktivität fast gleichmäßig in der Maus verteilt, nach 24 Stunden befand sich eine große Menge in der Schilddrüse, eine erhebliche Menge in Leber, Milz und Nieren. Im antigenen Tumor 15264 der Maus war die

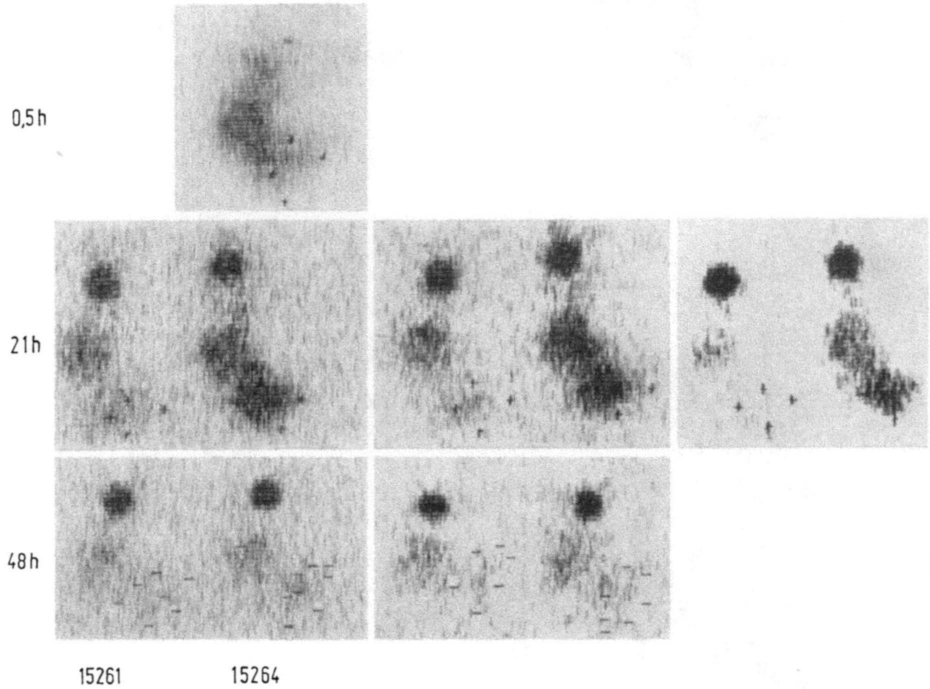

Abb. 9. Scan von Mäusen mit antigenem (15264) und nicht antigenem (15261) Tumor verschiedene Zeiten nach der i.v.-Injektion von ^{131}J-markierten Antikörpern gegen Tumorgewebe und unterschiedlicher Untergrundsubtraktion

Anreicherung 70mal höher als im nicht antigenen Tumor. Nach 48 Stunden war die Radioaktivität im antigenen Tumor sekundär freigesetzt, ähnlich wie in den Tierversuchen mit dem ^{131}J-HFAK.

Bei einer 50jährigen Patientin mit einem hypernephroiden Karzinom der re. Niere und intraoperativ gesicherten inoperablen Lymphdrüsenmetastasen erzeugten wir aus 100 g gewaschenem Tumorgewebe der exstirpierten Niere ein Antiserum beim Kaninchen. Die Reinigung des Antiserums ist in Abb. 10 dargestellt.

Im Ouchterlony-Test zeigte das Antiserum außer einer Reaktion gegen das Tumorgewebe auch Kreuzreaktionen gegen Serum- und Niereneiweiß. Nach erschöpfender Absorption an Zellulose-Serum-Proteinpartikel waren im Ouchterlony-Test keine Antigen-Antikörperreaktionen mehr gegen Serumeiweiß nachweisbar. Durch erschöpfende Absorption an Zellulose-Nierenproteinpartikel wurde das Antiserum völlig von kreuzreagierenden Antikörpern gegen Nieren-

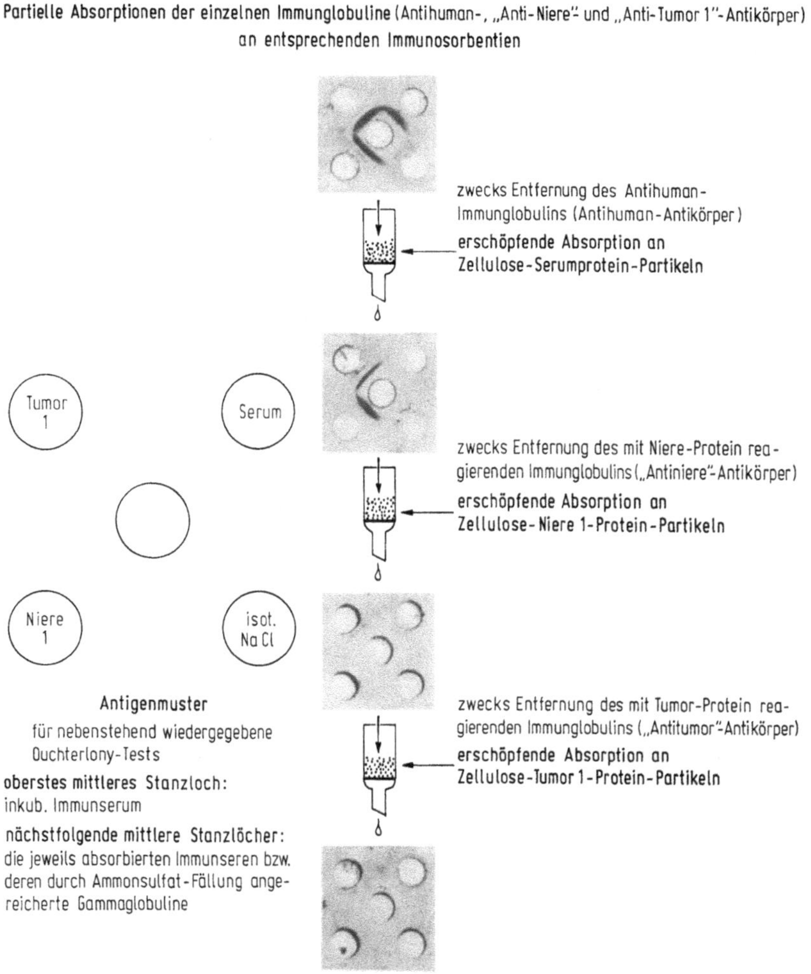

Abb. 10. Differenzierung des Immunserums

gewebe gereinigt. Das Antiserum zeigte jetzt nur noch eine schwache Reaktion gegen Tumoreiweiß. Nach Markierung von 2 mg gereinigten Antikörpern gegen Hypernephrom-Eiweiß mit 5,8 mCi ^{131}J injizierten wir die markierten Antikörper der Patientin und ermittelten die Blutaktivität.

Abb. 11 zeigt, daß die Blutaktivität mit 2 verschiedenen Geschwindigkeitskonstanten abfiel. $^2/_3$ der Aktivität hatte eine $T/2$ von 7 Stunden und $^1/_3$ eine $T/2$ von 52 Stunden. Durch Punktmessung und Szintigramme wurde bewiesen (Abb. 12), daß auch bei markierten Antikörpern gegen Tumorgewebe nur die Fraktion mit einer HWZ von etwa 10 h eine biologische Wirkung aufweist. Das Scan links zeigt 2 h nach der i.v.-Injektion von 5,8 mCi ^{131}J-Antikörpern gegen Tumorgewebe bei der auf dem Bauch liegenden Patientin eine erhebliche Radioaktivität in den zahlreichen bei der rechtsseitigen Nephrektomie zurückgebliebenen bis zu 2 cm großen Metastasen cranial des rechten Nierenlagers. Links ist über der

Milz und über dem Nierenbecken (Ausscheidung von Jodid) Radioaktivität nachweisbar. Die Nierenrinde weist keine ^{131}J-Aktivität auf — kreuzreagierende Antikörper gegen Nierengewebe waren nicht mehr nachweisbar. Nach 72 h (rechts) war die Radioaktivität in den Metastasen oberhalb des rechten Nierenlagers weitgehend verschwunden. Nur die Milz wies noch eine erhebliche ^{131}J-Aktivität auf, die wahrscheinlich durch die Fraktion mit der langen HWZ im Blut aufrechterhalten wird.

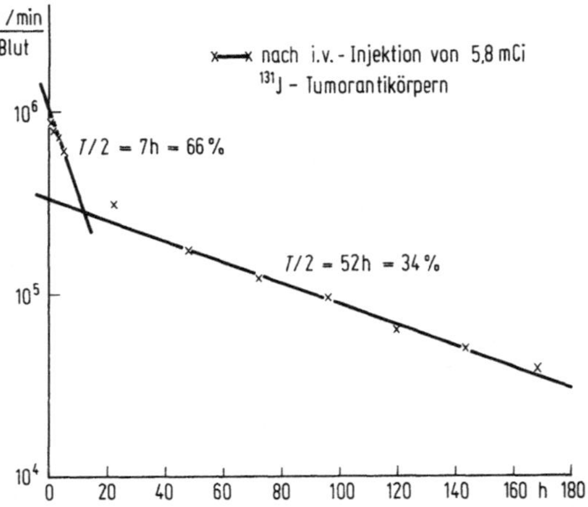

Abb. 11. Verhalten der Blutaktivität bei einer 50jährigen Patientin nach i.v.-Injektion von 5,8 mCi ^{131}J-Antikörper gegen Hypernephrom. Über 90% der Radioaktivität war am Antikörpereiweiß gebunden

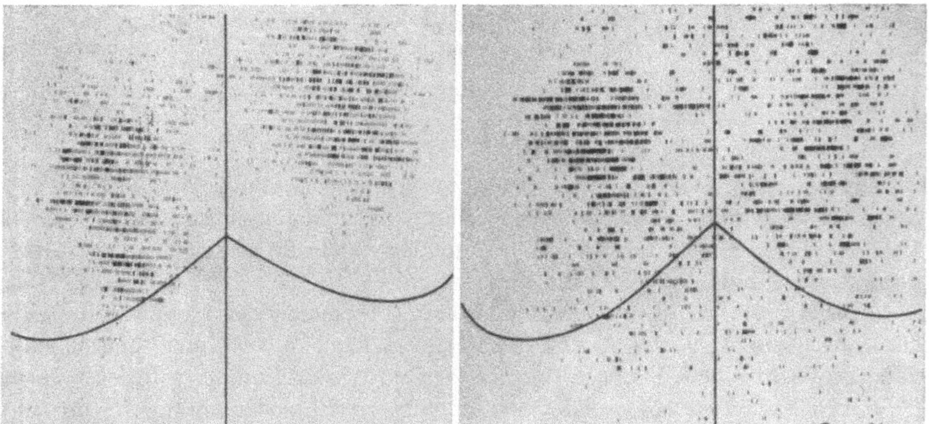

Abb. 12. p.a. Scan der Rücken- und Lendenregion einer Patientin mit Metastasen eines hypernephroiden Karzinoms oberhalb der resezierten rechten Niere nach i.v.-Injektion von 5,8 mCi ^{131}J-Antikörper gegen Tumorgewebe. Links: Scan 2 h p.i.; rechts: Scan 72 h p.i.

Diskussion und Zusammenfassung

Die von HAYES [15] und KAKEHI u. a. [16] in Salzburg gezeigten Szintigramme von positiv und negativ dargestellten Phantomen geben einen guten Eindruck von der Leistungsfähigkeit des positiven szintigraphischen Nachweises. Unser Anliegen war es, die Würfel- und Umgebungsaktivität sowohl bei negativer als auch bei positiver Darstellung zu bestimmen, um so die Möglichkeiten für die Erkennbarkeit von unterschiedlich großen Tumoren in verschiedener Tiefe und in unterschiedlich dicken Organen zu prüfen. Neben der prinzipiellen Überlegenheit der positiven Tumorszintigraphie über die negative sind in der Abb. 1 die Möglichkeiten und Grenzen beider Verfahren dargestellt. Der positive Nachweis ist selbst dann noch möglich, wenn im Organ eine erhebliche Untergrundstrahlung vorhanden ist.

Die hier zusammengefaßten Ergebnisse der Tierversuche zeigen, daß mit allen markierten Substanzen im Nukleinsäurestoffwechsel (^{32}P), Lipidstoffwechsel (^{131}J-Lezithin, ^{131}J-Stearat) oder Kohlehydratstoffwechsel (^{131}J-Monojodazetat) sowie mit Oxydationsmitteln (^{99m}Tc-Pertechnetat) und Schwermetallionen (^{203}Hg) lediglich eine unspezifische Anreicherung im Tumor erreicht wird, die zeitweise 15fach höher als im Muskelgewebe sein kann. Bei allen diesen Substanzen ist jedoch die Blutaktivität wie auch die Radioaktivität der parenchymatösen Organe im Vergleich zur Tumoraktivität hoch. Die Fixierung der Aktivität im Tumor ist ungenügend, so daß nach den Tierversuchen nur ein Einsatz zur Tumorszintigraphie an den Extremitäten sinnvoll erscheint.

Wesentlich günstigere Ergebnisse werden beim Y.Sa. mit artfremden ^{131}J-HF und ^{131}J-HFAK erzielt. Die Speicherung von HF und HFAK im Transplantationstumor ist durch die Untersuchungen von verschiedenen Autoren [11, 18] bekannt. Die vergleichende Kinetik der Substanzen zeigt darüber hinaus, daß beide Stoffe im gleichen Maße und in etwa gleicher Geschwindigkeit im Tumor gespeichert werden. Dabei ist bemerkenswert, daß die biologisch wirksamen artfremden Proteine und Antikörper mit relativ gleichen Halbwertzeiten aus dem Blut verschwinden und im Tumor angereichert werden. Während die Radioaktivität des Tumors nach ^{131}J-HFAK-Applikation sekundär wieder freigesetzt wird, verbleibt nach ^{131}J-HF-Injektion die einmal fixierte Radioaktivität im Tumor. ^{131}J-HF erweist sich somit beim Yoshida-Sa. der Ratte als ideale Substanz zur Tumorszintigraphie, da 24 Stunden und später die Radioaktivität im Tumor um Größenordnungen über der aller übrigen Gewebe lag. Da auch Ratten-Fibrinogen nicht im Tumor gespeichert wurde, besteht die Möglichkeit, daß artfremdes Fibrinogen eine uns nicht bekannte Beziehung zum Tumor hat. Diese Vermutung wird auch durch den deutlich schlechteren Tumornachweis mit ^{131}J-HF in der Tumordiagnostik beim Menschen unterstrichen.

Die Ergebnisse der Tumordiagnostik beim Menschen mit ^{131}J-HFAK entsprechen denen anderer Autoren [6, 19, 26].

Mit diesem unspezifischen Tumornachweis gelang es, bei 21 von 26 Patienten Malignome szintigraphisch darzustellen. Falsch positive Ergebnisse wurden bei entzündlichen Prozessen und bei kurz vorangegangenen Operationen beobachtet. Falsch negative Ergebnisse erhielten wir bei sehr kleinen Tumoren und einigen Mammakarzinomen.

Die interessantesten Aspekte für die positive Tumordiagnostik ergeben sich durch die orientierenden Versuche mit gereinigten Antikörpern gegen Tumorgewebe. Durch die primäre Abtrennung der kreuzreagierenden Antikörper war es möglich, aus dem Antiserum 90—95% der für die Diagnostik unerwünschten kreuzreagierenden Antikörper zu entfernen, die eine Affinität zum artfremden Eiweiß und zu anderen Organen haben. Mit Hilfe der anschließenden reversiblen Bindung der Antikörper gegen Tumorgewebe an diazotierte Zellulose-Tumorproteinpartikel war dann eine Absorption und Isolierung der mit dem Tumorgewebe reagierenden Antikörper möglich. Nach ^{131}J-Markierung konnten erstmals mit derartigen Antikörpern Anreicherungen im Tiertumor und in menschlichen Metastasen direkt szintigraphisch dargestellt werden. Die Anreicherung beim UV-induzierten Tumor war 70mal höher als in dem umgebenden Gewebe. Beim Menschen ermittelten wir in den 2—3 cm großen Lymphknotenmetastasen eine Anreicherung um einen Faktor von mindestens 30.

Für die Perspektive ergibt sich die Aufgabe, zu klären, ob es sich hier um spezifische oder unspezifische Antikörper handelt, und damit die Frage, ob diese Antikörper nur zur Metastasendiagnostik am gleichen Patienten oder auch bei anderen Patienten mit einem histologisch ähnlichen Tumor verwendet werden können. Von großem praktischem Interesse ist darüber hinaus die Frage, ob sich gammamarkierte Antikörper gegen Tumorgewebe auch zur Therapie von Metastasen bei Patienten mit Fernmetastasen eignen, wenn die chirurgische Therapie nicht mehr indiziert ist und die Chemotherapie nur einen geringen Erfolg bei erheblichen Nebenwirkungen erreicht. Durch eine gezielte Forschung besteht die begründete Aussicht, daß die Nuklearmedizin über die Ausnutzung immunologischer Prozesse einen wesentlichen Beitrag zur Diagnostik und Therapie maligner Geschwülste liefern kann.

Literatur

1. ALTENBRUNN, H.-J.: Positive Tumordarstellung mit radioaktiven Isotopen und markierten Verbindungen. Habilitationsschrift, Humboldt-Universität Berlin, 1969.
2. —: Tierexperimentelle und klinische Untersuchungen zur positiven Tumordarstellung mit radioaktiven Isotopen. Exp. Chirurgie **1**, 149 (1968).
3. — HERZMANN, H., PASTERNAK, G., STEENBECK, L., VOGT, R.: Möglichkeiten der Tumordiagnostik mit gammamarkierten Antikörpern. Radioisotope in Pharmakokinetik und klinischer Biochemie. 6. Jahrestagung der Gesellschaft für Nuclearmedizin,Wiesbaden 1968. Stuttgart: Schattauer (im Druck).
4. BANERJEE, R. N., GIBSON, K.: Preparation and purification of high specific activity insulin-131-iodine. J. Endocrinol. **25**, 145 (1962).
5. BASCHIERI, I., SCOGNAMIGLIO, A.: Perspectives d'emploi d'un nouveau système d'enregistrement à contraste accentué. Medical Radioisotope Scanning, International Atomic Energy Agency, Wien 1964, Band 1, S. 545.
6. BECCHINI, M. I., RICCIONI, N., MONASTERIO, G.: Le fibrigène radioactif pour le diagnostic des tumeurs chez l'homme. Radionuklide in der klinischen und experimentellen Onkologie. Stuttgart: Schattauer 1965, S. 337.
7. BLOMBÄCK, B., BLOMBÄCK, M.: Purification of human and bovine fibrinogen. Arkiv för Kemi **10**, 415 (1956).
8. CAMPBELL, D. H., LUESCHER, E., LERMAN, L. S.: Immunologic adsorbents. I. Isolation of antibody by means of a cellulose-protein antigen. Proc. Nat. Acad. Sci. **37**, 575 (1951).
9. CHRISTIE, J. H., TATSUNO, I., MACINTYRE, W. J.: The use of colloidal gold-198 and colloidal gold-199 for liver scanning. Radioaktive Isotope in Klinik und Forschung, Band VI, München-Berlin: Urban & Schwarzenberg 1965, S. 156.

10. DAY, E. D., LASSITER, S., MAHALEY, M. S.: The localization of radioantibodies in human brain tumors. III. Radioiodination of pre-purified localizing antibody. J. Nucl. Med. **6**, 38 (1965).
11. — PLANINSEK, J. A., PRESSMAN, D.: Localization of radioiodinated rat fibrinogen in transplanted rat tumors. J. Nat. Cancer Inst. **23**, 799 (1959).
12. — PRESSMAN, D.: Purification of tumor-localizing antibodies. Ann. New York Acad. Sci. **69**, 651 (1957).
13. DRISLICH, G. I., GURWITSCH, A. E.: Anwendung der Antikörper durch Fixierung an unlösliches Antigen zur spezifischen Isolierung verschiedener Eiweiße. Bichimia **29**, 1054 (1964) (russisch).
14. GENNA, S., FARMELANT, M. H., BURROWS, B. A.: Improved scintiscan resolution without sensitivity loss. Medical Radioisotope Scintigraphie, International Atomic Energy Agency, Wien 1969, Band 1, S. 561.
15. HAYES, M.: Comparison of gamma-ray imaging devices with a new three-dimensional phantom. Medical Radioisotope Scintigraphy, International Atomic Energy Agency, Wien 1969, Band 1, S. 523.
16. KAKEHI, H., ARIMIZU, N., UCHIYAMA, G., SAEGUSA, K., TATENO, Y.: Test patterns to determine sensivity and resolution of scintigraphy. Medical Radioisotope Scintigraphy, International Atomic Energy Agency, Wien 1969, Band 1, S. 509.
17. LAUGHLIN, J. S., KENNY, P. J., WEBER, D. A., DWYER, A. J., MAYER, K., GREENBERG, E. J., DIMICH, A. B., HASAN, S.: Methods and applications of quantitative computer-analysed scanning. Medical Radioisotope Scintigraphy, International Atomic Energy Agency, Wien 1969, Band 1, S. 633.
18. LISS, E., SCHMIDT, F., ERNST, H.: Versuche zur Tumorszintigraphie mit heterologem Fibrinogen nach Markierung mit ^{131}J. Strahlentherapie **117**, 223 (1962).
19. MARRACK, D., DEWEY, W. C., CORRY, P., SPAR, I. L., BALE, W. F.: ^{131}J-antihuman fibrinogen as an agent for the localisation of tumours in man — a comparison with other agents. Radioactive Isotopes in the Localization of Tumours, William Heinemann Medical Books Ltd. 1969, p. 73.
20. MCNEIL, C., LADLE, J. N., HELMICK, W. M., TRENTLEMAN, E.: A colon cancer antibody produced by ovarian mucinous cystadenoma. Proc. Amer. Ass. Cancer Res. **8**, Heft 3, 174 (1967).
21. PASTERNAK, G.: Immunologische Eigenschaften chemisch und physikalisch induzierter Tumoren. Arch. f. Geschwulstforsch. **29**, 113 (1967).
22. —: Immunologische Eigenschaften virusinduzierter Leukämien. Habilitationsschrift, Humboldt-Universität Berlin, 1965.
23. PRESSMAN, D., EISEN, H. N.: The zone of localization of antibodies. V. An attempt to saturate antibody-binding sites in mouse kidney. J. Immunology **64**, 273 (1950).
24. SODEE, D.: Delineation of anatomic structures and the detection of carcinoma utilizing low energy mercury labeled chlormerodrin. Radioaktive Isotope in Klinik und Forschung, Band VI, München-Berlin: Urban & Schwarzenberg 1965, S. 167.
25. SPAR, I. L., BALE, W. F., GOODLAND, R. L., IZZO, M. J.: Preparation and human tracer studies with ^{131}I-labeled antibody reacting with fibrin. The University of Rochester, Atomic Energy Project UR-614 (1962).
26. TAYLOR, D. M.: Radiopharmaceuticals for tumor localisation. Rapporteur report. Radioactive Isotopes in the Localization of Tumours. William Heinemann Medical Books Ltd. 1969, S. 73.
27. World Health Organization: Cancer treatment. Genf 1966, Technical Report, Series 322.

Integrierte Herz-Kreislaufanalyse mit Hilfe der quantitativen Funktionsszintigraphie*

Integrated Analysis of Cardiac Output by Means of Quantitative Scintigraphy of the Various Functions

F. BITTER, W. BESCH, N. SCHÄFER und E. SIGMUND

Sektion für Nuklearmedizin und Sektion für Cardiologie des Zentrums für Innere Medizin und Kinderheilkunde der Universität Ulm

Summary

In an attempt to improve on the poor evaluation of information obtainable by recording the passage of a radioactive tracer through heart and lungs, data recorded by a scintillation camera were stored on 3 tracks of 4-track magnetic tape, the 4th track being used for time signals. The data were then played through a selective photo-multiplier which enabled separate records to be obtained for left and right ventricles, aorta, etc. This gives a much narrower margin of error than conventional radiocardiography and furthermore provides important information which was not previously available, thus completing the integrated picture of cardiac output. Results of analyses carried out on 16 patients are reported.

Nachdem VEALL, PEARSON, HANLEY und LOWE gleichzeitig mit der Gruppe HUFF, FELLER und BOGARDUS (1954) eine klinikreife Methode der Radiocardiographie entwickelt hatten, konzentrierte sich das Interesse zahlreicher Untersucher auf die Verläßlichkeit des Verfahrens. Bei aller Anerkennung der Vorteile der Methodik — Ersatz der Arterienpunktion bei der Farbstoffverdünnungsmethode durch eine Außenkörpermessung über der Aorta direkt am Ausgang des Herzens — blieb das Urteil über die Verläßlichkeit dieses nuklearmedizinischen Verfahrens in der Cardiologie umstritten [ZIPF, MCGUIRE, WEBER und GROVE (1957), MCINTYRE, PRITCHARD und MOIR (1958), VAN DER FEER, DOUMA und KLIPP (1958), CARTER, JOHNSEN, LOEFFLER und SOUTHARD (1959), SCHREINER, LOVEJOY und YU (1959), GORTEN und GUNNELS (1960)]. Allgemein wurden technische Probleme der Methodik, so insbesondere die Ungenauigkeit der Detektorlokalisation über dem Herzen, für diskrepante Ergebnisse verantwortlich gemacht. In jüngerer Zeit richtete sich das Bemühen mehrerer Arbeitsgruppen einerseits auf eine Ausmerzung methodischer Fehlermöglichkeiten, andererseits auf eine bessere Ausschöpfung der in der Cardiographie enthaltenen Informationen. HUNDESHAGEN et al. lokalisierte mehrere Detektoren über Teilbereichen des Herzens und konnte gegenüber konventionellen Verfahren

* Durchgeführt mit Hilfe der Deutschen Forschungsgemeinschaft.

wesentlich detailliertere Aussagen gewinnen. DONATO et al. stellte die ganze Methodik auf eine breite mathematische Grundlage. KELERSOHN et al. richtete einen Detektor auf das Herz von ventral und einen zweiten von links-seitlich auf die Aorta. Daß trotz dieser zweifellosen Fortschritte das Problem der korrekten Detektorpositionierung und damit der exakten Datenerfassung nicht gelöst war, bewies der Verlauf der Tagung der Gesellschaft für Nuclearmedizin in Wien, 1967, wo gerade diese Probleme diskutiert wurden. Ganz neue Gesichtspunkte ergaben sich erst, als rasch registrierende Szintigraphiegeräte mit stehendem Detektor auf den Markt kamen. ANGER, POWELL, VAN DYKE, SCHAER, FAWWAZ und YANO (1966), SCHEER, JAHNS, KAZEM und GELINSKY (1966) sowie SCHENK, ZUM WINKEL, MAIER-BORST und SZEGVARI (1967) konnten den Durchfluß radioaktiver Substanzen durch das Herz mit Hilfe der Szintillationskamera kinematographisch verfolgen. Sie erhielten eine Serie von Bildern mit verschiedenen Füllungsphasen der Herzventrikel. Über eine *quantitative* Auswertung am menschlichen Herzen der mit der Szintillationskamera gewonnenen Daten berichteten erstmals 1967 ADAM, LORENZ und SCHEER. Die Autoren konnten mit Hilfe eines Computers den Durchfluß der radioaktiven Substanz durch das Herz in Form von Zeit-Aktivitätskurven getrennt für die rechte und linke Herzhälfte erfassen. Anhand von Herzmodellen konnten die Autoren darlegen, daß eine Fülle weiterer cardiologisch interessierender Parameter nach diesem Verfahren gewonnen werden kann.

Problemstellung

Die mit Hilfe der Tracer-Technik erzielbaren Informationen werden in der konventionellen Radiographie bei weitem nicht ausgenützt. Das ergibt folgende einfache Überlegung: Auf seinem Weg durch das Herz-Kreislaufsystem passiert der Tracer alle Herzbinnenräume, Arteria und Vena pulmonalis, die Lungen und die Aorta. Sofern es gelingt, jedes vom Tracer ausgesandte Quant einem einzelnen Compartment zuzuordnen, können nicht nur minimale und mittlere Kreislaufzeiten, sondern auch die durch die Füllungsschwankungen der einzelnen Herzbinnenräume bedingten Impulsschwankungen in Systole und Diastole registriert und zur Bestimmung von Auswurfsraten und Restvolumina benutzt werden. Eine exakte Zuordnung der im Detektor registrierten Impulse ist aber bei Szintigraphiegeräten mit stehendem Detektor in der Tat möglich, da im Gegensatz zu den normalen Detektoren jedem Impuls Ortskoordinaten beigegeben sind. Die Konsequenzen für die Herz-Kreislaufdiagnostik haben ADAM et al. dargelegt und im wesentlichen in Experimenten an Herzmodellen mit Hilfe eines Computers exemplifiziert. Nach den im Modell erzielbaren Daten kann die Herzdiagnostik wesentlich ausgebaut und methodisch vereinfacht werden. Unsere Untersuchungen sollten die Problematik der Übertragung der am Modell gewonnenen Erkenntnisse auf den Menschen abklären. Insbesondere interessierte die klinisch wichtige Frage, ob eingreifendere cardiologische Untersuchungen, z. B. ein Herzkatheterismus vermieden und andererseits Daten gewonnen werden können, die bis zum gegenwärtigen Zeitpunkt mit keiner anderen Methode erzielbar sind. Innerhalb einer integrierten Herz-Kreislaufanalyse käme einer solchen nuklearmedizinischen Methodik eine große Bedeutung zu.

Methodik der Patientenuntersuchung (Datengewinnung)

Sechzehn Patienten wurden einer quantitativen funktionsszintigraphischen Herzuntersuchung, davon acht zusätzlich einer Herz-Minutenvolumen-Bestimmung mit Hilfe der Thermodilutionsmethode zugeführt. Ein Intracath-Katheter wurde in den rechten Ventrikel eingeführt. 0,2 ml ^{99m}Tc waren in einem der Injektionsspritze vorgelagerten Zwischenstück enthalten. Die Injektionsspritze enthielt 5 ml physiologischer Kochsalzlösung bzw. Kältelösung. Die Injektion erfolgte als single-shot-Injektion EKG-gesteuert in der Phase der Diastole. Der gesamte Spritzeninhalt wurde in weniger als 1 Sekunde injiziert. Damit war die Gewähr gegeben, daß die an der Spitze der Injektionssäule befindliche Tracersubstanz sich vollständig und in einem Zeitraum kleiner als 0,2 sec im rechten Ventrikel verteilte, so daß bei Beginn der nachfolgenden Systole eine weitgehend homogene Verteilung der Tracer-Substanz im Volumen des rechten Ventrikels unterstellt werden konnte. Die Aufnahme des Herzdurchflusses erfolgte im zweiten schrägen Durchmesser. Zu diesem Zwecke wurde der präcordial auf den liegenden Patienten eingestellte Kopf der Szintillationskamera um 45° seitlich geneigt. Vorausgehende Untersuchungen (Schenk et al., Adam et al.) hatten ergeben, daß in dieser Aufnahmestellung linker und rechter Herzventrikel am besten differenziert werden können (siehe Abb. 1). Die von der Szintillationskamera kommenden x-, y- und z-Informationen wurden auf drei Spuren eines CEC-Magnetbandes analog gespeichert (AM-Verfahren, maximale Frequenz 300 kHz, Impuls derzeit auf 20 μsec verlängert). Auf dem vierten Kanal wird ein von der R-Zacke abgeleiteter Impuls von 3 V Amplitude und 1 msec Länge gespeichert. Spätestens 60 sec nach Injektion wird das Magnetband gestoppt. Bei einem Teil der Patienten erfolgte gleichzeitig die Bestimmung des Schlagvolumens mit der Thermodilutionsmethode.

Methodik der Magnetbandauswertung (Datenverarbeitung)

Da uns z. Z. in Ulm ein Computer noch nicht zur Verfügung steht, wandten wir zur Gewinnung von Zeit-Aktivitätskurven aus Herzteilbereichen das wenig aufwendige von Adam und Lorenz inaugurierte Photomultiplier-Verfahren an. Anhand einer photographischen Aufnahme bei beleuchtetem Raster wird die dem interessierenden Herzteilbereich (etwa rechter, linker Ventrikel, Arteria pulmonalis, Aorta) entsprechende Fläche des Oszillographenschirms mit Hilfe einer Schablone, die alle übrigen Teile abdeckt, ausgespart (Abb. 2). Ein auf den Schirm aufgesetzter Photomultiplier sieht lediglich den interessierenden Teilbereich, so daß beim Abspielen des Bandes lediglich die in diesem Teilbereich erscheinenden Lichtblitze registriert werden. Ein im Multiscalingbetrieb arbeitender 400-Kanal-Analysator erlaubt die Fortschaltung von Kanal zu Kanal in Zeitschritten von 10 bis 1000 msec, und damit die Gewinnung von Zeit-Aktivitätskurven über 4 bis 400 sec (Abb. 3). Der Photomultiplier arbeitet streng linear bis zu einer Frequenz über 50 kHz. Um die Zeit-Aktivitätskurven übersichtlich einzelnen Herzphasen zuordnen zu können, wurde der durch die R-Zacke ausgelöste Impuls dazu verwendet, den der R-Zacke folgenden Kanal (100 msec) zu sperren (Abb. 4 I).

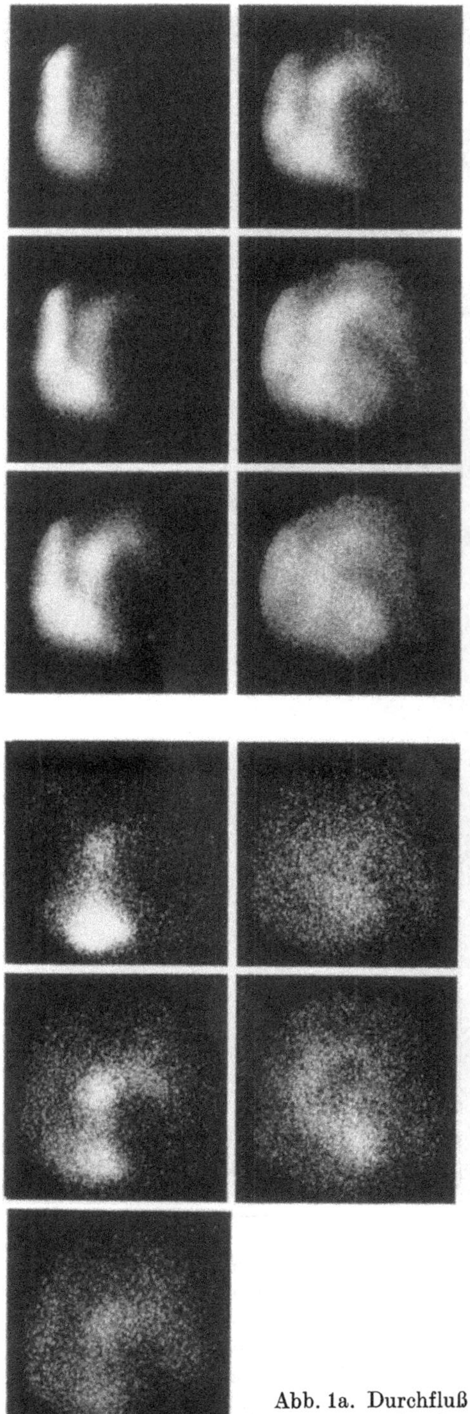

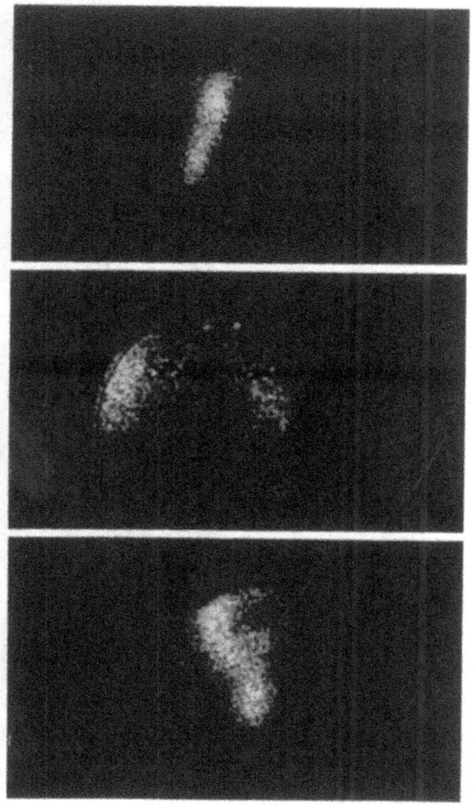

Abb. 1b. Ankunft des Aktivitätsembolus im rechten Ventrikel (oben) und in der Arteria pulmonalis, in der Lunge (Mitte) und im linken Ventrikel und in der Aorta (unten)

Abb. 1a. Durchfluß eines Embolus durch das Herz. Zwei Beispiele der von der Anger-Kamera gewonnenen Originalaufnahmen

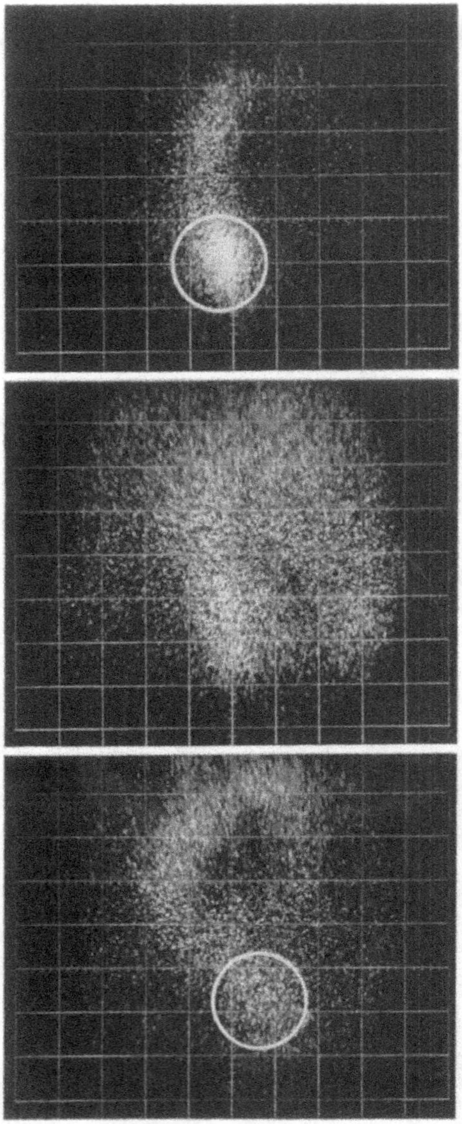

Abb. 2. Durchfluß eines Embolus durch das Herz. Reproduktion vom Magnetband auf einen Oszillographen. Eingezeichnet sind die repräsentativen Teilbereiche, oben rechter und unten linker Ventrikel

---

→

Abb. 3. Zeit-Aktivitätskurven. a Gewonnen am 400-Kanal-Analysator. I linker Ventrikel; II Lungenperipherie; III A. pulmonalis und Aorta; IV rechter Ventrikel. Abszisse: Zeit = 100 msec/Kanal; Ordinate: Aktivität (relativ). — b Gewonnen an einem ausgeprägt insuffizienten Herzen. I rechter Ventrikel; II Lungenperipherie; III linker Ventrikel

Abb. 4. Zeit-Aktivitätskurve des rechten Ventrikels. I Jeder der *R*-Zacke folgende Kanal wird gesperrt, so daß ein Ausfall in diesen Kanälen resultiert. Diese Kurve dient ausschließlich der Zuordnung des Elektrokardiogramms; II Beispiel der Zuordnung des Elektrokardiogramms zur „reinen" Kurve; III Kurve II, gemittelt nach dem Verfahren des gleitenden 3-Werte-Durchschnitts; IV Kurve II, gemittelt nach dem Verfahren des gewichteten 3-Werte-Durchschnitts

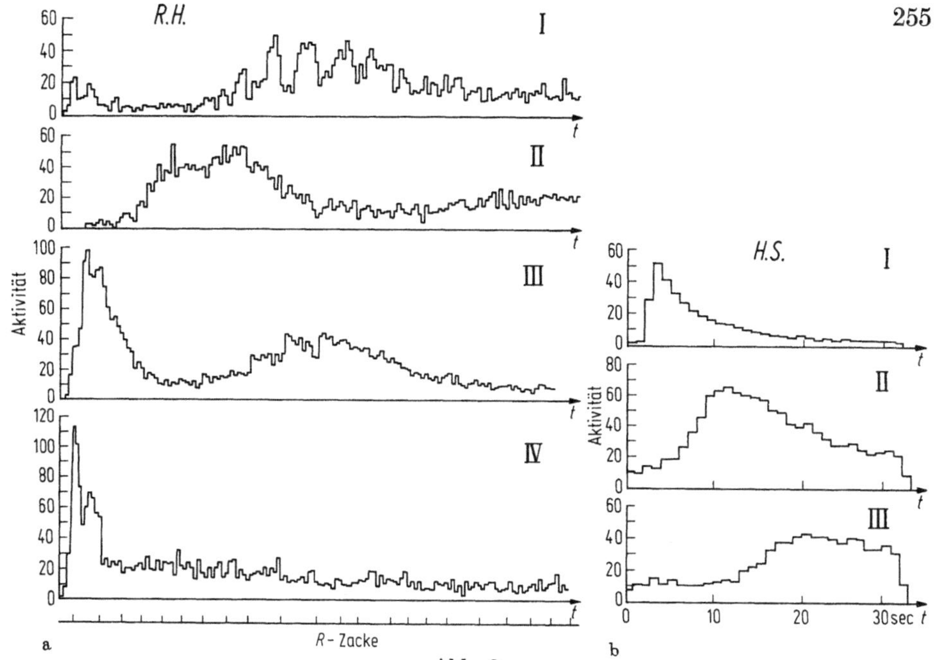
R.H.
I
II
III
IV
Aktivität
t
R-Zacke
a
H.S.
I
II
III
Aktivität
0
10
20
30 sec
t
b

Abb. 3

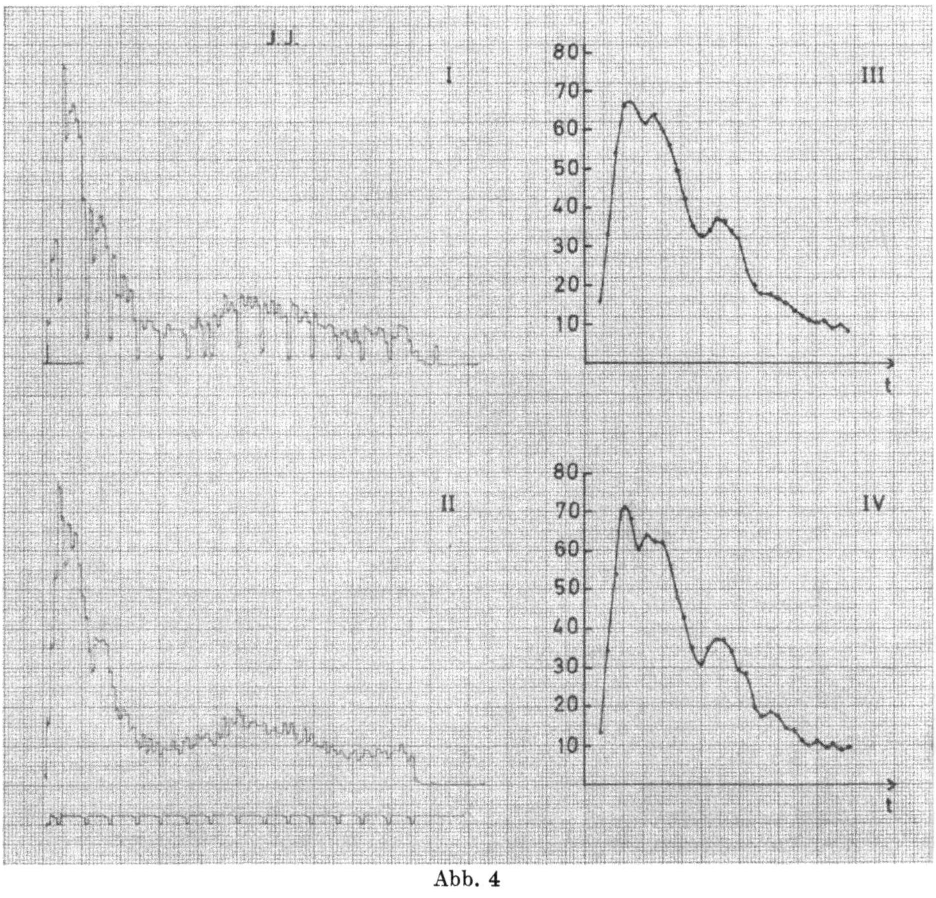
J.J.
I
II
III
IV
t

Abb. 4

Ermittlung der Herz-Kreislauf-Größen aus den funktionsszintigraphisch gewonnenen Zeit-Aktivitätskurven

1. Kreislaufzeiten

Mittlere Durchflußzeiten können allgemein nach folgender Formel berechnet werden:

$$\bar{t}_i = \frac{\int_0^\infty t\, R_i(t)\, dt}{\int_0^\infty R_i(t)\, dt}, \tag{1}$$

wobei die Zeit t gerechnet wird vom Zeitpunkt der Aktivitätsankunft in dem entsprechenden Compartment. $R_i(t)$ ist die Aktivität im Compartment als Funktion der Zeit t. Neben der Bestimmung der mittleren pulmonalen Durchflußzeit nach (1) verwenden wir die sogenannte left-peak-Methode nach DONATO: hierbei errechnet sich die mittlere Durchflußzeit als Differenz aus der Zeit des Aktivitätsmaximums im linken Ventrikel und der mittleren Durchflußzeit durch den rechten Ventrikel (Tab. 1).

Tabelle 1. *Kreislaufzeiten*

$$\bar{t}_i = \frac{\int_0^\infty t\, R_i(t)\, dt}{\int_0^\infty R_i(t)\, dt}\ *$$

Patient	Herzfrequenz $[min^{-1}]$	Rechter Ventrikel $\bar{t}_{RV}$ [sec]	Lunge $\bar{t}_L$ [sec]	Linker Ventrikel $\bar{t}_{LV}$ [sec]	Lunge nach DONATO $\bar{t}_L$ [sec]	Druck Art.pulm. [mm Hg]
B.K.	101	3,35	4,08	7,89	5,3	31/15
H.S.	120	10,40	16,40	14,20	10,6	59/20
H.M.	68	1,86	8,79	6,99	8,5	28/15
R.H.	103	1,11	3,14	4,18	6,0	28/15
W.M.	75	1,26	4,59	4,28	4,4	21/13
G.S.	97	1,80	3,45	3,53	3,5	21/11
H.F.	88	1,15	2,83	3,53	4,1	22/11
L.W.	124	1,61	3,16	4,48	4,2	30/19
F.J.	100	1,61	3,33	4,69	5,7	28/11
J.G.	90	1,66	4,08	6,31	7,0	23/12
J.J.	80	2,12	4,56	3,43	4,1	26/13
T.H.	—	1,30	3,33	3,22	3,5	20/10
F.D.	—	2,52	5,48	5,57	6,0	37/21
H.W.	95	1,01	3,32	3,26	4,0	29/15
J.G.	100	1,34	—	2,48	3,5	21/13
F.B.	—	—	2,72	—	3,9	37/19
$\bar{t}$		1,54	3,66	4,08	4,61	
s		0,45	0,83	1,08	1,16	
SEM		0,13	0,24	0,31	0,32	
N		12	12	12	13	

* Als Durchflußzeit wurde also nicht die Differenz der Zirkulationszeiten zweier aufeinanderfolgender Compartments definiert. Sie ist hier also abhängig vom Injektionsort.

2. Berechnung der Volumina

a) Die Bestimmung der relativen Restvolumina beider Ventrikel erfolgt aus den Zeit-Aktivitätskurven. Das Verhältnis des enddiastolischen Füllungsvolumens zum Restvolumen ist gleich dem Verhältnis des Aktivitätsmaximums zum Aktivitätsminimum *einer* Herzrevolution.

In Abb. 4 III, IV wird die Zeit-Aktivitätskurve II geglättet mit Hilfe des gleitenden 3-Werte-Durchschnitts (III ungewichtet, IV gewichtet).

Diese Mittelung ist notwendig, um ein möglichst exaktes Verhältnis bilden zu können. Eine weitere Methode wird in Abb. 5 dargestellt: hierbei startet jede *R*-Zacke den Vielkanal-Analysator im Multiscalingbetrieb, so daß sich Aktivitätsschwankungen, die Volumenschwankungen entsprechen, phasengleich aufaddieren. Mit einer Fortschaltzeit von 10 msec wurden die Kurven in Abb. 5a, b gewonnen.

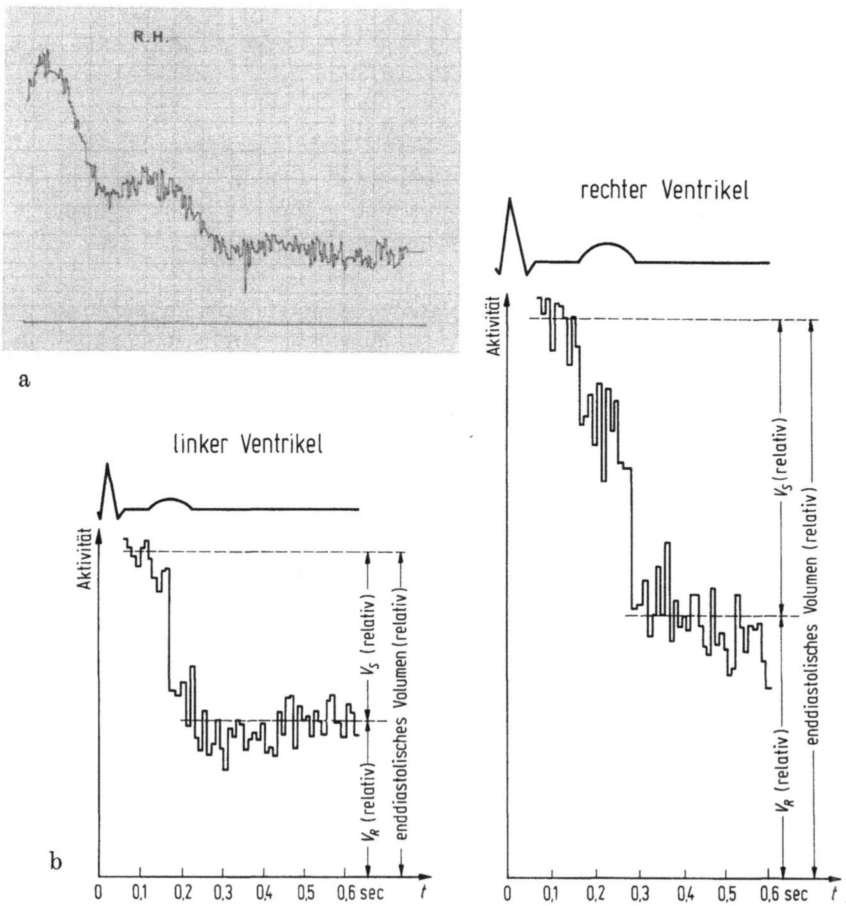

Abb. 5. *R*-Zacken-gesteuerte Summations-Zeit-Aktivitätskurve. a vom rechten Ventrikel, b vom linken und rechten Ventrikel. (Die relativen Volumina sind eingezeichnet)

Aus den so errechneten relativen Volumina der Ventrikel und den mit Hilfe der Thermodilutionsmethode gemessenen absoluten Werten der Herz-Zeit-Volumina werden die absoluten Volumina der Herzkammern bestimmt.

b) Das pulmonale Blutvolumen wird als Produkt aus mittlerer pulmonaler Durchflußzeit und Schlagvolumen des rechten Ventrikels errechnet.

Ergebnisse

1. Kreislaufzeiten

Wie unsere Zeit-Aktivitätskurven (Abb. 3) zeigen, können mit dem Verfahren der quantitativen Funktionsszintigraphie verschiedene funktionale Einheiten (Compartments) exakt und ohne wesentliche Überlagerung von anderen Compartments erfaßt werden (Abb. 2). Nahezu alle Zeit-Aktivitätskurven zeigen einen idealen Verlauf, wie wir es von der Simulation der Herzpassage am Analog-Computer her kennen. Die Ankunft des radioaktiven Embolus in einem Compartment markiert sich als Aktivitätsanstieg von der Null-Linie aus, so daß exakte Kreislaufzeitbestimmungen möglich sind. Die konventionell radiocardiographisch notwendigen Extrapolationsmanöver bei den üblicherweise resultierenden Summationskurven entfallen völlig. Die Kreislaufzeiten lassen sich — wie Tab. 1 zeigt — exakt festlegen.

Bei unseren Patienten ergeben sich mittlere Durchflußzeiten von 1,54 $\pm$ 0,13 sec durch den rechten, von 4,08 $\pm$ 0,31 sec durch den linken Ventrikel. Die mittlere Durchflußzeit durch die Lunge mit 3,66 $\pm$ 0,24 sec ist signifikant vermindert gegenüber der nach DONATO bestimmten Durchflußzeit von 4,61 $\pm$ 0,32 sec.

Zur Mittelwertsbildung werden die Daten der ersten drei Patienten in der Tabelle nicht herangezogen, da sie sich in einzelnen Parametern zu stark von der Hauptgruppe unterscheiden.

Wie Tab. 2 zeigt, besteht eine enge Korrelation zwischen den mittleren Durchflußzeiten des rechten Ventrikels, der Lunge und des linken Ventrikels. Für die Berechnung der Korrelationskoeffizienten wurden die Daten des gesamten Patientengutes verwendet.

Tabelle 2. *Korrelationskoeffizienten*

$$\bar{t}_i = \frac{\int_0^\infty t\, R_i(t)\, dt}{\int_0^\infty R_i(t)\, dt}, \quad r = \frac{N\, \Sigma\, \bar{t}_i \bar{t}_k - (\Sigma \bar{t}_i) \cdot (\Sigma \bar{t}_k)}{[N\, \Sigma(\bar{t}_i)^2 - (\Sigma \bar{t}_i)^2]\, [N\, \Sigma(\bar{t}_k)^2 - (\Sigma \bar{t}_k)^2]}$$

	t_{RV}	t_{Lunge^*}	t_{LV}
t_{RV}	1	0,7327	0,9273
t_{Lunge}	0,7327	1	0,8897
t_{LV}	0,9273	0,8897	1

* t_{Lunge} bestimmt nach der left-peak-Methode von DONATO et al.

2. Berechnung der Volumina

Aus den Zeit-Aktivitätskurven des *rechten Ventrikels* werden relative Restvolumina zwischen 34% und 67% des enddiastolischen Füllungsvolumens errechnet. Der Mittelwert liegt bei 51,6 ± 3,6%. Die absoluten Werte für die Restvolumina liegen zwischen 21 und 134 ml, für die Gesamtvolumina zwischen 52 und 203 ml.

Mit Hilfe des bekannten Schlagvolumens des rechten Ventrikels, bei Unterstellung gleicher Schlagvolumina für beide Ventrikel, läßt sich das enddiastolische Füllungs- und Restvolumen des *linken Ventrikels* bestimmen. Zwei Beispiele sind in Tab. 3 angeführt.

Für das *pulmonale Blutvolumen* lassen sich nach DONATO Werte zwischen 316 und 541 ml angeben, nach der funktionsszintigraphischen Methode Werte zwischen 165 und 565 ml.

Tabelle 3. *Bestimmung der Herzvolumina*

Patient	Rechter Ventrikel			Lunge		Linker Ventrikel		
	V_{RV} [ml]	HZV [ml/min]	V_R [ml]	PBV [ml] 1	2	V_{LV} [ml]	HZV [ml/min]	V_R [ml]
R.H.	52	3160	21	316	165	44	3160	13
B.K.	108	3620	73	320	246			
W.M.	196	7380	98	541	565			
G.S.	138	6830	70	398	393			
H.F.	203	6080	134	415	286	165	6080	95
L.W.	78	5590	33	391	294			
F.J.	86	4920	37	467	273			
J.G.	92	4550	41	531	309			

[1] Berechnet mit $\bar{t}_L$ nach DONATO et al. [2] Berechnet mit $\bar{t}_L = \frac{\int_0^\infty t\,R(t)\,dt}{\int_0^\infty R(t)\,dt}$.

Diskussion

Die Ergebnisse belegen in schlüssiger Form die Überlegenheit quantitativ-funktionsszintigraphischer Verfahren gegenüber der konventionellen Radiocardiographie. Da die Zeit-Aktivitätskurven unter Sicht am Oszillographenschirm gewonnen werden, lassen sich untersuchte Organe und Organteilbereiche exakt abgrenzen. Daraus ergeben sich folgende Vorteile:

Das zu untersuchende Organ kann vollständig unter Elimination des umgebenden Gewebes zur Gewinnung der Zeit-Aktivitätskurve herangezogen werden.

Bei teilweiser Überlagerung durch benachbarte Organe läßt sich ein überlagerungsfreier Teilbereich zur Gewinnung der Zeit-Aktivitätskurven heranziehen, der das Gesamtcompartment repräsentiert. In einem solchen Falle läßt sich also eine reine Zeit-Aktivitätskurve, z. B. eine reine Durchflußkurve des linken Ventrikels, gewinnen (Abb. 3a I).

Es steht also außer Frage, daß das Verfahren gegenüber der konventionellen Radiocardiographie eine deutliche Präzisierung der Meßwerte erbringt.

Der Einwand, daß zur Erzielung exakter Ergebnisse auch bei unseren Untersuchungen ein Katheter eingelegt wurde, ist nicht stichhaltig. Dabei handelt es sich ja lediglich um einen Intracath-Katheter, der in der Cardiologie routinemäßig zur Bestimmung von Druckgrößen in den rechten Ventrikel und die Arteria pulmonalis eingeführt wird. Sein Durchmesser ist zu klein, als daß damit Kontrastuntersuchungen der Herzbinnenräume durchgeführt werden könnten.

Der weitere unter dem Abschnitt „Problematik" gemachte Einwand gegen die konventionelle Radiocardiographie betraf die mangelhafte Ausschöpfung der Information, die von einem das Herz und die Lunge passierenden Aktivitätsembolus geliefert wird. Daß auch dieser tatsächlich berechtigte Einwand bei der quantitativen Funktionsszintigraphie vollkommen entfällt, belegen die Herz-Kreislaufdaten, die wir an unseren Patienten mit Hilfe des neuen Verfahrens gewonnen haben. Dabei ist es sogar in einigen Fällen möglich, einzelne Parameter nach mehreren Verfahren zu bestimmen (z. B. Restvolumen des rechten Ventrikels, pulmonales Blutvolumen). Im folgenden sollen nur noch Größen erwähnt werden, die bis jetzt auch mit eingreifenden cardiologischen Verfahren unter keinen Umständen gewonnen werden können. In erster Linie handelt es sich dabei um den linken Ventrikel betreffende Parameter, im einzelnen um das enddiastolische Füllungsvolumen und das Restvolumen dieses Ventrikels. Das dabei angewandte Verfahren der Mittelung mehrerer Herzaktionen während der Passage des Tracers durch den linken Ventrikel soll in späteren Untersuchungen auch im steady-state angewandt werden. HOFFMANN und KLEINE führten eine Mittelung der Herzbewegung mit einem blind aufgesetzten Detektor durch. Nach dem hier vorgelegten Verfahren können Mittelwertskurven der Füllungsschwankungen von Herzteilbereichen, von Gefäßabschnitten und auch von Lungenteilbereichen durchgeführt werden, sofern eine im Blutpool verbleibende Aktivität angewandt wird. Untersuchungen dieser Art haben wir noch nicht durchgeführt, da uns eine in der Blutbahn verbleibende Substanz sehr hoher spezifischer Aktivität (2 mCi in 0,2 ml) nicht zur Verfügung stand. Das ist auch der Grund dafür, daß wir auf die übliche Herzminutenvolumenbestimmung mit Hilfe der Stewart-Hamilton-Gleichung verzichten und die Thermodilutionsmethode im gleichen Untersuchungsgang anwenden.

Abschließend soll aus den vorgelegten Daten ein Punkt von Interesse gesondert diskutiert werden.

Eine sehr enge Korrelation fand sich zwischen der pulmonalen Passagezeit, bestimmt nach DONATO und bestimmt nach der direkten funktionsszintigraphischen Methode. Allerdings zeigte sich, daß unsere Werte systematisch und signifikant unter denen von DONATO liegen. Die Erklärung ergibt sich aus der Analyse der unterschiedlichen methodischen Verfahrensweise: DONATO berücksichtigt die großen Lungengefäße und deren große Verzweigungen, während wir bei der Auswahl eines repräsentativen Teilbereiches der Lungenperipherie lediglich die kleineren Äste der Arteria und Vena pulmonalis sowie die Kapillaren in Rechnung stellen.

Zusammenfassung

Die konventionelle Radiocardiographie wertet die während der Passage einer Tracer-Substanz durch das Herz und die Lungen anfallenden Informationen nur mangelhaft aus. Mit Hilfe einer Szintillationskamera, eines Analog-Magnetbandes, eines Photomultipliers und nachfolgender Registriereinrichtungen lassen sich Herz-Kreislaufgrößen gewinnen, die nuklearmedizinisch bisher lediglich mit großen Fehlerbreiten erzielbar waren. Darüber hinaus gelingt es, wichtige Herz-Kreislaufgrößen zu bestimmen, die bisher weder nuklearmedizinisch noch mit eingreifenden cardiologischen Verfahren zu gewinnen waren. Es wird über Untersuchungen an 16 Patienten berichtet. In einer integrierten Herz-Kreislaufanalyse nehmen die auf diese Weise funktionsszintigraphisch bestimmten Parameter einen wichtigen Platz ein.

Literatur

ADAM, W. E., LORENZ, W. J., SCHEER, K. E.: Quantitative radiocardiographische Untersuchungen mit der Szintillationskamera. Vortrag auf der 5. Jahrestagung der Ges. f. Nuclearmedizin, Wien 1967.

— —: Ein einfaches Verfahren zur analogen Auswertung funktionsszintigraphischer Messungen mit der Szintillationskamera. Atompraxis **5,** 67 (1967), Direktinformation.

ANGER, H. O., POWELL, M. R., VAN DYKE, D. C., SCHAER, L. R., FAWWAZ, R., YANO, Y.: Recent Applications of the Scintillation Camera. In: Radioaktive Isotope in Klinik und Forschung, Band VII. München-Berlin-Wien: Urban & Schwarzenberg 1967.

CARTER, B. L., JOHNSEN, S. E., LOEFFLER, R. K., SOUTHARD, M. E.: A critical evaluation of external body surface counting in the determination of cardiac output with radioactive isotopes. Amer. J. Roentgen. **82,** 618 (1959).

DONATO, L., GIUNTINI, C., BIANCHI, R., MASERI, A.: Quantitative Radiocardiography for the Measurement of Pulmonary Blood Volume. In: Dynamic Clinical Studies with Radioisotopes. US Atomic Energy Commission, Division of Technical Information, June 1964.

HOFFMANN, G., KLEINE, N.: Eine neue Methode zur unblutigen Messung des Schlagvolumens am Menschen über viele Tage mit Hilfe von radioaktiven Isotopen. In: Verh. d. Dtsch. Ges. f. Kreislaufforschung, 31. Tagung. Darmstadt: Steinkopf 1965.

HUFF, R. L., FELLER, D. D., BOGARDUS, G.: Cardiac output by body surface counting of ^{131}I-human serum albumine. J. Clin. Invest. (Abstract) **33,** 944 (1954).

HUNDESHAGEN, H.: Die Bestimmung der Coronardurchblutung nach Applikation von radioaktiven Indikatoren und Körperoberflächenmessung. Vortrag auf der 5. Jahrestagung der Ges. f. Nuclearmedizin, Wien 1967.

MCINTYRE, W. J., PRITCHARD, W. H., MOIR, T. W.: The determination of cardiac output by the dilution method without arterial sampling. II. Validation of precordial recording. Circulation **18,** 1147 (1958).

SCHEER, K. E., JAHNS, E., KAZEM, I., GELINSKY, P.: Klinische Untersuchungen mit der Szintillationskamera und ^{99m}Tc. In: Radioaktive Isotope in Klinik und Forschung, Band VII. München-Berlin-Wien: Urban & Schwarzenberg 1967.

SCHENK, W., ZUM WINKEL, K., MAIER-BORST, W., SZEGVARI, M.: Angiokardiographie mit der Szintillationskamera. Vortrag auf der 5. Jahrestagung der Ges. f. Nuclearmedizin, Wien 1967.

SCHREINER, B. F. J., LOVEJOY, F. W. J., YU, P. N.: Estimation of cardiac output from precordial dilution curves in patients with cardiopulmonary disease. Circ. Res. **7,** 595 (1959).

VAN DER FEER, Y., DOUMA, J. H., KLIPP, W.: Cardiac output measurement by the injection method without arterial sampling. Amer. Heart J. **56,** 642 (1958).

VEALL, N., PEARSON, J. D., HANLEY, T., LOWE, A. E.: A method for the determination of cardiac output *k* (preliminary report). Proc. Second Radioisot. Conf. Oxford, 1954. London: Butterworths 1954, p. 183.

ZIPF, R. E., MCGUIRE, R. F., WEBER, J. M., GROVE, G. R.: Determination of cardiac output by means of external monitoring of radioisotope injected intravenously. Am. J. Clin. Path. **28,** 134 (1957).

Blood Flow through Arteriovenous Anastomoses

B. A. RHODES

Division of Nuclear Medicine, The Johns Hopkins Medical Institutions,
Baltimore, Maryland, U.S.A.

Arteriovenous anastomoses or communications between arteries and veins larger in calibre than capillaries have been reported by anatomists in a number of tissues [1]. Their function has been observed by microscopic observations of the microcirculation of living tissues [2]. In order to evaluate the physiological significance of this phenomenon on the overall circulatory system, it has been necessary to estimate the relative magnitude of blood flow through these anastomoses compared to flow through the capillaries. This has been accomplished with arterial injections of particles. When particles too large to pass through capillaries are injected into an artery, only the particles carried by the blood flowing through the anastomoses will pass to the venous side of the circulation. Thus, the fraction of injected particles flowing out of the perfused tissues estimates shunting, or the relative blood flow through the anastomoses. Using this principle, significant shunt blood flow was demonstrated in the hindlegs of anesthetized dogs fifteen years ago by PIIPER and SCHOEDEL [3]. Recently we repeated and verified the results of these earlier investigators [4] using a new technique developed by the author. In this method, ^{99m}Tc-labeled serum albumin microspheres are injected into an artery and the shunted microspheres detected by external radiation counting after they are trapped in the lungs. As in the previous study, we found that a significant and variable fraction of extremity blood flow was shunted. With this new technique, however, it has been possible to make the measurements without interfering with the normal circulation and the measurements can be made repeatedly in the same subject. This allows one to study changes in shunting in response to the various parameters of blood flow. Because the particles are made from human serum albumin the method is applicable to man as well as experimental animals.

This is a report of experiments which demonstrate that shunting in extremities of both man and dog can be relatively significant and variable within the same individual. Acetylcholine was administered to the dogs in order to alter shunting. This substance was chosen because previous studies suggest that the arteriovenous anastomoses are responsive to this neurohormone [1, 2, 5, 6].

Materials and Methods

Microspheres: Human serum albumin microspheres were manufactured using a modification of the method of ZOLLE et al. [7] by The 3M Company, St. Paul, Minnesota, U.S.A. Two sizes of microspheres were used: 40 micron mean diameter (range 25 to 50) and 20 micron mean diameter (range 10 to 30). Twenty milligrams of the microspheres were labeled just prior to injection with 20 or more millicuries of ^{99m}Tc by the method of RHODES et al. [8]. The labeled microspheres are suspended in 15 cc of saline containing 0.1% Tween 80 and $0.8 \times 10^{-4} N$ HCl. The vial used has a porous glass frit (10—15 micron) at one end so that liquid can be withdrawn from both ends. At one end saline free of microspheres can be withdrawn and at the other end saline containing microspheres can be withdrawn.

Acetylcholine cloride (Merck) was dissolved in physiological saline just prior to use in a concentration of 1 mg/cc.

Procedure in Man: The patients were placed in the supine position and a $3'' \times 3''$ collimated NaI scintillation detector was placed over the upper right lung field. The spectrometer was calibrated to register pulses 140—190 keV. The data was recorded with an automatic digital printer at 6-second intervals. One milliliter of the saline used to suspend the microspheres is injected into one of the femoral arteries at the level of the inguinal ligament. This allowed measurement of background due to unbound technetium (0.5 to 1.5% of the technetium in the radiopharmaceutical is not bound to the microspheres). Two minutes later, 1—3 ml of the microsphere suspension is injected through the same needle. Approximately five minutes later, a repeat injection of microspheres in the contralateral artery was made if required. After approximately another five minutes, one ml of the microsphere suspension was injected into a superficial vein of the arm and the radioactivity recorded for another 2—3 minutes. The radioactivity of each syringe is assayed before and after each injection to determine the exact amount of activity injected. The per cent shunt was calculated by the following equation:

$$S = \frac{[(A_2 - A_1) - NB]/a}{(V_2 - V_1)/v}$$

where: S = fractional shunting of the microspheres
A_1 = cpm before arterial injection
A_2 = cpm after arterial injection
N = ml of microsphere suspension injected intra-arterially
B = cpm after injection of supernate
V_1 = cpm before intravenous injection of microspheres
V_2 = cpm after intravenous injection of microspheres
a = mCi ^{99m}Tc injected intra-arterially
v = mCi ^{99m}Tc injected intravenously.

Procedure in Dogs: Mongrel dogs weighing 30—35 pounds were initially anesthetized with sodium pentabarbital (30 mg/kg) and a solution of 1.4 mg/ml of sodium pentabarbital in 0.9% saline was administered by intravenous drip (5—15 drops/min). A respiration pump was used to assure constant and normal respiration throughout the procedure. After surgical cutdown just below the inguinal ligament, Cournand needles were inserted into each femoral artery so that the needles pointed upstream. To inject, the stylet was removed and a syringe containing the suspended microspheres was connected to the needle. The mixture was then injected in about 10 seconds and the syringe flushed with blood three times. The stylet was then replaced until the next injection.

The measurements in the dogs were made as described previously for man, except they were usually repeated many times over a period of several hours. For repeat measurements lower doses of microspheres (0.1 cc) were used. In the acetylcholine studies, the drug was also administered through the Cournand needle, usually 15 seconds before injecting a dose of microspheres to measure shunting. In order to measure blood pressure, a catheter leading to a Hg manometer was inserted into the inferior mesenteric artery in some of the dogs. A screw clamp was placed around the aorta one to two inches superior to the inferior mesenteric artery. This was used to mechanically lower the blood pressure to the lower extremities. In some dogs the circulation to the paw was occluded at the time of injection of microspheres so that shunting was measured without paw circulation. In other dogs the skin was removed so that shunting was measured without skin circulation.

At the conclusion of the studies the dogs were scanned with a 3″ rectilinear scanner to determine the pattern of distribution of the microspheres trapped in the legs.

Results of Human Studies

The shunting of 40 micron microspheres has been determined in thirteen individuals. The results are given in Table 1. The values ranged from 0 to 41%. This high value was found in a man suspected of having arteriovenous fistulation

Table 1. *Summary of results of leg shunt studies in man*

Subject	Disease	% shunting of 40 μ microspheres
L.H. (female, age 25)	varicose veins	13
O.P. (female, age 54)	varicose veins	3
		1 (same patient, two weeks later)
A.D. (female, age 64)	varicose veins	1
B.R. (male, age 33)	normal	0
C.B. (male, age 40)	tuberculosis, with clubbing	4
T.Z. (male, age 50)	tuberculosis, normal extremity circulation	3
R.A. (male, age 56)	tuberculosis, normal extremity circulation	2
K.M. (male, age 20)	venous blockage, lower leg	1
B.B. (female, age 25)	livedo reticularis	37 (leg with sympathetic block)
		0 (other leg)
J.A. (female, age 18)	a.v. fistula in lower leg	7
R.J. (female, age 32)	undiagnosed leg pain	0
		0 (repeat measurement, same day, same leg)
H.B. (male, age 60)	question of a.v. fistulation in left leg	41
E.S. (male, age 21)	question of a.v. fistulation in right leg	5

in the left leg. Another patient with high shunting, B.B., had cold, mottled lower extremities on standing and was diagnosed as having a variant of hemato-dyskinesia with prominent livedo reticularis. One and one-half hours before the study an anesthetic was administered to produce temporary sympathetic nerve block to one leg. This resulted in relief of symptoms in that leg at the time of the study. Thirty-seven per cent shunting was found in this leg, while shunting was not detected in the untreated leg. Shunting in three subjects with no known circulatory disease ranged from 0 to 3%.

Results of Dog Studies

Fig. 1 shows the per cent shunting of 20 micron microspheres as a function of time in a hindleg of three different dogs. In the first dog, the flow through the shunts remained fairly constant for a period of two and one-half hours. In the other two dogs, shunting tended to decrease with time. The third dog was given additional sodium pentabarbital at 2.5 hours. This was followed by a rise in shunting.

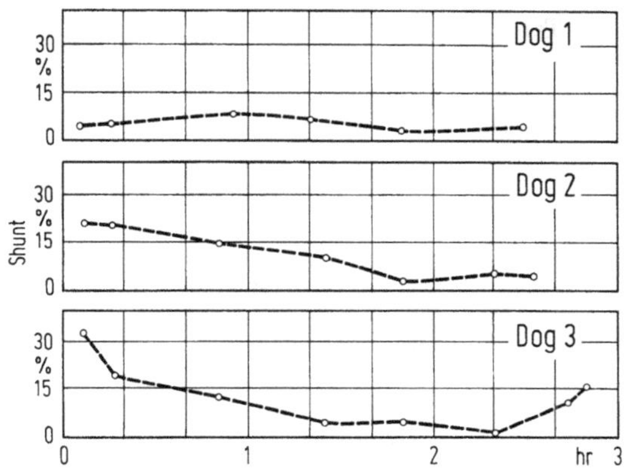

Fig. 1. Results of multiple measurements of shunting of 20 micron microspheres in the hindleg of an anesthetized dog

The effect of acetylcholine on the blood pressure is shown in Tables 2a and 2b. In the range of 0.1 to 4 mg, acetylcholine produced a drop in blood pressure which began immediately. The maximum lowering of the blood pressure occurred 35 seconds after the injection. The pressure returned to pre-injection values within five minutes after a 1 mg dose. These results are the averages of four determinations.

Table 3 lists the results of shunt measurements made before, 15 seconds after and five minutes after acetylcholine. In every case a significant decrease occurred immediately following the acetylcholine injection. Fig. 2 shows graphically the effect of acetylcholine on shunting. Since the shunting was measured at the time the acetylcholine was causing a 10 to 15% reduction in blood pressure, shunting was also measured when the blood pressure was mechanically reduced

Table 2a. *Effect of acetylcholine on blood pressure*

Dose (mg)	% of original pressure
0	100
0.001	100
0.01	100
0.1	79
1.0	52
4.0	43

Table 2b. *Time course of the effect of 1 mg of acetylcholine on blood pressure*

Time (sec)	% of original pressure
0	100
5	97.6
10	94.7
15	88.8
20	73.8
25	55.8
30	48.4
35	43.3
40	49.3
50	58.4
70	66.7
100	76.9
150	86.4
300	98.2

Table 3. *Effect of acetylcholine on shunting*

Expt. No.	Dose (mg)	% shunting		
		before dose	15 sec after dose	5 min after dose
24	0.01	8.5	2.4	6.5
	0.05	6.5	2.3	6.9
	0.15	6.9	4.4	13.7
32	4	61.3	0	42.8
	4	52.3	6.8	48.7
	4	65.9	0	60.5
34	1	57.3	7.7	n. d.
35	1	2.0	0	3.6
	1	13.9	5.1	12.4
	1	17.4	4.2	17.0
36	1	35.2	0.2	38.1
	1	39.6	1.3	26.6
	1	39.7	0	45.9
	1	31.6	0	29.2
	1	35.9	0	28.1
	1	26.5	0	18.4

to the same extent. This was accomplished by tightening the screw clamp on the aorta prior to injection of the microspheres. This also produced a decrease in shunting although not as much as did the acetylcholine.

Figs. 3 and 4 show how 1 mg of acetylcholine alters the distribution of blood flow in the hindlegs of dogs. The left side of the picture shows the distribu-

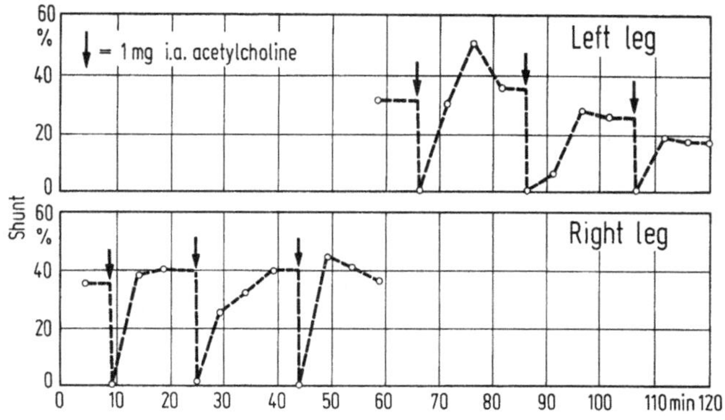

Fig. 2. Shunting of 20 micron microspheres in the hindlegs of a dog showing that 15 seconds after administration of 1 mg of acetylcholine into the femoral artery, shunting was repeatedly blocked

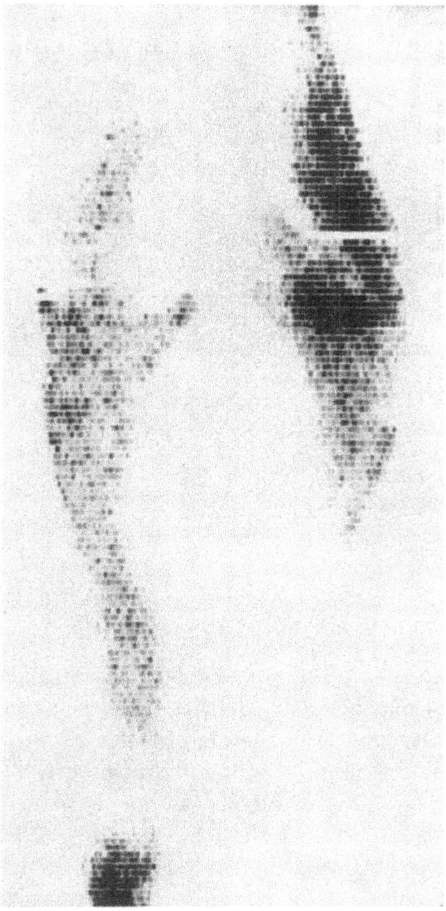

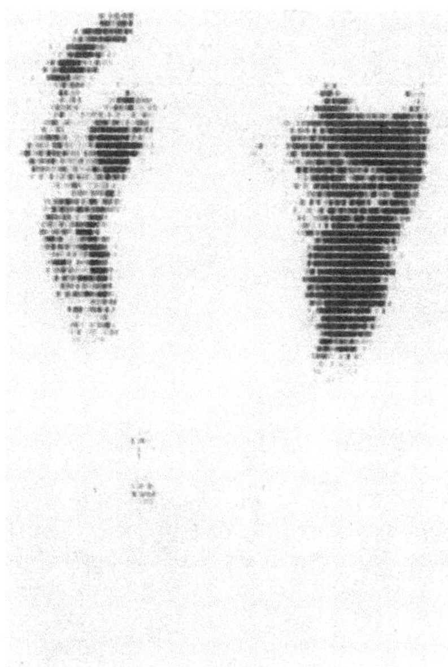

Fig. 4. The left side of the picture shows the distribution of radioactive microspheres in the hindleg of a dog in which 57% of the microspheres were shunted. The right side shows the distribution when the microspheres were injected while the shunts were closed by the action of acetylcholine

Fig. 3. The left side of the picture shows the distribution of radioactive microspheres in the hindleg of a dog in which 45% of the microspheres were shunted. The right side shows the distribution when the microspheres were injected while the shunts were closed by the action of acetylcholine

tion when the shunts are open; the right side shows the distribution when the shunts are closed as a result of prior administration of acetylcholine. Apparently the action of the acetylcholine is to increase blood flow to the muscles of the upper extremity at the expense of flow to the lower extremity, especially the paw.

Table 4 shows the result of the experiments in which the skin was removed or in which the circulation to the paw was occluded. This data shows that the shunts are not located primarily in the skin or paw. Fig. 5 shows a scan of a dog in which shunting was measured during occlusion of flow to the right paw. This dog shunted 19.5% of microspheres through the right leg, compared to 10.4% through the left or control leg.

Table 4. *Effect of skin and paw circulation on shunting*

Procedure	% shunting	
	before	after
Skin removed from leg	0.8	1.4
	20.8	34.2
Occluded circulation to paw	4.9	3.7
	5.0	9.6
	7.3	10.4
	10.9	11.8
	9.1	16.4
	10.4	19.5

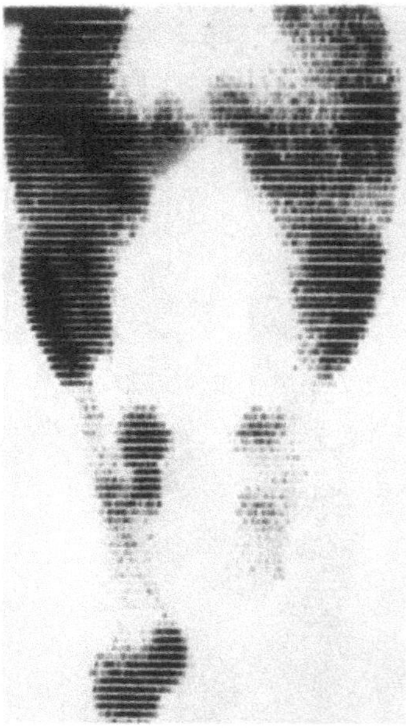

Fig. 5. The left side of the picture shows the distribution of radioactive microspheres in the hindleg of a dog in which 10.4% of the microspheres were shunted. The right side shows the distribution when the microspheres were injected during occlusion of paw circulation. 19.5% of the microspheres were shunted

Discussion

These experiments in man and dog demonstrate the physiological significance of the passage of blood through communications in the extremities that are larger than capillaries. Previously it has been thought that the significance of such shunting was primarily in the control of the distribution of skin and fingertip or paw blood flow [2,5,9]; however, since the shunting of blood was not decreased when the paw or skin circulation was removed, it appears that the overall circulation to the extremity is involved. This is also suggested by the magnitude of the shunted blood flow [3, 4].

GRANT [2] observed that acetylcholine opened the anastomosis in the ears of rabbits and UNVÄS [6] observed that the stimulation of vasodilator nerves in the hypothalamus increased venous outflow from skeletal muscle of the hindleg of the dog by a factor of 4 to 5, but oxygen consumption was decreased, suggesting that shunts were being opened. The transmitter was thought to be acetylcholine [10] because the effect was blocked by atropine [6]. The present results, like the earlier findings of BOSTROEM and SCHOEDEL [5], do not support the hypothesis that acetylcholine causes shunts to open, at least in the hindlegs of dogs. In fact, acetylcholine always reduced shunting. The mechanism by which this occurs is not yet known; however, the data suggest an explanation: the vasodilation action of the acetylcholine greatly increases muscle blood flow, probably by opening up the muscle capillaries. Normally, most of the capillaries would be closed during periods of inactivity. The diversion of the blood into capillaries together with the decreased blood pressure may indirectly limit the flow through the shunts.

Previous studies of extremity blood flow in man with ^{131}I-labeled macroaggregated albumin by WAGNER and his coworkers [11] suggested that the study of shunting in the extremities of man was of potential value in the assessment of peripheral vascular disease. They found that the uptake of the macroaggregates in the leg after a femoral artery injection was less in normal resting subjects than in normal subjects with experimentally induced reactive hyperemia. On the other hand, the uptake in the leg was not increased by reactive hyperemia in patients with arteriosclerosis. These results have suggested that shunts can be open in normal resting subjects but closed as required to meet metabolic demands; however, the shunts did not appear to the responsive in the arteriosclerotic patients. The data in the present study shows that the measurement of shunting in the extremities of man is feasible and that the resultant information is also potentially useful in the diagnosis and management of patients with peripheral vascular disease. Further studies in normal subjects and further understanding of the basic mechanism of the opening and closing of the shunts is needed, however. All subjects tolerated the test without difficulty. Also, the albumin microspheres have been used safely and effectively as a lung scanning agent in a larger series of patients [12].

Limitations of the accuracy of this method arise from the basic assumptions. These are: 1. uniform mixing of the microspheres with the blood perfusing the tissues; 2. all particles are too large to pass through capillaries yet small enough to pass through all arteriovenous anastomoses; and 3. all shunted particles are carried by the blood stream to the lungs where they are completely and uniformly trapped.

The method of injection is designed to maximize mixing. The microspheres are mixed with blood in the syringe, then injected countercurrent so as to create turbulence and mixing at the point of injection. The rheology of microspheres in blood indicated a uniform distribution in arteries [13]; however, a tendency of the microspheres to pass preferentially to larger channels could result in overestimations of shunting.

The ideal size microsphere for shunt studies is yet to be determined. Since capillaries are usually less than 10 microns in diameter, it is probable that 10 to 15 micron microspheres would be nearly ideal. In previous studies [3, 4] it has been found that shunting of larger particles is less than that of smaller particles but the difference between 20 and 40 microns was not great. Because the microspheres used in this study may be slightly larger than required to be stopped by capillaries, shunting may have been underestimated.

Whole body scans of dogs obtained after the shunt studies show that the radioactivity is localized in the legs and lungs. Measurement of transpulmonary shunting in man [14] and animal distribution studies [8] have shown the lungs effectively remove approximately 99% of these microspheres from the venous circulation and that the microspheres are distributed uniformly in the lungs of healthy subjects. If the detector were placed over an area of non-perfused lung, underestimates of shunting could result. Hence it has become a part of the routine to scan following each shunt study.

Webber et al. [15] have shown that albumin particles are trapped in areas of veins with thrombophlebitis. This has also been observed with albumin microspheres. Even a venipuncture in the leg will cause a local accumulation of radioactivity following microsphere injection. When this occurs, shunted microspheres fail to reach the lungs and shunting is underestimated. Again, this problem can be detected by scanning following the shunt study.

Conclusions

These studies demonstrate that ^{99m}Tc-labeled albumin microspheres can be used to estimate relative blood flow through non-capillary arteriovenous communications in the extremities of man and dogs. Shunting measured in this manner ranged from zero to more than 50 per cent and a wide difference could be found in the same individual. The studies in dogs show that these shunts are not located primarily in the skin or paw and that they are closed rather than opened by acetylcholine as had been previously thought.

The method has been safely and effectively used in man. These results indicate that the method is potentially useful in the diagnosis and management of patients with peripheral vascular disease.

References

1. Liebow, A. A.: Situations which lead to changes in vascular patterns. In: Handbook of Physiology, Vol. II, Section 2, W. F. Hamilton and P. Dow (Eds.). Washington, D.C.: American Physiological Society 1963, p. 1251.
2. Grant, R. T.: Observations on direct communications between arteries and veins in the rabbit's ear. Heart **15,** 281 (1929—1931).
3. Piiper, J., Schoedel, J. W.: Untersuchungen über die Durchblutung der arteriovenösen Anastomosen in der hinteren Extremität des Hundes mit Hilfe von Kugeln verschiedener Größe. Pflügers Arch. Ges. Physiol. **258,** 489 (1954).
4. Lopez-Majano, V., Rhodes, B. A., Wagner, H. N., Jr.: Arteriovenous shunting in extremities. J. Appl. Physiol. **27,** 782 (1969).
5. Bostroem, B., Schoedel, J. W.: Über die Durchblutung der arteriovenösen Anastomosen in der hinteren Extremität des Hundes. Pflügers Arch. Ges. Physiol. **256,** 371 (1953).
6. Unväs, B.: Sympathetic vasodilator system and blood flow. Physiol. Rev. **40,** Suppl. **4,** 69 (1960).
7. Zolle, I., Rhodes, B. A., Wagner, H. N., Jr.: Preparation of metabolizable radioactive human serum albumin microspheres for studies of the circulation. Int. J. Appl. Rad. Isotopes **21,** 155 (1970).

8. RHODES, B. A., ZOLLE, I., BUCHANAN, J. W., WAGNER, H. N., Jr.: Radioactive albumin microspheres for studies of the pulmonary circulation. Radiology **92,** 1453 (1969).
9. MAGGIO, E.: Microhemocirculation. Springfield, Illinois: C. C. Thomas 1965, p. 92.
10. GREGG, D. E.: Regulation of pressure and flow in the systemic and pulmonary circulation. In: The Physiological Basis of Medical Practice, C. H. BEST and N. B. TAYLOR (Eds.). Baltimore, Maryland: Williams & Wilkins 1951, p. 256.
11. WAGNER, H. N., Jr.: Principles of Nuclear Medicine. Philadelphia: W. B. Saunders 1968, p. 577.
12. RHODES, B. A., WAGNER, H. N., Jr.: Improved lung scans using ^{99m}Tc-albumin microspheres. J. Nucl. Med. **10,** 432 (1969).
13. PHIBBS, R. H., WYLER, F., WENTZE, J.: Rheology of microspheres injected into circulation of rabbits. Nature **216,** 1339 (1967).
14. HURLEY, P. J., STRAUSS, H. W., RHODES, B. A., WAGNER, H. N., Jr.: Quantification of transpulmonary shunting in man by measuring labeled particle distribution. J. Nucl. Med. **10,** 344 (1969).
15. WEBBER, M. M., BENNETT, L. R., CRAGIN, M., WEBB, R., Jr.: Thrombophlebitis — demonstration by scintiscanning. Radiology **92,** 620 (1969).

Unilateral Renal Clearances by External Counting in the Study of Hypertension

C. Bianchi, A. Coli, R. Palla and P. Giannotti

Medical Clinic, C.N.R. Laboratory of Clinical Physiology* and Surgical Clinic**, University of Pisa, Italy

Introduction

The quantitative evaluation of unilateral renal function in hypertensive subjects is of considerable usefulness to demonstrate an eventual unilateral renal disease responsible for hypertension. It is essential before any operation either of nephrectomy or of vascular reconstruction to establish the function of the injured kidney with the other one most probably normal. This is done for the purpose of evaluating the possibilities of curing hypertension and of foreseeing the entity of residual renal function in case of nephrectomy.

The available methods of measuring divided renal functions are based on the separated clearance measurement (of inulin, PAH, or other suitable tracers), performed by urine collection through double ureteral catheterization. Nevertheless these methods are not satisfactory, as the ureteral catheterization often involves some nasty effects, such as pain, hematuria and infection. Furthermore, in a relevant number of cases there is leakage of urine into the bladder causing unreliable results. Finally, ureteral catheterization could actually alter renal function in a significant manner.

A method has been set up for the measurement of unilateral renal clearances by external counting, thus avoiding ureteral catheterization. At first it has been applied to measure the effective renal plasma flow (RPF) of each kidney, by using ^{131}I-hippuran as a tracer [1—4]. More recently, also the unilateral glomerular filtration rate (GFR) has been measured by ^{131}I-hypaque as a tracer. This method and its possibilities in the study of hypertensive subjects are described and discussed.

Method and Materials

The method is based on the measurement of the radioactivity increase recorded on each kidney during a venous perfusion of a gamma-emitting tracer eliminated through the kidney, when the urine-flow into the bladder is stopped. This method is represented in Fig. 1 where the plasma, kidney and bladder radioactivity levels are shown. When constant plasma levels are attained, renal activity remains

* C. Bianchi, A. Coli, R. Palla.

** P. Giannotti.

constant while the bladder activity increases linearly in time. If urine-flow into the bladder is stopped (indicated by the broken line in the figure), a linear increase of radioactivity of each kidney occurs while the bladder and plasma activities

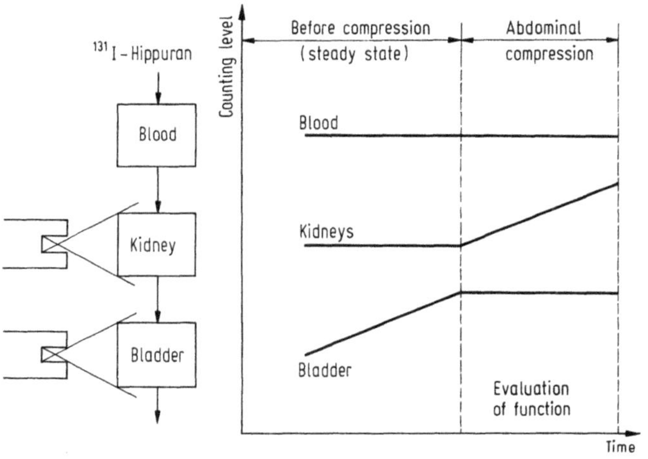

Fig. 1. Schema of the external counting method to measure unilateral RPF. The vertical broken line indicates the beginning of the abdominal compression

naturally remain constant. It is assumed that the radioactivity build-up of the two kidneys, caused by the accumulated tracer in the pelves, be proportional to the relative clearances. That is:

$$\Delta L : \Delta R = \mathrm{Cl}\ L : \mathrm{Cl}\ R$$

where ΔL and ΔR are the radioactivity increases over the left and right kidneys respectively and Cl L and Cl R left and right clearances.

The unilateral clearances are thus obtained:

$$\mathrm{Cl}\ L = \frac{\Delta L}{\Delta L + \Delta R} \cdot \text{total clearance.}$$

The percent difference between the clearances of the two sides is calculated dividing the difference D (of the two clearances) by the average clearance of the two kidneys as following:

$$\%\ \text{difference} = \frac{2\,D}{\mathrm{Cl}\ L + \mathrm{Cl}\ R} \cdot 100.$$

Urine-flow was stopped by means of abdominal compression with an apparatus already described [5]. This apparatus consists of air pressure operated pistons with two wooden wedges placed on the abdomen about 2 cm below the transverse umbilical line. In order to measure the radioactivity of the two kidneys and of the bladder three scintillation counters, whose characteristics have already been described [4, 6], were used.

It is important to note that each renal counter is recording solely the radioactivity of one kidney and not of the other one or of the bladder. The relative counting efficiency of the renal counters on a plane situated 5 cm from the opening of the collimators is shown in Fig. 2. The complete apparatus (the abdominal compressors, the bladder counter and the renal counters, placed under the bed) are visible in Fig. 3.

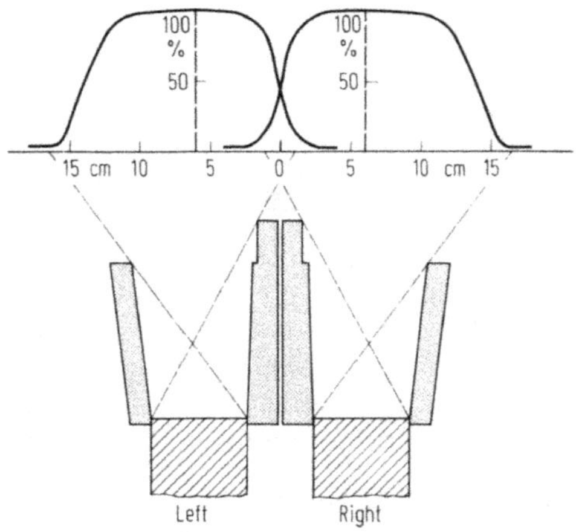

Fig. 2. Counting efficiency of the renal counters

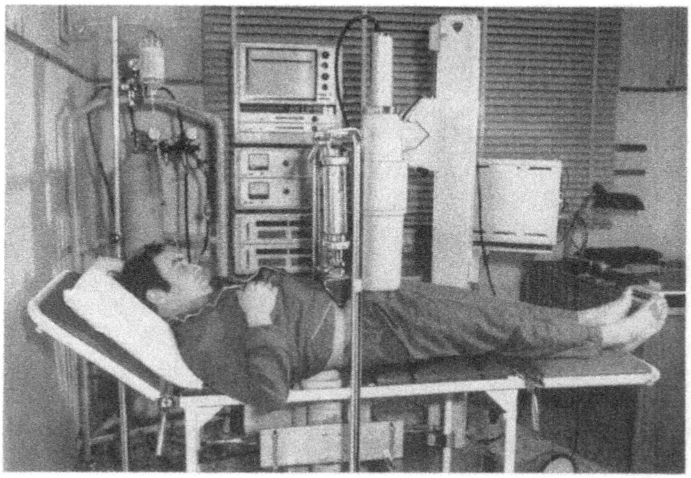

Fig. 3. A patient during unilateral RPF measurement. The vesical counter and the device for abdominal compression are above the patient; the two renal counters beneath

In order to obtain a correct measurement of the radioactivity increases it is necessary that both pelves are found completely in the visual angle of the renal counters. For this purpose the renal collimators are furnished with two lead diaphragms with a central opening of 2.5 cm in diameter. At a certain time, when total renal clearances are measured, during the perfusion with radioactive tracers, the urine-flow is stopped. The two renal counters with the diaphragms in place are moved on their supports under the bed to look for the areas of highest activity. This research is facilitated by an acoustic and visual monitoring system. After the two highest radioactivity areas are localized, the two diaphragms are taken off. It is assumed that in these conditions the renal counters see completely the pelves. The renal counters are shown in detail in Figs. 4 and 5.

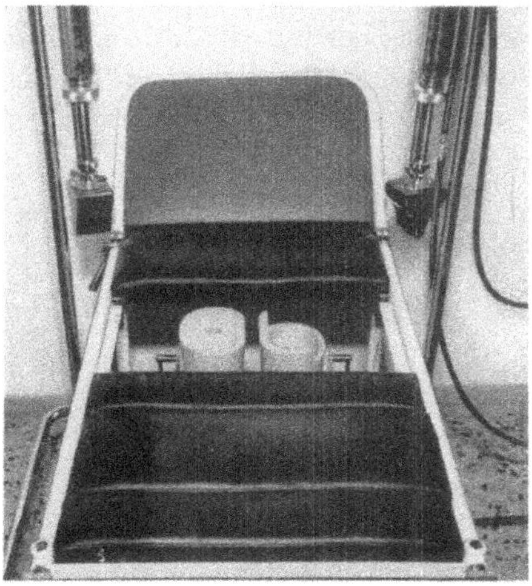

Fig. 4. The two renal counters with (on the left) and without the lead diaphragm. Abdominal compressors can be seen

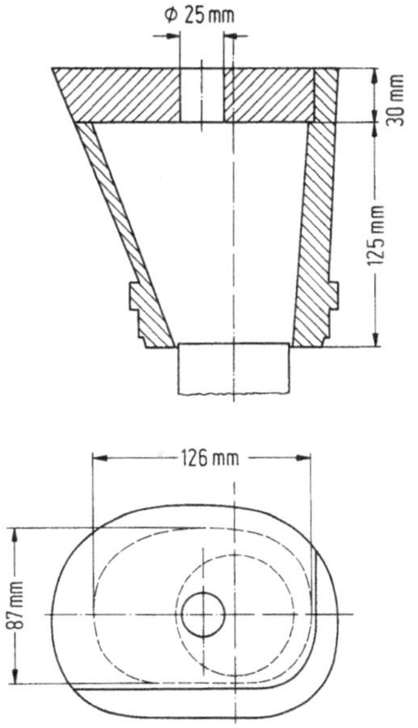

Fig. 5. Schema of a renal counter. Sagittal section (above); viewed from above (below)

In order to evaluate the effectiveness of this method the unilateral RPF was measured in 15 subjects apparently normal with regard to their renal function (6 males and 9 females, ranging from 18 to 44 years of age) who were hospitalized for minimal extrarenal diseases. Afterwards the unilateral RPF measurement with the above described method was applied also in 28 hypertensive subjects, 12 males and 16 females ranging from 22 to 56 years of age, in whom the total renal function was sufficiently preserved (GFR superior to 50 ml/min). In 17 of these subjects the unilateral RPF was also measured using ureteral catheterization under mild sedation with diazepam; No. 6 Ch catheters were used. In these cases, all the other tests, necessary to demonstrate the characteristics of the unilateral renal disease, were carried out: rapid sequence pyelography, renography, renal scan and aortography, when necessary. Two typical cases of this group are described in detail in the Results.

Recently the above described method of external counting was applied in unilateral GFR measurement. Twelve subjects presumably non-nephropathic (except one with chronic glomerulonephritis) were examined: 5 males and 7 females, age ranging from 19 to 60, hospitalized for minimal extrarenal diseases.

Divided RPF was measured using ^{131}I-hippuran and divided GFR was measured with ^{131}I-hypaque. Doses of about 2 microcuries per kg of body weight were dissolved in 500 ml of saline; 10 ml of PAH at 20% were added as a carrier in the solution for RPF measurement. In the cases without renal involvement a priming dose of 100 drops per minute for 10 min and after a sustaining dose of 30 drops per minute till the end of the examination was administered i.v. When the equilibrium was obtained (as judged from the precordial counting) a total clearance was measured with the standard technique of urine collection in some cases and with the external counting method [6] in the remaining ones.

In each case two or more clearance periods during compression were performed recording the kidney and bladder radioactivity increases. The typical behaviour of a normal case is shown in Fig. 6. The clearance periods lasted 7 to 15 minutes. In 40 subjects at the end of the first or second period, the highest radioactive points were again checked to evaluate the exactness of the first collimation.

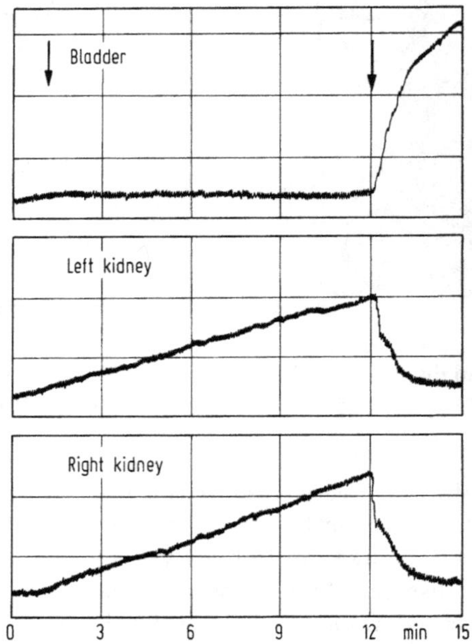

Fig. 6. Unilateral RPF measurement. Recording of vesical and renal curves during urine stop-flow in a normal case. The arrows on the left and on the right indicate respectively the beginning and the end of the abdominal compression

Results

The results are reported in Table 1, 6, 7 and shown in Figs. 7, 18, 19.

The unilateral RPF values in normal subjects with the external counting method (Table 1, Fig. 7) are quite similar to those obtained by ureteral catheterization. The mean RPF ratio between the left and right kidney is 1.01 $\pm$ 0.07 S.D.

and the mean percent difference 5.04 ± 4.35 S.D. As one can observe in Fig. 7, 13 of 15 experimental points are situated in the range of ± 10% from the identity line.

Table 1. *Total and unilateral RPF in normal subjects*

Subject	Sex	Age (years)	Total RPF (ml/min)	Left RPF (ml/min)	Right RPF (ml/min)	Left RPF / Right RPF	% Difference
F.V.	F	36	396.17	197.24	198.93	0.99	0.85
G.G.	F	18	430.06	212.30	217.76	0.97	2.53
M.V.	M	44	440.05	228.17	211.88	1.07	7.40
S.A.	M	43	539.93	256.24	283.69	0.90	10.16
D.D.M.	F	26	468.22	235.42	232.80	1.01	1.11
C.A.	F	19	577.99	286.78	291.21	0.98	1.53
S.A.	F	19	442.17	237.51	204.66	1.16	14.85
R.L.	F	37	662.57	330.80	331.77	1.00	0.29
B.B.	M	33	564.28	297.59	266.69	1.11	10.95
R.D.	F	52	366.66	189.75	176.91	1.07	7.00
C.M.	F	41	470.48	241.47	229.01	1.05	5.00
B.M.	M	24	627.95	315.97	311.95	1.01	1.28
D.E.M.	M	23	447.49	220.33	227.15	0.96	2.38
B.F.	M	26	530.01	256.73	273.28	0.93	6.24
T.L.	F	35	501.46	245.74	255.71	0.96	3.97
Mean						1.01 ± 0.07	5.04 ± 4.35

No value is corrected to 1.73 sq m body surface area.

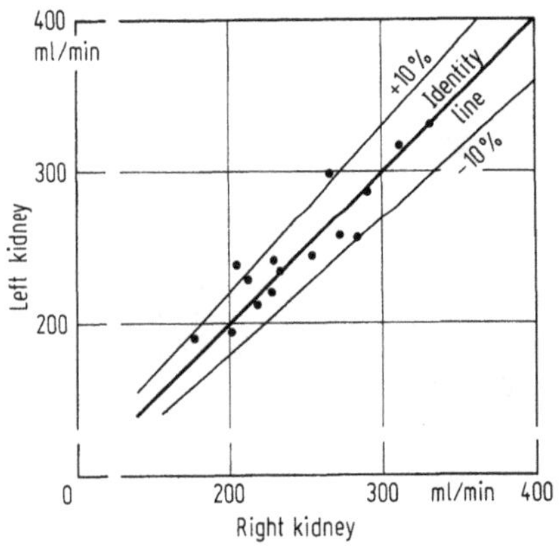

Fig. 7. Unilateral RPF measurement by external counting in 15 normal subjects. The values of the right kidney are on the abscissa, those of the left kidney on the ordinate

In the 17 hypertensive subjects in which unilateral RPF was also measured using the ureteral catheterization the results were reliable in 10 of them. In these cases the urine leakage was practically nothing or even lacking, except in one case in which it was significant, but the values of the urine concentration in sodium, creatinine and ^{131}I-hippuran permitted us to locate exactly the side of urine leakage. The data obtained by ureteral catheterization (urine-flow, urine concentration in sodium, creatinine and ^{131}I-hippuran) plus clinical and laboratory data of each case permitted diagnosis of the nature of the renal disease. In the 10 cases presented we came across 3 unilateral pyelonephritis, 3 renovascular hypertension, 1 renal tuberculosis, while in the other 3 cases the diagnosis was not ascertained.

In order to demonstrate the importance of the unilateral clearance measurement (and hence of our method) in comparison to the other clinical and laboratory data, in the study of hypertension, two cases (renovascular hypertension and unilateral pyelonephritis) are reported in detail.

Main Left Renal Artery Obstruction (Atherosclerotic)

C.G., a 53 year old man, was admitted to the Hospital in October 1967 complaining of headaches of 4 months' duration. There was no family history of hypertension. His systolic blood pressure was 200 in June 1967. On admission to the Hospital the blood pressure was 230/130 lying and standing. Auscultation of the abdomen was negative. Funduscopic examination revealed signs of early hypertensive retinopathy. The heart was not clearly enlarged. The ECG showed left ventricular hypertrophy. Urinalysis showed a trace of proteinuria without significant sediment. The urine culture was sterile. BUN was 30 mg% and serum creatinine 1.38 mg%. The total serum cholesterol was 245 mg%. An intravenous pyelogram at rapid sequence showed a delay in the appearance of the contrast medium on the left side (Fig. 8) and a greater concentration on the same side (Fig. 9). The two kidney's sizes did not seem evidently different. The renogram (^{131}I-hippuran) in hydration state, showed a prolonged accumulation to a peak at about five minutes and a marked decreased slope of the excretory phase on the left side (Fig. 10). The renal scan (^{197}Hg-chlormerodrin) demonstrated

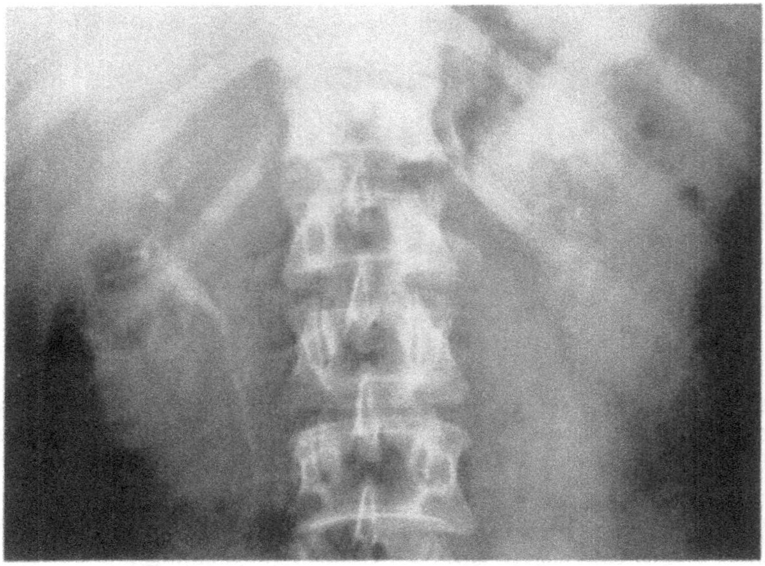

Fig. 8. C.G., 3-min intravenous pyelogram. There is a delay in the appearance of the contrast medium on the left side

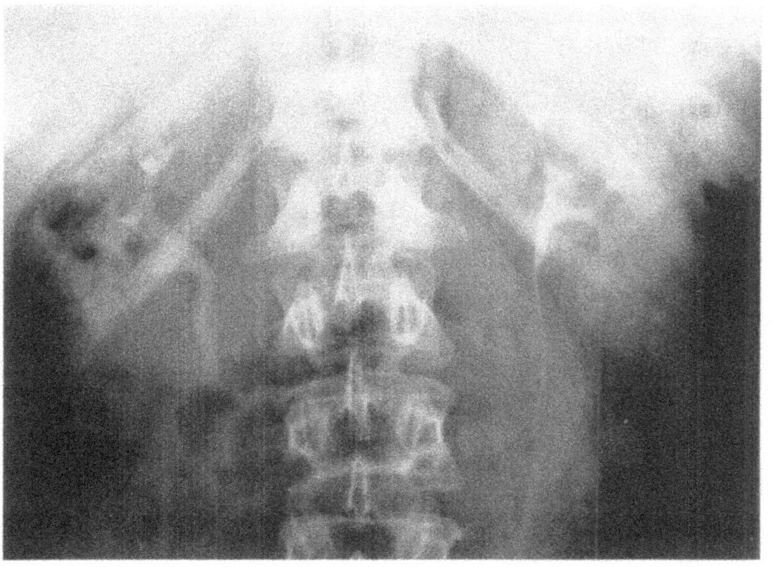

Fig. 9. C.G., 10-min intravenous pyelogram. Hyperconcentration is prominent in the left kidney

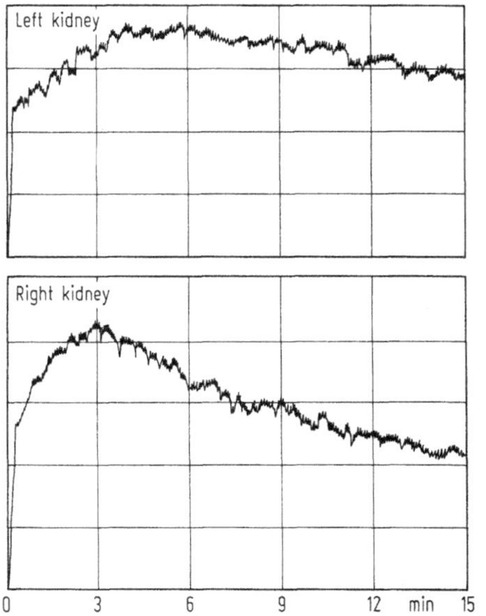

Fig. 10. C.G., renogram. There is a prolonged accumulation and a decreased slope of the excretory phase on the left side

a reduced left kidney uptake of the tracer (Fig. 11). During hospitalization unilateral RPF has been measured by external counting and by ureteral catheterization at a 28 days' interval. In this case RPF by ureteral catheterization has been performed by employing PAH (instead of ^{131}I-hippuran) as a tracer. The results, reported in Table 2, show similar percent differences between the two sides, although the absolute values obtained by the two methods are different.

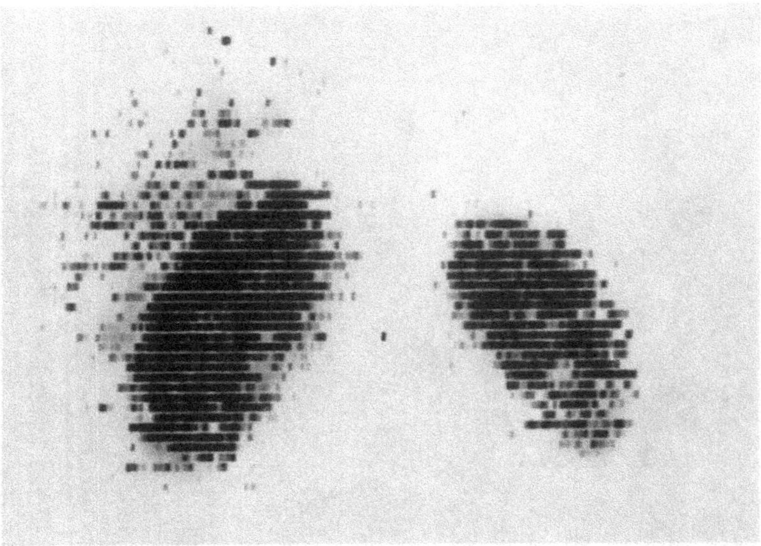

Fig. 11. C.G., renal scan. Left kidney is smaller than the right

Table 2

Subject: C.G.	By external counting	By ureteral catheterization
Left kidney RPF (ml/min)	55.20	103.33
Right kidney RPF (ml/min)	237.53	330.33
% difference	124.58	104.69

Ureteral catheterization permitted us to measure unilateral urine volumes and concentrations of PAH, creatinine and sodium. The results of these measurements are reported in Table 3.

Table 3

Clearance No.		Urine-flow (ml/min)	PAH concentration (mg%)	Creatinine concentration (mg%)	Sodium concentration (mg%)
1	Left	0.45	95	40	135
	Right	3.54	60	13	262
2	Left	0.45	135	35	92
	Right	4.00	40	11	250
3	Left	0.53	125	32	113
	Right	4.93	35	10	230
Mean	Left	0.47	118	35.66	113
	Right	4.15	45	11.33	247

The above reported data were quite suggestive for the diagnosis of left renal ischemia. The retrograde femoral arteriogram confirmed this evidence, showing a severe stenosis in the proximal portion of the left renal artery with poststenotic dilatation (Fig. 12). This was definitively confirmed by surgery which demonstrated an atherosclerotic stenosis of the first portion of the artery. A biopsy specimen, taken during surgery, showed a prominent macula densa and hyperplasia of the juxtaglomerular cells (Fig. 13). After left aortorenal by-pass (dacron) blood pressure fell gradually to normal values.

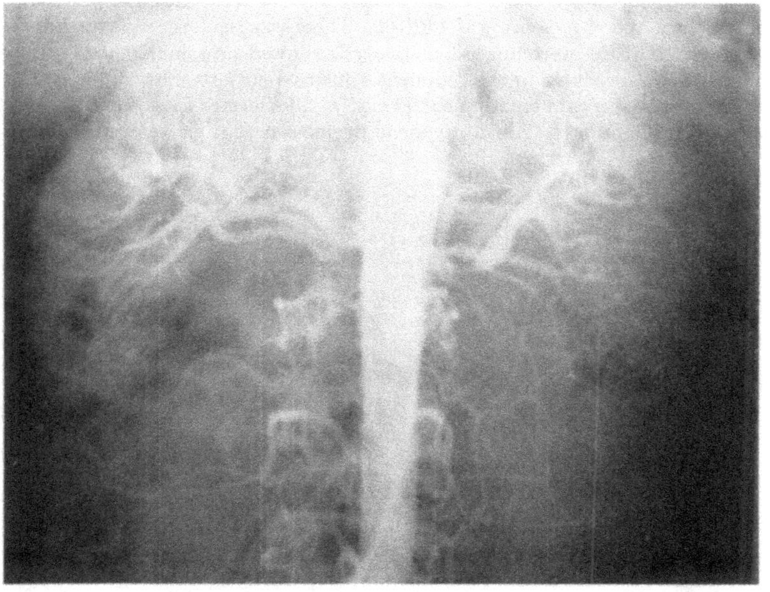

Fig. 12. C.G., transfemoral aortcgram. A severe stenosis in the proximal portion of the left renal artery with poststenotic dilatation is evident

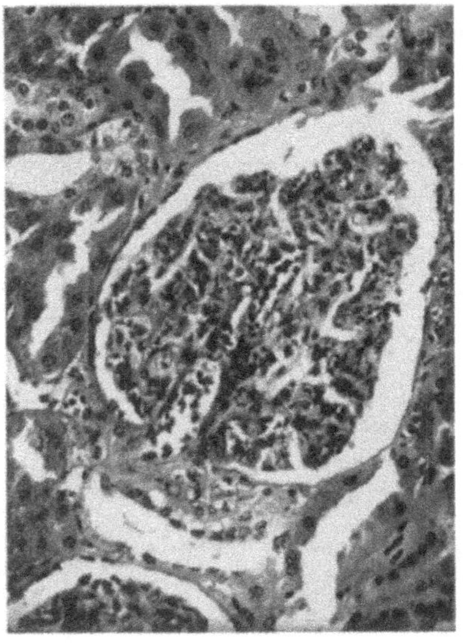

Fig. 13. C.G., left kidney biopsy. Microscopic section of a typical glomerulus shows a prominent macula densa and hyperplasia of the juxtaglomerular cells

Left Kidney Pyelonephritis

V.D., a 22 year old man, was admitted to the Hospital in August 1968 for persistent mild pain in both flanks of 1 month's duration, accompanied by tiredness, headache and intermittent nausea and vomiting. Nothing in family history. In March 1967, he experienced severe pain in the left flank considered as a renal colic. On this occasion an intravenous pyelogram was negative and the blood pressure was normal. He was diagnosed as having "nephritis". Afterwards urinalysis always demonstrated proteinuria. A second pyelography performed in February 1968 showed the left kidney smaller than the right one. On admission to the Hospital the blood pressure was 160/105. There was tenderness in the left flank. The chest X-ray and the ECG were normal. Urinalysis showed proteinuria (0,25 g/L) and a few red blood cells and leucocytes in the sediment. A urine culture showed 100 000 E. Coli/ml, BUN was 32 mg% and serum creatinine 0.90 mg%. Glomerular filtration rate (^{131}I-hypaque clearance) resulted 76 ml/min. An intravenous pyelogram at rapid sequence showed prompt

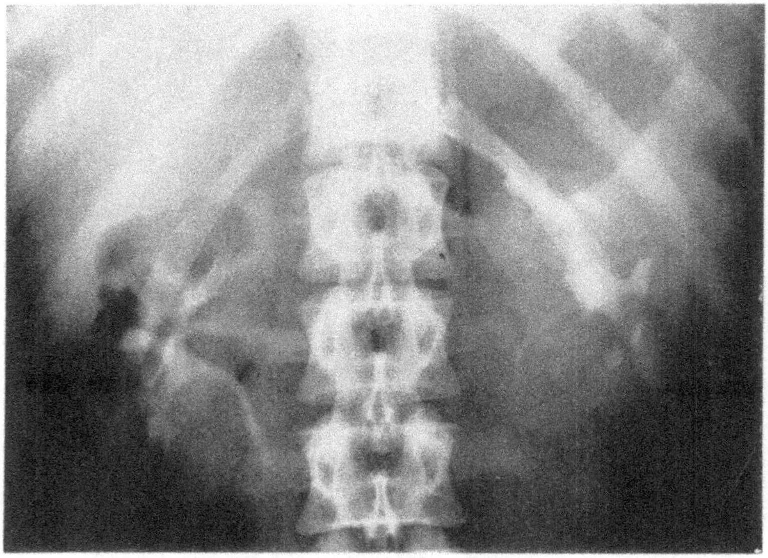

Fig. 14. V.D., 7-min intravenous pyelogram. Some typical features of the pyelonephritis can be seen on the left side

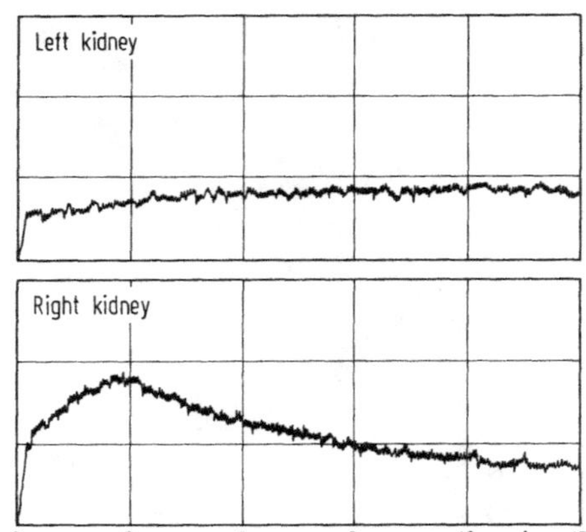

Fig. 15. V.D., renogram Left kidney non-functioning

excretion bilaterally without differences of concentration between the two sides; on the left side calices were distorted and their margins were irregular (Fig. 14). These findings were suggestive of a left kidney pyelonephritis. The renogram (^{131}I-hippuran) in hydration state showed no evidence of renal activity on the left side (Fig. 15). The renal scan (^{197}Hg-chlormerodrin) demonstrated an uptake of the tracer by the left kidney markedly reduced (Fig. 16). The measurement of unilateral renal plasma flow by means of ^{131}I-hippuran has been performed by external counting and by ureteral catheterization at a 3 days' interval. The results by the two methods, reported in Table 4, are very similar.

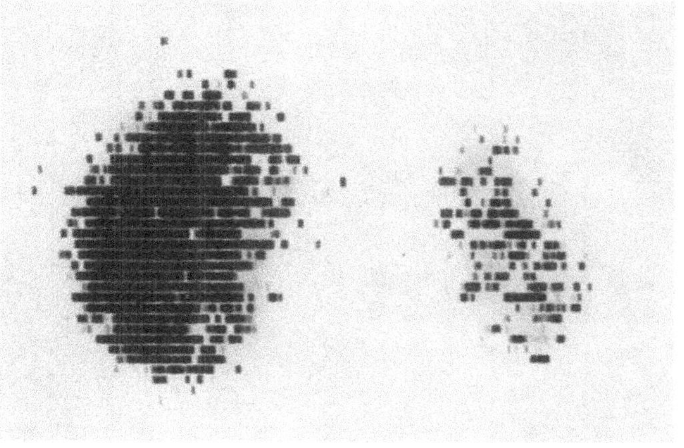

Fig. 16. V.D., renal scan. Left kidney is smaller than the right; activity is irregularly distributed in the left kidney

Table 4

Subject: V.D.	By external counting	By ureteral catheterization
Left kidney RPF (ml/min)	159.81	167.66
Right kidney RPF (ml/min)	353.56	359.49
% difference	75.48	72.78

The values of unilateral urine volumes and concentrations of ^{131}I-hippuran, creatinine and sodium are reported in Table 5.

Table 5

Clearance No.		Urine-flow (ml/min)	^{131}I-hippuran concentration (cpm)	Creatinine concentration (mg%)	Sodium concentration (mg%)
1	Left	1.19	152 942	30	178
	Right	2.25	181 040	34	225
2	Left	1.80	138 466	23	158
	Right	3.27	152 576	26	170
3	Left	2.80	82 870	14	96
	Right	5.80	88 758	15	110
Mean	Left	1.93	124 759	22	144
	Right	3.77	140 791	25	168

A retrograde femoral arteriogram showed a normal filling of the main renal vessels on both sides. The left kidney is significantly smaller than the right; on the left side the cortex is reduced in thickness and the cortical margin is irregular (Fig. 17).

All the above reported results are consistent with the diagnosis of pyelonephritis of the left kidney.

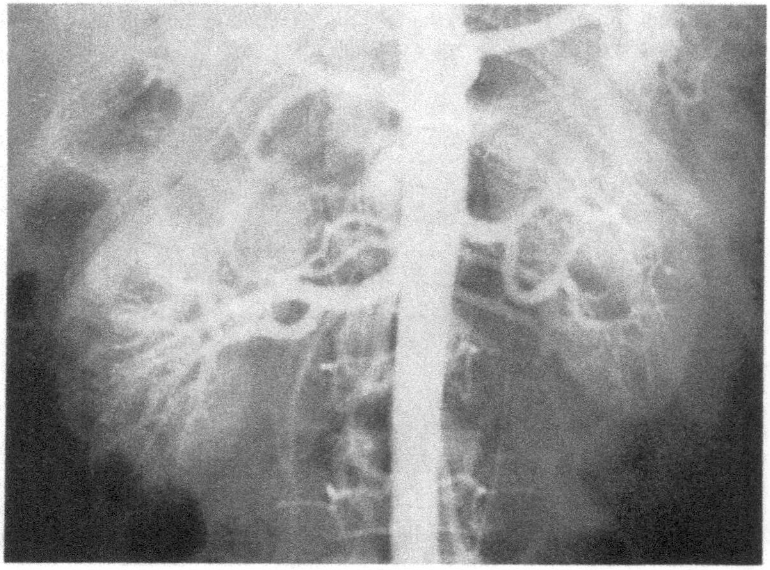

Fig. 17. V.D., transfemoral aortogram. There are no vascular obstructions. Left kidney is smaller than the right; its cortex is reduced in thickness and the cortical margin is irregular

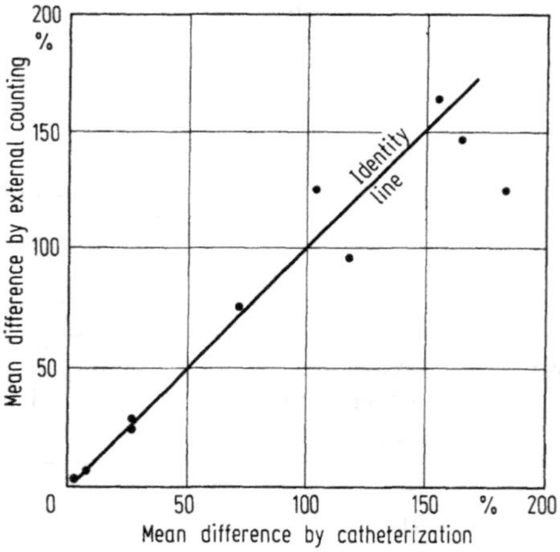

Fig. 18. Unilateral RPF measurement in 10 hypertensive subjects. On the abscissa the mean percent differences between the two sides obtained by catheterization, on the ordinate those obtained by external counting

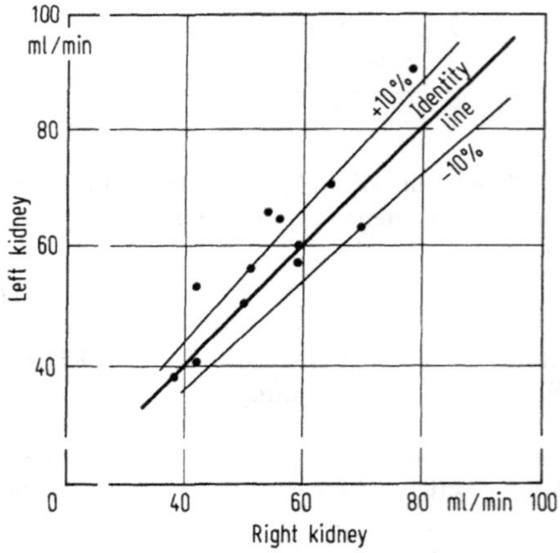

Fig. 19. Unilateral GFR measurement by external counting in 12 normal subjects. The values of the right kidney are on the abscissa, those of the left kidney on the ordinate

Table 6. *Unilateral RPF in hypertensive subjects*

Subject	Diagnosis	Sex	Age (years)	Blood pressure	Renal plasma flow					
					By external counting			By catheterization		
					(ml/min)		% Diff.	(ml/min)		% Diff.
L.A.	left kidney tuberculosis	M	35	240/170	Left Right	62.63 177.03	95.47	Left Right	45.37 177.75	118.66
C.G.	left renal artery stenosis (atherosclerotic)	M	53	220/135	Left Right	55.20 237.53	124.58	Left Right	103.33 330.33*	104.69
C.G.	right kidney pyelonephritis	M	42	240/145	Left Right	217.73 34.09	145.85	Left Right	251.15 22.99	166.45
V.D.	left kidney pyelonephritis	M	22	160/105	Left Right	159.81 353.56	75.48	Left Right	167.66 359.49	72.78
C.A.	unknown	F	31	160/100	Left Right	259.25 202.83	24.42	Left Right	203.24 154.54	27.22
B.C.	right renal artery stenosis (fibromuscular dysplasia)	F	24	190/125	Left Right	216.26 51.51	123.05	Left Right	281.91 11.36	184.50
B.G.	unknown	M	25	160/120	Left Right	210.53 157.07	29.08	Left Right	261.54 197.98	27.66
C.R.	left kidney pyelonephritis	M	56	205/120	Left Right	21.95 214.78	162.90	Left Right	19.84 161.34	156.19
V.F.	unknown	F	32	190/110	Left Right	315.13 303.22	3.85	Left Right	281.48 275.52	2.14
A.C.	left renal artery stenosis (external constriction)	F	48	210/125	Left Right	188.35 203.05	7.51	Left Right	172.08 186.71	8.15

No value is corrected to 1.73 sq m body surface area.

* Performed with PAH.

Apart from these cases even in the remaining 8 hypertensive subjects a satisfactory agreement between the values obtained with the described external counting method and that of the ureteral catheterization was obtained (Table 6, Fig. 18). The mean percent difference between the two sides in RPF was 79.22% with external counting method and 86.84% with the ureteral catheterization.

With regard to the external counting method in the case of unilateral GFR measurement in normal subjects, till now examined (Table 7, Fig. 19), the results agree satisfactorily with those obtained with the standard technique of urine collection via ureteral catheterization. The mean ratio between GFR of the left kidney with that of the right is 1.07 ± 0.11 S.D. and the mean percent difference 8.89 ± 7.44 S.D. In this case 8 experimental points out of 12 are situated in the range of ± 10% from the identity line.

Table 7. *Total and unilateral GFR in normal subjects*

Subject	Sex	Age (years)	Total GFR (ml/min)	Left GFR (ml/min)	Right GFR (ml/min)	Left GFR / Right GFR	% Difference
G.E.	F	43	82.60	40.58	42.01	0.97	3.46
L.B.	M	44	135.41	70.62	64.69	1.09	8.75
N.G.P.	M	28	116.34	57.37	58.97	0.97	3.36
G.M.C.	F	26	132.80	62.95	69.85	0.90	10.39
G.L.*	F	42	76.83	38.41	38.42	1.00	0
M.B.	F	46	107.57	56.16	51.39	1.09	8.87
S.A.	F	29	100.09	50.26	49.82	1.01	0.87
L.R.	F	25	119.24	60.03	59.21	1.01	1.37
S.E.	M	20	119.63	65.53	54.10	1.21	19.10
M.G.C.	M	19	169.06	90.51	78.55	1.15	14.13
B.B.	F	36	120.88	64.55	56.33	1.15	13.60
S.P.	M	60	94.86	52.83	42.03	1.26	22.77
Mean						1.07 ± 0.11	8.89 ± 7.44

No value is corrected to 1.73 sq m body surface area.
* Chronic glomerulonephritis.

Discussion

To compare correctly the external counting method for split function evaluation with the standard technique of ureteral catheterization, one must point out that it is necessary to avoid the leakage of urine outside of the catheter in the latter technique, while the external counting method requires a complete urine stop-flow. This fact is actually obtained in the majority of cases with the apparatus for abdominal compression. In fact in 101 cases examined up to now (of both sexes and various physical make-ups) we obtained a complete arrest of urine-flow (demonstrated with external counting over the bladder) in 87 subjects. In the cases where divided RPF was measured with ureteral catheterization, a significant leakage of urine into the bladder occurred in 30% of cases. These results are similar to those reported by other authors, who studied a wider number of cases [7]. It therefore seems feasible to stop the urine-flow into the bladder by the apparatus of abdominal compression rather than collecting the urines completely with ureteral catheterization. It is noteworthy to point out that abdominal compression causes slight pain, usually well tolerated for a period between 7 to 15 min, and moreover, in our opinion, always less than with ureteral catheterization. Furthermore with regard to eventual modifications induced by the methods of divided function evaluation, it has been demonstrated that ureteral catheterization involves significant variations in PAH and inulin clearances [7]. There is no evidence in humans that the urine stop-flow obtained by abdominal compression induces variation of the renal function at least during the time of our observation. Our direct experience seems to demonstrate the opposite thesis. In fact, during venous perfusion of both ^{131}I-hippuran and ^{131}I-hypaque the radioactivity increases linearly over each kidney during the stop-flow till the end of the abdominal compression period (Fig. 6). It seems rather difficult that a variation in renal function occurs just at the beginning of the examination. If, on the other hand, one admits that the increase of pressure in the pelves (caused by the urine stop-flow) modifies the renal function after a certain time, the radioactivity increases recorded over each kidney would not be linear. One must note also that ureteral catheterization can cause infection of the urinary tract. This naturally does not occur with the external counting method, a factor which presents a real advantage over the former technique.

An indispensable condition for this technique is that both pelves be in the visual angle of the collimators in order to exactly measure the radioactivity increases over both kidneys. To demonstrate the validity of this condition, we measured divided RPF in normal subjects. The results, reported here, are almost similar to those obtained by others using the catheterization method [8]. Moreover the reproducibility of the unilateral RPF values with external counting was satisfactory. In fact the repetition of renal counter placement in the same subjects showed a mean percent variation of 7.9%.

With reference to hypertensive subjects, the results obtained in the ten reported cases, studied both with the external counting method and with the ureteral catheterization, show that the former method seems adequate to evaluate functional inequalities between the two sides even of slight degree. It should be noted that the results of the two methods may differ because they are not carried out simultaneously and therefore are not perfectly comparable. The two types of measurements were usually carried out in the same subject within a month. During the measurement with the two methods, the blood pressure was sometimes different for different salt intake and therapy which was unavoidable. One must also remember that there could be variations in RPF due to ureteral catheterization [7].

For its simplicity and non-traumatic effects, the external counting method permits the extension of divided RPF measurement to a greater number of hypertensive subjects, till now not possible with the catheterization method.

The importance of unilateral RPF evaluation in the study of hypertension appears clearly in the two cases reported in detail. The knowledge of this parameter is useful not so much to demonstrate an unilateral renal disease (for this purpose pyelography, renography and renal scan are generally sufficient), but as to evaluate quantitatively the function of both, injured and considered normal, kidneys. This information is indispensable to determine the prognosis and to decide the method of medical or surgical treatment.

It is however evident that unilateral RPF values alone can not be sufficient for the differential diagnosis among various unilateral kidney diseases responsible for hypertension. These, commonly represented by unilateral pyelonephritis and occlusive diseases of the renal arteries, demonstrate a particular pattern of the various solute concentrations (sodium, creatinine, PAH etc.) in the urines of the two kidneys collected via bilateral ureteral catheterization. This peculiar pathway depends on the glomerular-tubular imbalance, prevailing tubular function in the ischemic kidney and filtration function in the pyelonephritic one. Thus the filtration fraction (FF) is reduced if renal ischemia is present, while it may be normal or even increased in the pyelonephritic kidney [9—11].

For the purpose of measuring unilateral FF, the external counting method was later applied in unilateral GFR measurement. This measurement is however in itself interesting in both physiological and pathological renal studies. The unilateral GFR values in normal subjects with the external counting method were not dissimilar to those obtained by ureteral catheterization [8].

For unilateral GFR measurement ^{131}I-hypaque was used. This tracer, which is eliminated via the kidneys by glomerular filtration only, is employed routinely in our clinic for the determination of total glomerular filtration rate with external counting over the bladder [12,13]. In the near future unilateral RPF and GFR will be simultaneously determined in order to measure the FF of each kidney. For this purpose two gamma-emitting tracers of different energies will be employed: ^{131}I-hippuran and ^{140}La-DTPA (diethylene-triamine-pentacetic acid) for RPF and GFR respectively. The latter tracer seems to be eliminated through the kidneys by glomerular filtration only [14, 15]. The simultaneous use of these tracers could thus permit a differential diagnosis between unilateral renal diseases (namely renal artery diseases and pyelonephritis) without bilateral ureteral catheterization.

Summary

A new method for the measurement of unilateral renal clearances by external counting without ureteral catheterization is described. This is based on recording the radioactivity increases of each kidney, after intravenous injection of appropriate tracers, when urine-flow into the bladder is stopped by a suitable apparatus. It is assumed that the radioactivity build-up of each kidney be proportional to the respective clearance. To test the reliability of this technique, unilateral renal plasma flow (RPF) was determined in fifteen normal subjects, using ^{131}I-hippuran as a tracer. The mean ratio between plasma flow of the left kidney and of the right kidney is 1.01 ± 0.07 S.D. and the mean percent difference is 5.04 ± 4.35 S.D. These results agree with those obtained by others using ureteral catheterization. Ten hypertensive subjects suspected of unilateral renal diseases were studied with the external counting method and also by ureteral catheterization. The two methods provided similar results; the mean percent difference between the two sides being 79.22 by external counting and 86.84 by ureteral catheterization. The external counting method during urine stop-flow would then seem adequate for the determination of unilateral RPF. The same technique is now under investigation for the measurement of unilateral glomerular filtration rate (GFR) using ^{131}I-hypaque as a tracer. Twelve normal subjects have been so far examined. The mean ratio between GFR of the left kidney and of the right kidney is 1.07 $\pm$ 0.11 S.D. and the mean percent difference 8.89 $\pm$ 7.44 S.D. If these results of divided GFR determination are confirmed, the simultaneous measurement of unilateral RPF and GFR would become possible using two different gamma-emitting tracers; thus simplifying the differential diagnosis between unilateral pyelonephritis and renovascular hypertension.

References

1. Bianchi, C., Colizzi, C., Meozzi, A., Pedetta, G. P.: La determinazione separata della portata plasmatica dei due reni mediante l'uso di indicatori radioattivi e del conteggio esterno. Minerva Nefrol. **12,** 122 (1965).
2. — Giannotti, P., Coli, A.: Studio preliminare sulla determinazione mediante radiohippuran-^{131}I della portata plasmatica di ciascun rene nell'uomo. Minerva Med. **56,** 4224 (1965).
3. — Coli, A., Giannotti, P., Lenaers, A.: A new approach to the measurement of divided renal plasma flow by ^{131}I-hippuran and external counting in humans. Radioaktive Isotope in Klinik und Forschung. Strahlentherapie Sonderbd. **65,** 136 (1967).
4. — — Palla, R., Rindi, P.: Divided renal plasma flow measurement: the improvement of a new technique. Proceedings 1st Int. Symp. on the use of radioisotopes in the study of renal function, Liège, 1967. Excerpta Medica. Int. Congress Series **178,** 273.
5. — —: A device for the study of divided renal plasma flow. J. Nucl. Biol. Med. **10,** 76 (1966).
6. — Toni, P.: The measurement of the renal clearance of γ-emitting tracers by the use of external counting. Experientia **20,** 148 (1964).
7. Fritjofsson, A., Aurell, M.: Bilateral renal clearance. Technical problems and the effect of ureteric catheterization on renal function. Scand. J. Urol. Nephrol. **2,** 103 (1968).
8. Hulet, W. H., Baldwin, D. S., Biggs, A. W., Gombos, E. A., Chasis, H.: Renal function in the separate kidneys of man. I. Hemodynamics and excretion of solute and water in normal subjects. J. Clin. Invest. **39,** 389 (1960).
9. Dustan, H. P., Poutasse, E. F., Corcoran, A. C., Page, I. H.: Separated renal functions in patients with renal arterial disease, pyelonephritis, and essential hypertension. Circulation **23,** 34 (1961).
10. Stamey, T. A.: Alcune osservazioni sulla frazione di filtrazione, sul trasporto del sodio e dell'acqua nel rene ischemico e sull'importanza prognostica del F.P.R. del rene controlaterale nell'ipertensione renovascolare. Proceedings of the CIBA Symposium "Antihypertensive therapy", Siena, 1965.
11. DePalma, J. R., Cutler, R. E.: Simplified bilateral differential function studies in hypertensive patients utilizing radioactive clearance techniques. J. Urol. **97,** 559 (1967).
12. Bianchi, C., Toni, P.: Mesure du taux de filtration glomérulaire chez l'homme par comptage externe, avec un indicateur émetteur de rayons gamma (hypaque-I^{131}). Proceedings 2nd Int. Congress of Nephrology, Prague, 1963. Excerpta Medica, Int. Congress Series **78,** 517 (1963).
13. — Coli, A., Gallucci, L., Paci, A., Palla, R., Rindi, P.: The measurement of glomerular filtration rate in children by ^{131}I-hypaque and external counting. J. Nucl. Biol. Med. **11,** 144 (1967).
14. Funck-Brentano, J. L., Lellouch, J., Leski, M.: Nouvelle méthode de mesure de la filtration glomérulaire sans prélèvement d'urine. Mesure de la clearance du DTPA Lanthane 140 par enregistrement de la décroissance de la radioactivité recueillie par détection externe. Rev. Franç. Etud. Clin. Biol. **12,** 790 (1967).
15. Bianchi, C., Blaufox, M. D.: ^{131}I-hypaque and ^{140}La-DTPA for the measurement of glomerular filtration rate in dog. J. Nucl. Biol. Med. **12,** 117 (1968).

Experiences with Lung Scanning on Patients Suffering from Boeck's Sarcoidosis

KATALIN MÁRTONFFY, TIBOR HERNÁDY and ÉVA TÖRÖK

Radiological Clinic of the Medical University of Budapest, Hungary

The Boeck's sarcoidosis has formerly been regarded as a very rare disease. Since the fifties of this century, however, numerous international congresses and symposia have dealt with it. This may be understood by that screening tests have been made, from the late forties, all over the world and thus the population of whole countries have practically been controlled. As an extra result of this, it could be stated the relative frequency of the lung sarcoidoses.

According to LOEFGREN'S data, in Sweden, there occurred 40—50 sarcoidoses among 100 000 inhabitants. In the U.S.A., in hospitals of the Veterans Administration 1194 cases were diagnosed in five years. In Switzerland, as a result of the military screening tests, 13 diseases have fallen to 100 000 individuals. As reported by MÁNDI, the disease is relatively frequent in Hungary, too [8].

The earlier data have reported serious disease-processes. According to these statistics, the mortality rate was 20—25 per cent. The duration of the illness was estimated to be 6 years, however, there were statements about a possible duration of 30 years [8].

According to more recent data, the prognosis is much more favourable, mainly in the acute phase of the disease.

92 per cent of the Loefgren's cases, having started with erythema nodosum, recovered in two years without trace or with scar; only 8 per cent turned into the stage II. 73 per cent of his cases diagnosed by screening or by examining some complaints have shown themselves as recovered in two years of observation, that is, only in 27 per cent has evolved the chronic form. As it is shown by WURM'S data, the Middle European average concerning the recovery within two years of observation period amounts to 42 per cent. These statistics include also the cases of the so-called Loefgren syndrom. KUMSCHICK et al. have observed that in cases of the hilus sarcoidosis starting without any signs the recovery within two years amounts to about the half of that showing itself in the Loefgren syndrom [3, 8].

The less serious disease-process observed lately may be interpreted, according to MÁNDI, by that the earlier data were mainly reflecting the sarcoidosis material of the dermatological and ophthalmological institutes concerning predominantly the chronic diseases. One of the manifestation forms of that disease of unknown aetiology, which affects numerous organs and produces the most various symptoms, has first been described by HUTCHINSON as the "Mortimer disease". Similar dermatological case was described by BESNIER having termed it as "lupus pernio". In 1897 BOECK had reported such dermatological signs which he has later termed — on the basis of the histological picture — as "skin sarcoid" and, more later, as "miliar lupoid". Similar histological changes have been found by HEERFORDT in diseases of the eye, by MIKULITZ in the salivary glands and by MOROSOFF in the phalanges of the hand and the foot.

SCHAUMANN has first recognized that all these diseases termed variously and held as distinct ones are but representing special manifestations of the same general disease entity which is mainly localized at the lymph nodes and—in most of cases—affects the thoracal lymph nodes. He has termed the disease as lymphogranulomatosis benigna.

Aetiology and pathology of that disease were investigated in the last ten years by many authors. A lot of papers were published concerning the alterations to be due to sarcoidosis and affecting almost all organs (e.g. the central nervous system, the kidneys, the heart, the tubular bones).

According to the literary data known by us, it seems that the digestive system, the adrenal glands and the serous membranes are not affected by the disease.

The aetiology of the Boeck's sarcoidosis remained unclarified up to the present [2]. Significant part of the German literature means—on the basis of the pathohistological picture mainly—that it may have a tuberculous origin. It is supposed by numerous authors that the disease may be caused by a non-acid fast type of the Mycobacterium tuberculosis. As for ZETTERGREN, this is a special type of the tuberculosis. According to this concept, the Ziegler's macrocellular form would represent a transition between the sarcoidosis and a classic manifestation of the tuberculosis. The possible virus-origin, too, has been suggested. According to MITCHELSON, OSBORNE and KALKOFF, it may be a polyaetiological syndrom.

Because of its morphological similarity to the periarteritis nodosa, RIJSELL has proposed the possibility of its allergic origin. His view is supported by the theory of TEILUM according to which the alterations start with a globulin precipitation in the RES caused by an allergic hyperglobulinaemia. In accordance with this, the inclusions described by SCHAUMANN within the cytoplasm of the RES cells would represent hyperglobulinaemic precipitates. It is without doubt that also the so-called asteroid bodies of FRIEDMANN are similar to that described by TEILUM [2]. The observations (WALDGREN, ISRAEL et al.) reporting that the sarcoidosis does occur conspicuously frequent at districts rich in pine-forests, also could support the allergic origin. In the opinion of CUMMINGS and HUDGINS, namely, the pine-pollen, as an allergen, might play a significant role. This hypothesis gets some support by the observation of ISRAEL reporting that inhalation of hairspray causes a roentgenmorphological alteration quite similar to the sarcoidosis [3].

The fact that the frequency of the sarcoidosis is not higher neither in tuberculous families, nor in populations strongly infected with tuberculosis runs counter to the tuberculous origin of the disease. As a counterargument can serve also the observation that the lymph nodes affected by sarcoidosis do never break into the bronchi, respectively, one never could observe caseation [8].

Consequently, the statement accepted by the Conference of Washington in 1956, according to which "the sarcoidosis represents a system-disease, or disease-group of an unknown aetiology and pathogenesis" remains valid up to the present [8].

Also the various methods of the medical treatment applied since that the disease had been described, point to the incertain nature of the aetiology. The spontaneous recovery occurs in a relatively high percentage, thus, during all therapeutic procedures one always could observe several favourable cases.

The antituberculotics failed to produce any medicating effect. Recently, the application of the steroids, too, was naturally tested. The actual principle of the treatment is that in the stage I steroid will only be given if the disease starts with acute signs and fiver, or, no spontaneous melioration shows itself after an observation period of 4—8 months, respectively, the disease is getting worse. In the stage II, one immediately starts with the dosing of steroids under protection of antibiotics. In the stage III of the disease, when having a sure diagnosis and no counterindication for steroid therapy, the latter must be applied [3, 8].

Thus, in our actual knowledge, this disease shows spontaneous recovery in a high percentage. The most recent therapeutic results render the disease-process more favourable. The recovery, in this disease, is shown by not the melioration of the patient's general state, because, as it is well known, the patients have, in most of cases, no complaints. No complaints are indicating for the progression yet, but the transition from the stage I into the stage II is "silent", and, generally, it is diagnosed accidentally [3].

As a consequence of this, in controlling the patients, the X-ray examination proves to be the most reliable. It is important to know that in patients declared, roentgenmorphologically, as recovered, what kind of lung functioning can be found, or, in cases recovered with residuum, of what an extent is the decrease in the respiratory function.

The spirometry gives data about the common function of the two lungs only. The bronchospirometry can though deliver data about function of isolated lung-parts, however, the indication of such examinations is restricted.

DRESSLER and LEITNER have described the decrease of the oxygen concentration in the arterial blood. American authors have registered some diffusion trouble in the diffuse form of the disease having shown itself as significant at overstraining and as non-significant at rest. The cortison treatment has not influenced this state [10].

We have found, in our treated patients, that there is no concordance between the subjective complaints and the results of the X-ray and respiratory-function examinations. We have tried to resolve this seeming discrepancy with the aid of the scanning.

In the last years, the lung scanning has become a valuable functional examining method and its indication is significantly widened. As it is well supported by many literary data, by using the perfusion scanning, the measurable activity above certain lung parts is in direct ratio to the blood-quantity streaming through them [1, 11].

This statement has turned the attention to the functional examinations which, naturally, has not decreased the importance of the scanning in the diagnosis of pulmonal embolism and central lung tumors.

As for developmental anomalies, tuberculosis, patients suffering from emphysema and silicosis we have found, in accordance with the literary data, that the scans say more about the functional state of the lungs than the more detailed X-ray pictures or the examination of the respiratory function [4, 6, 9]. This latter statement is more effective for the cases when one cannot perform the bronchospirometry for that the bronchial system has some pathologic deformity or for other reasons. The scanning is favoured by the facts that it is practically free from risk, does not represent some non-physiological stimulus or a load for the patient. While the result of the spirometry is highly depending on the patient's cooperation, that of the scanning is not influenced by that factor.

We have made scanning on 13 patients suffering from Boeck's sarcoidosis. 9 histologically verified cases were evaluated. The right diagnosis was assured by mediastinoscopy, thoracoscopy and by the histological examination of the lymph node excised from the axilla. In three further cases, the diagnosis could only be rendered probable on the basis of the clinical picture and the X-ray examinations. One of our cases shows signs of the starting disease; here, naturally, one cannot speak of recovery. The latter four cases are not included in our analysis.

Case	Sex	Age (years)	Spirometry	Complaint	X-ray	Scan
1. B.M.	♀	31	∅	∅	∅	+
2. B.S.	♂	26	∅	∅	+	+
3. B.J.	♂	40	+	∅	+	++
4. Cs.L.	♀	18	∅	+	+	++
5. Cz.Gy.	♂	37	∅	∅	+	++
6. K.E.	♀	34	++	+	+	++
7. M.V.	♀	26	+	+	∅	+
8. N.J.	♀	35	+	+	+	+
9. S.D.	♂	39	∅	∅	+	+

Our patients were examined after being treated with antituberculotics and steroids in the stage of the regression or the recovery. In each of our patients we have made summation and stratigraphied pictures as well as lung scans; also the bronchoscopy and the spirometry were performed (VC, MBC, Tiffeneau).

The scanning was performed by injecting 120—300 μCi iodine-131-labeled human serum albumin macroaggregatum into patients lying on their back, with the Picker Magnascanner III apparatus and using the coloured dot technique.

Our case reports are the following:

Case 1. B.M., a 31-year-old woman. She was admitted to the hospital by having on both legs an erythema nodosum and in the hili strongly enlarged lymph nodes (Loefgren's syndrom). In the beginning of the disease she had a dyspnea on exertion being ceased during the treatment. The spirometric values are normal. In spite of this, above the right apex and the basis of the left lower lobe, a decreased activity can be registered which cannot be interpreted based on roentgenogram (Fig. 1a, b).

Fig. 1a

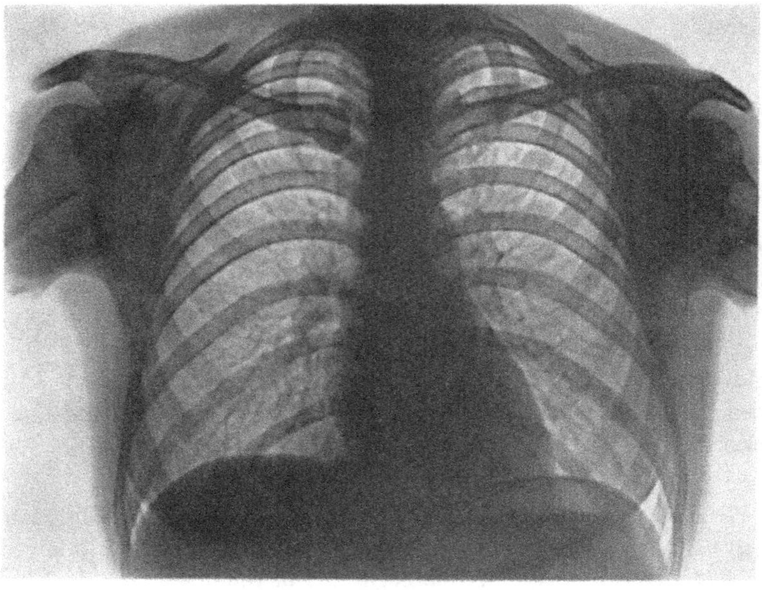

Fig. 1b

Case 2. B.S., a 35-year-old man. He was admitted to the hospital in 1960 with the suspicion of having tuberculous changes in the hilar lymph nodes. The Scalenus biopsy has revealed the Boeck's sarcoidosis. After therapy, a regression of high degree has shown itself. The roentgenogram made after 9 years shows that the right hilus has remained more massive. At the right side, in the heart-diaphragma angle, a well expressed network-like picture can be seen. In the little fissure a fine pleural shadow-streak is observable. The medial half of the right apex is covered. The scan corresponds with the X-ray examination, the level of activity being lower at the right lower lung region. The patient has, at present, no complaints, also his spirometric values are normal.

Case 3. B.J., a 42-year-old man. He was sent to the hospital after screening. After the finished therapy, two years later, the spirometric values show a ventilatory disorder of a mediate seriousness (VC: 3030 ml −37%; MBC: 81 l −17%; Tiffeneau: 89%). The change on the roentgenogram does not account for the degree of the ventilatory disorder. Nevertheless, the scan shows a significantly lower activity at the projections of both apexes, as well as above the bases.

Case 4. Cs.L., a 28-year-old woman. She was examined because having fiver and tussis. After the treatment, the values of spirometry show no significant changes. The patient suffers continuously from dyspnea. During the disease, the seriousness of the roentgen-changes at the left lower lobe has intensified when compared with the right one. The scan shows, at about the projection of the left lower lobe, a strongly decreased activity giving an explanation for the patient's complaints (Fig. 2a, b).

Case 5. Cz.Gy., a 39-year-old man. His disease has been recognized by screening. After therapy, the subjective complaints of the patient ceased. The spirometric values are found to be normal. The roentgenogram shows discrete changes

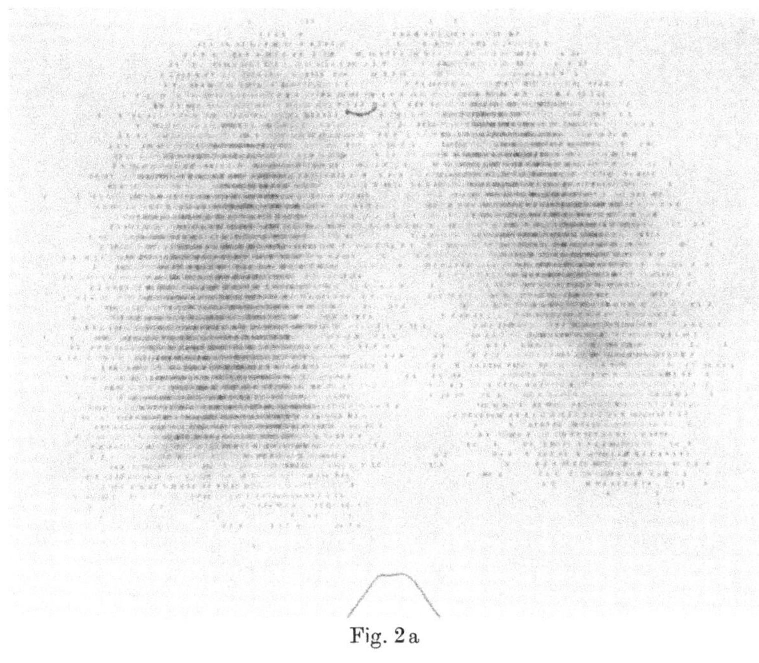

Fig. 2a

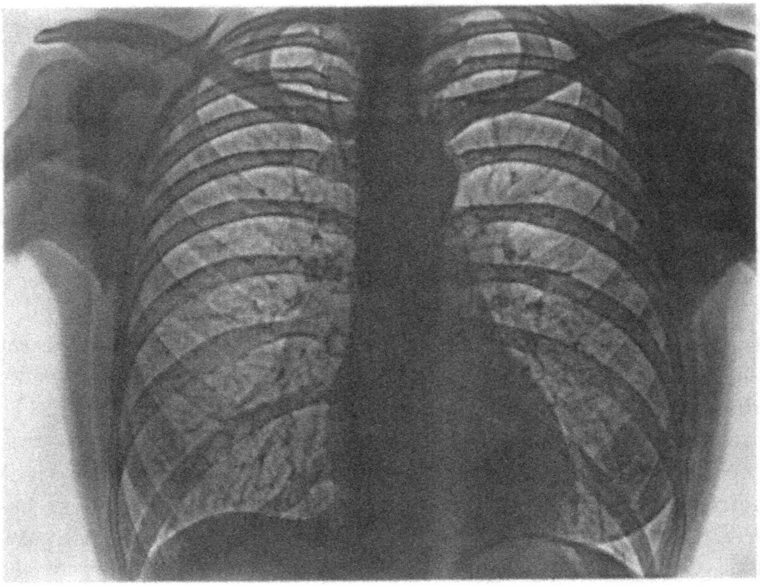

Fig. 2b

at the right lower lobe of the lung. On the scan, at the projection of the left lung-base, one can find a low activity, which cannot be explained by the roentgenogram. It is noteworthy that on the finger V of the patient's left hand the base-phalange shows cystic lesion (Fig. 3a, b, c).

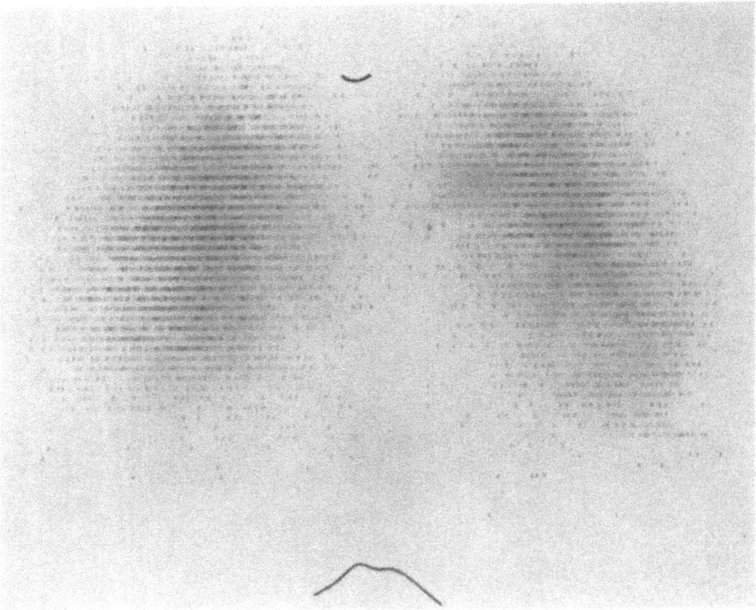

Fig. 3a

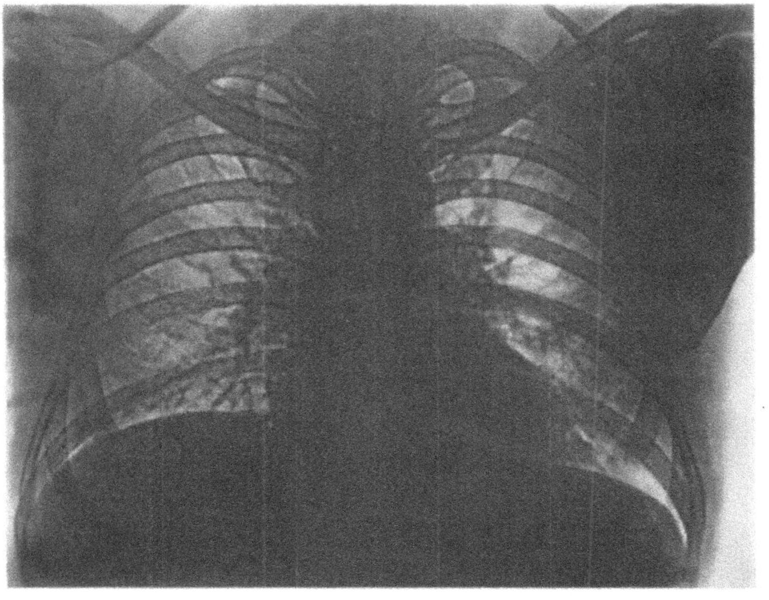

Fig. 3b

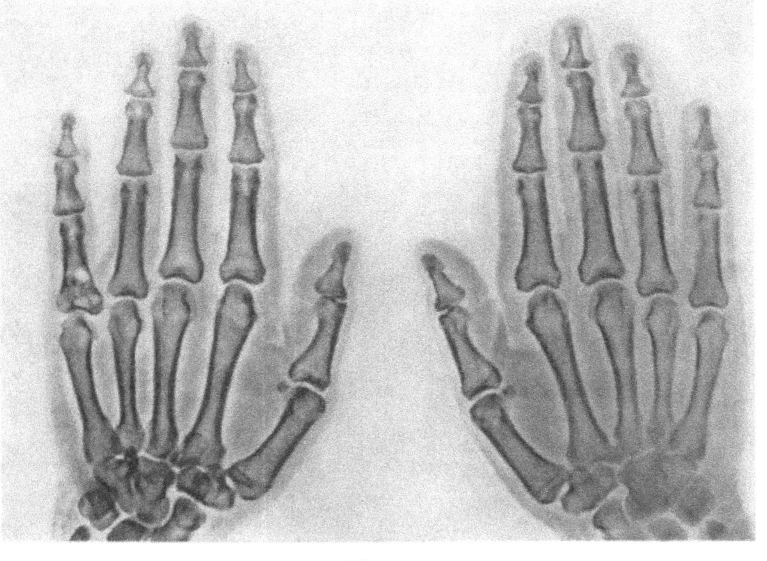

Fig. 3c

Case 6. K.E., a 38-year-old woman. She was admitted to the hospital in an extremely bad state. After therapy, on the roentgenogram we observed a regression of high degree. In spite of this, the spirometry shows a serious restrictive ventilatory disorder (VC: 1860 ml −48%; MBC: 38,7 l −45%; Tiffeneau: 76%). The patient has, accordingly, dyspnea. As to the ventilatory disorder, the scan gives a more convincing explanation than the roentgenogram. On the scan, at the projection of the lower segments of both lobes and in the neighbourhood of the right hilus a decreased activity can be found (Fig. 4a, b).

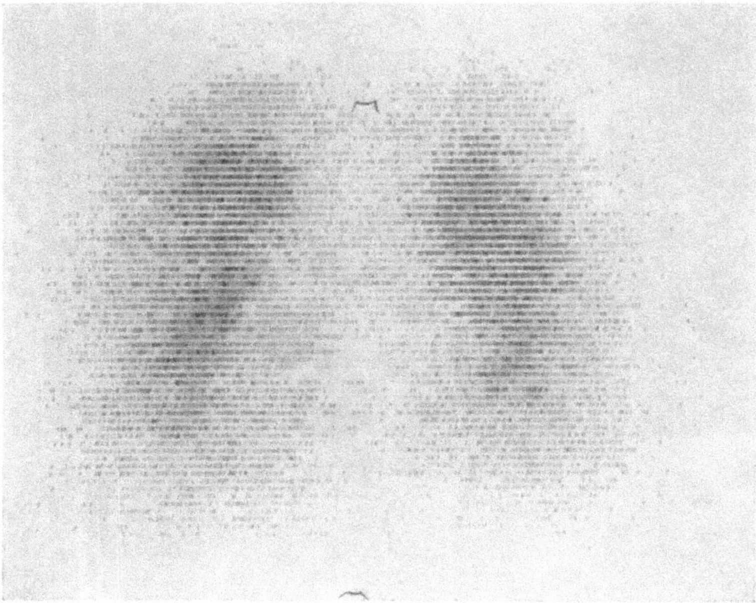

Fig. 4a

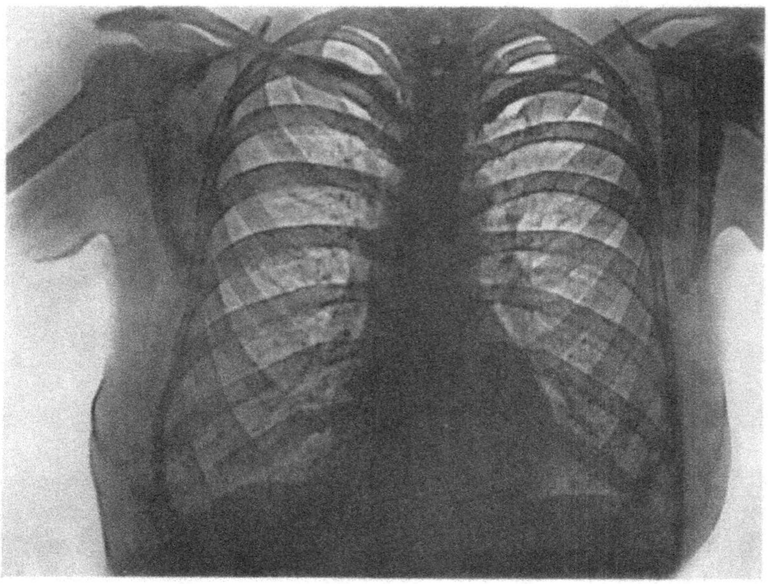

Fig. 4b

Case 7. M.V., a 30-year-old woman. She was admitted with dyspneal complaints. According to the spirometry, the patient has a moderate ventilatory disorder of mixed character (VC: 2880 ml -17%; MBC: 54,9 l-28%; Tiffeneau: 63%). On the roentgenogram, at the region of the right lower lobe, a more intensive bronchovascular picture can be seen. On the scan, at the projection of the right apex, no activity can practically be found. Above the left lung-base, too, a lower activity shows itself. The scan supports the spirometric data, in spite of that the patient has no subjective complaints. The roentgenogram shows no significant changes.

Case 8. N.J., a 40-year-old woman. She was admitted with a Loefgren's syndrom. After a therapy lasting two years the patient has dyspnea. The spirometry has diagnosed an obstructive ventilatory disorder (VC: 2520 ml -12%; MBC: 45 l-26%; Tiffeneau: 59%). The roentgenogram shows circumscribed focal changes at both sides perihilary. The scan, too, is pathologic. In the right lung, the distribution of activity is inhomogeneous; the deposition of MAA in the right side is generally lower than in the left lung.

Case 9. S.D., a 49-year-old man. 13 years ago, the sagittal roentgenogram showed a wide-spread disseminated change of sarcoid character. The present spirometric values reveal no significant changes; the patient has no dyspneal complaints. On the sagittal roentgenogram, as a residuum of the process, mainly at the right side, a well expressed fibrotic reticular tracery is seen in the neighbourhood of the callous thickening of the mediastinal pleura. On the scan, corresponding to the right apical region, there is a total absence of activity; above the right lung, the level of activity is lower when compared with the counter-side.

Comprehensive Analysis of our Cases

The breakdown by sex of our patients suffering from Boeck's sarcoidosis was the following: 5 women, 4 men. At the diagnoses they showed the following breakdown of age: between 10 and 20 years: 1, 20 and 30 years: 2, 30 and 40 years: 6. Each of our patients was medicinally treated. Seeing that the treatment of the patients was performed in different hospitals, there was no uniform therapy. At the diagnosis, 3 of them were in the stage I, 5 in the stage II and 1 in the stage III. (The stages correspond to those of WURM.) The period of observation varied from several months to 10 years and averaged 5,5 years. During this period none of our patients has shown aggravation. After having compared the spirometric results and the subjective complaints with the roentgenograms and the scans, we have observed the following.

In the case of 2 patients, the roentgenogram practically points to total recovery, while their scans proved to be positive. One of the patients has dyspneal complaints, too, with the objective signs of a ventilatory disorder. In our opinion, it is important that in each of our cases (in 5 patients without any complaints, too) the scans pointed to a perfusion disorder varying in degree. In the case of 4 patients, the disorder of perfusion demonstrated by the scan has proved to be of a higher degree than what one could expect based on X-ray pictures. This can be interpreted by that on the roentgen-films one cannot visualize the fine, microscopic changes. The scan, in spite of having a resolution power much more low as to the details, demonstrates, nevertheless, sensibly these destructions. At the scarred places, namely, the MAA deposition is naturally less.

It seems that in Boeck's sarcoidosis the cases showing themselves as recovered or stagnant by X-ray control, on the basis of lung scanning—concerning the pulmonary perfusion— cannot be regarded as having a restitutio ad integrum.

In spite of the low number of our cases we, nevertheless, mean that our results, too, support the view pointing to that in the estimation of the lung's functional state the scanning cannot be disregarded. Consequently, our data, too, are documenting the functional significance of the lung scanning.

Summary

The authors have analysed the lung scans and roentgenograms of 9 recovered patients suffered from Boeck's sarcoidosis. The scans proved, in all cases, to be positive. In the case of 4 patients, the perfusion disorder detected by the scan has proved to be of a higher degree than what could be expected when based on roentgen pictures. In cases regarded as recovered or stagnant from the point of view of X-ray controlling, one cannot speak of a restitutio ad integrum when the pulmonary perfusion is taken into consideration.

References

1. Chernick, V., Lopez-Majano, V., Wagner jr., H. N., Dutton jr., R. E.: Am. Rev. Resp. Diseas. **92,** 958 (1965).
2. Kelemen, J. T.: Orvosi Hetilap **102,** 1116 (1966).
3. Kumschick, H., Siegenthaler, Rhomberg, F.: Schweiz. med. Wschr. **97,** 1407 (1967).
4. Lopez-Majano, V., Wagner jr., H. N.: Dis. Chest **54,** 46 (1968).
5. — Chernick, V., Wagner jr., H. N., Dutton jr., R. E.: Radiology **83,** 697 (1964).
6. — Kandel, M. G. E., Wagner jr., H. N.: Bull. Physiopath. Resp. **5,** 39 (1969).
7. — Kieffer, R. F., Marine, D. N., Garcia, D. D., Wagner jr., H. N.: Am. Rev. Resp. Diseas. **99,** 554 (1969).
8. Mándi, L.: Orvosi Hetilap **102,** 1105 (1961).
9. Oeser, H., Schumacher, W., Ernst, H., Frost, D.: Atlas der Szintigraphie. Kapitel VI. Berlin: de Gruyter 1969.
10. Purjesz, B.: A belgyógyászat és határterületeinek syndromái. Budapest: Medicina 1965, p. 139.
11. Wagner jr., H. N.: Principles of Nuclear Medicine. Philadelphia: W. B. Saunders 1969, p. 472.
12. Wurm, K., Reindell, H., Heilmeyer, L.: Der Lungenboeck im Röntgenbild. Stuttgart: G. Thieme 1958.

Druck: Mercedes-Druck, Berlin 61
Einband: Konrad Triltsch, Graphischer Betrieb, Würzburg

Subject Index — Sachverzeichnis

GPSR Compliance
The European Union's (EU) General Product Safety Regulation (GPSR) is a set of rules that requires consumer products to be safe and our obligations to ensure this.

If you have any concerns about our products, you can contact us on

ProductSafety@springernature.com

In case Publisher is established outside the EU, the EU authorized representative is:

Springer Nature Customer Service Center GmbH
Europaplatz 3
69115 Heidelberg, Germany

www.ingramcontent.com/pod-product-compliance
Lightning Source LLC
Chambersburg PA
CBHW082005130926
51338CB00021B/46
* 9 7 8 3 6 4 2 6 5 0 9 4 9 *